W0116045

Quality Management of Cement Concrete Construction

PS Gahlot ME (Str. Engg), Grad. Cert. Ed. (UK), FIE

Director General
Yagyavalkya Institute of Technology, Jaipur

Ex-Professor (Civil Engg)
National Institute of Technical Teachers' Training and Research, Chandigarh

Deep Gehlot BSc, BE (Civil Engg.)

Manager (Civil)
TCE Consulting Engineers Ltd, Pune

CBS

CBS Publishers & Distributors Pvt. Ltd.

New Delhi • Bengaluru • Chennai • Kochi • Kolkata • Mumbai
Hyderabad • Nagpur • Patna • Pune • Vijayawada

Disclaimer
The authors and publisher have tried their best to incorporate all the latest available information in this book. This book is published with the understanding that authors, publisher and printer are neither responsible for the result of action taken on the basis of this work nor for any errors or omissions.

ISBN: 978-81-239-1698-9 (PB)
ISBN: 978-81-239-1684-2 (HB)

First Edition: 2009
Reprint: 2010, 2013, 2016

Copyright © Authors & Publisher

All rights reserved. No part of this book may be reproduced or transmitted in any form or by any means, electronic or mechanical, including photocopying, recording, or any information storage and retrieval system without permission, in writing, from the publisher.

Published by:

Satish Kumar Jain for CBS Publishers & Distributors Pvt. Ltd.,
4819/XI Prahlad Street, 24 Ansari Road, Daryaganj, New Delhi - 110002
delhi@cbspd.com, cbspubs@airtelmail.in • www.cbspd.com
Ph.: 23289259, 23266861, 23266867 • Fax: 011-23243014

Corporate Office: 204 FIE, Industrial Area, Patparganj, Delhi - 110 092
Ph: 49344934 • Fax: 011-49344935
E-mail: publishing@cbspd.com • publicity@cbspd.com

Branches:
• *Bengaluru:* 2975, 17th Cross, K.R. Road, Bansankari 2nd Stage,
  Bengaluru - 70 • Ph: +91-80-26771678/79 • Fax: +91-80-26771680
  E-mail: cbsbng@gmail.com, bangalore@cbspd.com
• *Chennai:* No. 7, Subbaraya Street, Shenoy Nagar, Chennai - 600030
  Ph: +91-44-26681266, 26680620 • Fax: +91-44-42032115
  E-mail: chennai@cbspd.com
• *Kochi:* Ashana House, 39/1904, A.M. Thomas Road, Valanjambalam,
  Ernakulum, Kochi • Ph: +91-484-4059061-65
  Fax: +91-484-4059065 • E-mail: cochin@cbspd.com
• *Kolkata:* 6-B, Ground Floor, Rameshwar Shaw Road, Kolkata - 700014
  Ph: +91-33-22891126/7/8 • E-mail: kolkata@cbspd.com
• *Mumbai:* 83-C, Dr. E. Moses Road, Worli, Mumbai - 400018
  Ph: +91-9833017933, 022-24902340/41 • E-mail: mumbai@cbspd.com

Representatives:

• Hyderabad: 0-9885175004   • Nagpur: 0-9021734563
• Patna: 0-9334159340   • Pune: 0-9623451994
• Vijayawada: 0-9000660880

Printed at:
India Binding House, Noida (UP)

Quality Management of Cement Concrete Construction

Foreword

At the very outset I must congratulate the authors for bringing out an excellent volume on **Quality Management of Cement Concrete Construction**, a subject of great relevance and importance to all those engaged in the study or practice related to managing cement concrete constructions. Managing quality of cement concrete at all stages, viz. design, selection of ingredients and their testing, mixing, placing, compacting and curing, are extremely important to attain the desired criterions of workability, safety, strength and durability. Mismanagement by way of any wrong decision, practice or action taken at any of the stages mentioned above could prove uneconomical and damaging to the quality, strength and durability of concrete. It is, therefore, important that the students of civil engineering and practising engineers engaged in managing cement concrete constructions are equipped with all the knowledge, competence and best practices for managing cement concrete constructions.

The book deals exhaustively with all the topics related to managing cement concrete constructions for achieving the desired quality effectively. Starting with an overview of the subject and explaining in general the problems and thus the need and importance of quality management of cement concrete constructions, the book covers the various elements and aspects of quality management of cement concrete constructions such as basic properties of cement concrete; influence of basic ingredients and special admixtures on the quality of concrete; influence of concreting operations on quality; management, statistical control and acceptance criteria of quality; cement concrete mix design procedures; concrete mix quality assurance trials; testing of ingredients and concrete, repair maintenance and protection of concrete, and managing quality of concrete under special conditions of placement. The book makes use of the latest information, knowledge, techniques, best practices and standards available on the subject while dealing with each of the above topics.

The authors, in making use of their long experience of teaching and training have structured and sequenced the content of the subject by following principles of learning hierarchy. This will facilitate step-by-step acquisition, comprehension, application and transfer of learning of the important concepts and principles of quality management of cement concrete constructions under different site conditions and usage.

The quality of cement concrete materials and admixtures has undergone a sea change during the last three decades causing a big change in the technology of construction. I feel that the writing of such a comprehensive book on the subject is very timely. It will go a long way in facilitating and promoting the use of new materials and technology available for quality cement concrete constructions.

The book will prove very useful to the students who are undergoing courses of study in concrete technology and managing concrete constructions and more so to the practising engineers who are engaged in managing and assuring quality of cement concrete constructions under different conditions and usage.

M.M. MALHOTRA

Ex-Principal
National Institute of Technical Teachers' Training and Research
Chandigarh

Foreword

Concrete is the most widely used construction material in all types of structures due to its strength, mouldability and durability. Construction of buildings and other structures have been the area of great concern in all development projects in the country. The cement concrete construction subject is of vital importance to the practising as well as budding engineers to achieve economy, quality and speed to ensure optimal functionality, durability and longevity of such structures. To realize this, there is an urgent need to strengthen the curricula of civil engineering as well as to offer specialized education and training programs to practising engineers on the subject of quality management of cement concrete construction, the use of correct practices on quality management of cement concrete construction can result in saving of crores of rupees from various projects.

Concrete structures are susceptible to cracking, while steel gets corroded. Many a times deterioration occurs due to lack of adequate care, knowledge and coordination at the time of construction to enforce primary objectives of quality, safety, durability and serviceability of structures.

Construction of cement concrete is required to ensure that all the operations are conducted as per the requirements of users to eliminate failure at inconvenient moments. It is the responsibility of construction engineers that the buildings are constructed and maintained as per the standard practices dominated by the functional requirements. There are many factors such as aging, destructive action of natural forces, poor selection of materials, occupational human factors that could contribute to the deterioration and malfunctioning of the various sub-systems of building structures.

The book *Quality Management of Cement Concrete Construction* provides the latest information on methods/procedures, techniques, materials and systematic planning. The authors have rich experience of field practices and researches in cement concrete. The book also provides an extensive treatment on the use of the latest available materials and techniques of cement concrete construction.

The book meets the long-standing need of a comprehensive treatment on the subject of quality management of cement concrete construction. I hope that it will prove very useful for the students of civil engineering at degree as well as diploma levels, and for AMIE examinations, UPSC Engineering Service examinations, and practising field engineers and supervisors at various levels.

Dr BRAHAM DATT
Ex-President of BAI
Vice-Chairman, Som Datt Builders Ltd
56-58, Community Centre, East of Kailash
New Delhi 110 065

Preface

Cement concrete is the most important modern construction material used in almost all type of structures. Considering the importance and knowing it well that no good books are available on the subject of "quality management of cement concrete construction", the authors have tried to bring out a comprehensive book on the subject generally neglected in the curricula of civil engineering. The major purpose of undertaking this venture has been to make available a textbook in "quality management of cement concrete construction" for civil engineering students. Understanding, investigating and providing suitable approach to quality construction is the most essential for all level of civil engineering professionals. Many universities and boards have introduced this subject at all levels of programmes (MTech, BTech and Diploma). While writing this book, the authors have considered and focused on the needs of construction industry as a whole to suit all levels of engineers and technologists. The book is written considering the correct engineering practices and is most suitable for working professionals associated with construction industry as contractors, consultants, engineers, and supervisors. The book provides description of innovative techniques and new materials.

Each chapter of the book starts with the expected learning outcomes followed by inputs or related introduction, description of materials, techniques and illustrative applications in specific situations. The purpose of providing learning objectives at the beginning of the chapter is to raise expectations of what the learner would achieve at the end of study of the chapter. The illustrations are followed by questions related to construction practice to develop and assess learner's professional competence. The book is divided in four units which are sequenced according to the theory of learning (concepts, principles and applications) for managing quality. Chapters of Unit I, mostly deal with concepts, basic materials and fundamental aspects of understanding correctly. Unit II is based on the influence of operations and correct techniques for achieving quality. Unit III deals with principles required for understanding quality of cement concrete. Unit IV deals with applications of principles and practices to achieve quality concrete construction. The Unit IV also provides field practices and tests for the control of quality. The sample examples are also given for the mix design based on real life situations drawn from the field. Many of the special solutions are based on chemical stability of materials under the environmental conditions.

System International (SI) units of measurement have been used wherever necessary. Teachers of engineering colleges and polytechnics, and practising engineers have provided information on the problems, which has helped the authors to write this book incorporating field problems and solutions.

PS GAHLOT

DEEP GEHLOT

Contents

Unit II

Chapter 5 **Influence of Concreting Operations on Quality** **97**

Unit III

Unit IV

UNIT I

Introduction to Cement Concrete and Quality Management

LEARNING OBJECTIVES

The learner understands the importance of cement concrete **and** quality management **in** construction industry and will be able to:

- Define **cement concrete**;
- Describe **historical development** of modern cement concrete;
- Describe **importance** of cement concrete in construction industry;
- Explain **failures** of concrete;
- Explain quality with reference to the concept of **concrete chain**;
- Explain the **importance of quality** management in construction;
- Explain the importance of studying the text in a particular **sequence**.

1.1 DEFINITION

Concrete may be defined as a **mixture of cement** (binding material), **aggregates** (inert materials-both coarse and fine), and **water**, which when mixed and placed in suitable forms and allowed to **cure** under appropriate conditions, **hardens like stone**. The chemical **reaction between cement and water** (hydration) results into **binding** of aggregates to form "**concrete**"- a superior construction material of great potential. **Cement, water** and **sand** forms **mortar** and fills the voids in coarse aggregate particles. The concrete gains strength and other related ·properties with curing and age. The process of hardening by curing results in gain of strength of concrete.

1.2 HISTORY

It may be noticed from the findings of ancient civilizations of Mohenjodaro, and Greek, that concrete and mortars were used in one or the other form in the construction. Later on Romans developed mortar and concrete leading to the present mortars and concrete. These construction materials came to a sound footing after the **advent of portland cement** by an English brick layer named **Joseph Aspdin in 1824**. With the research and scientific development, the cement concrete became one of the most important and **versatile construction material** being used all over the world. The present day cement concrete is a highly developed construction material and is used in almost all type of structures. With the advancement in technology, it is now possible to obtain cement concrete of wide range of properties (such as **strength, durability** and **resistance** to certain specific forces).

A few years ago when concrete technology was not understood well by engineers, technicians and contractors, cement concrete was being used as a simple construction material without exercising scientific principles for its quality control. Different ingredients were mixed in an adhoc manner without knowing and predicting the accurate behaviour of concrete so obtained. **It was Duff Abram who, by his finding of water-cement ratio law in 1919** changed the concept of concrete as a simple adhoc construction material to concrete as a **scientific construction material** with possibility of control on its **quality**. Since Duff Abram's statement of water-cement ratio law, lot of research has been carried out on the behaviour of concrete, its quality and ingredients in different situations.

As a result of these developments, the cement concrete construction was better understood and carried out based on sound scientific principles. Availability of scientific knowledge and data made it possible to **use cement concrete** extensively in variety of structures under **different environmental conditions** and for various uses. Cement concrete is used with proper **mix design**, provides a number of advantages over other construction materials in most of modern structures.

1.3 IMPORTANCE

Initially in India, the construction industry and the engineers have not taken advantages of research and scientific developments in concrete construction to its fullest extent due to lack of education and training facilities in concrete technology. Due to lack of knowledge and skill of supervisors and construction contractors have **wrong notion that greater the quantity of cement in concrete better will be the quality and strength of concrete**. For desired quality, strength and economy in concrete construction, it is necessary to study various aspects of concrete on scientific basis.

India is a developing country and it has launched five-year plans for multipurpose development projects in Agriculture, Industry, Mining, Transport, Power, etc. All these five year plans involve large construction of roads, bridges, dams, industries, dock and harbours, power houses, nuclear power plants, aerodromes, irrigation schemes, public health engineering schemes, educational buildings, residential building schemes and other public structures. All these constructions/schemes demand optimum and efficient use of construction resources. Most of modern heavy constructions require huge quantity of cement concrete. Thus for efficient and optimum use of construction resources, it is most important to study properties and behaviour of cement concrete. Proper knowledge and skill of cement concrete shall lead to lot of economy and quality in construction.

The good quality cement concrete construction has many advantages over other construction materials. Cement concrete can be produced for very high strength (as high as 40 N/mm^2) with ordinary portland cement of 33 grade and 80 N/mm^2 with other high strength cements of 43 and 53 grades and use of admixtures. Cement concrete provides possibilities of creating aesthetically beautiful structures (shells, folded plates). Concrete can be used for construction in very wide range of environments (extreme cold/hot weather, under water, chemically surcharged, and atomic reactors) with the use of **special admixtures** and special cements. Cement concrete can also be used in combination with other materials (**RCC** and **prestressed** cement concrete) for structures subjected to different type of forces. Cement concrete can also be advantageously used for **precast construction** to replace many conventional scarce and costly materials such as timber. Cement concrete constructions are used for **covering very large spans**. Due to its **impermeability** cement concrete construction is used for **water retaining structures**. Cement concrete constructions are **economical** in comparison to many other construction materials.

The civil engineers and technicians having complete knowledge, competency and skill in concrete construction stand good job opportunities in construction industries in public and private sectors. They also have good scope of starting and successfully running their own industry in precast concrete constructions and allied industries.

1.4 FAILURE OF CONCRETE

Many times structures fail due to failure of concrete. Such failures occur in following ways:

a) **Overstressing** of weak and/or porous concrete leading to structural cracks;
b) Improper location and formation of **construction and expansion** joints in concrete structures leading to cracks;
c) Failure of **reinforcement due to corrosion** caused by **insufficient cover** and **porous concrete**; and
d) Improper **form work fixing** and removal.

The main causes of failure of concrete are attributed to its poor **quality of ingredients** and/or **poor concrete practices**. Both these reasons are caused due to **ignorance and negligence** on the parts of supervisors and engineers. To avoid concrete failures, care has to be taken right from the **selection of ingredients** till hardened concrete is put to service. The following is the list of factors/causes mainly responsible for the failure of concrete constructions:

i. Use of poor quality and deteriorated cement;
ii. Use of poor quality aggregate;
iii. Use of poor quality and contaminated mixing water;
iv. Using too much quantity of mixing water for ease of compaction;
v. Lack of proper rigid and water proof formwork;
vi. Inefficient batching, mixing, transporting and placing practices;
vii. Improper and insufficient compaction;
viii. Improper formation and location of construction and expansion joints;
ix. Poor surface finishing and bleeding on the surface;
x. Lack of proper curing; and
xi. Fixing and improper removal of form work.

The failure of concrete can be understood easily through a concept of "**Concrete Chain**" shown in Fig. 1.1. Quality of a chain is attributed to the quality of its poorest or **weakest link**. Thus considering concrete as a chain, its **quality shall be determined by its weakest link** in the concrete chain. For achieving good quality of concrete, every link and stage shall be equally important and must be understood well.

To avoid failures of concrete, the engineers and supervisors/technicians should understand the behaviour of cement concrete and develop appropriate competencies in testing its quality. Properly constructed and supervised concrete structures serve functionally well for the designed life span of 50 to 100 years. It is, therefore, essential for engineers and technicians to undergo education and training in concrete construction for:

i. **Understanding the principles** of concrete technology and apply them during construction, supervision and testing;
ii. **Supervising and managing** concrete manufacture and construction;
iii. **Developing skills in selecting and testing** concrete ingredients and concrete for certain specific requirements;
iv. **Interpreting the test result** and accordingly modifying the concrete proportions or ingredients to suit the required quality and specifications; and
v. **Planning for quality and economy** in concrete construction.

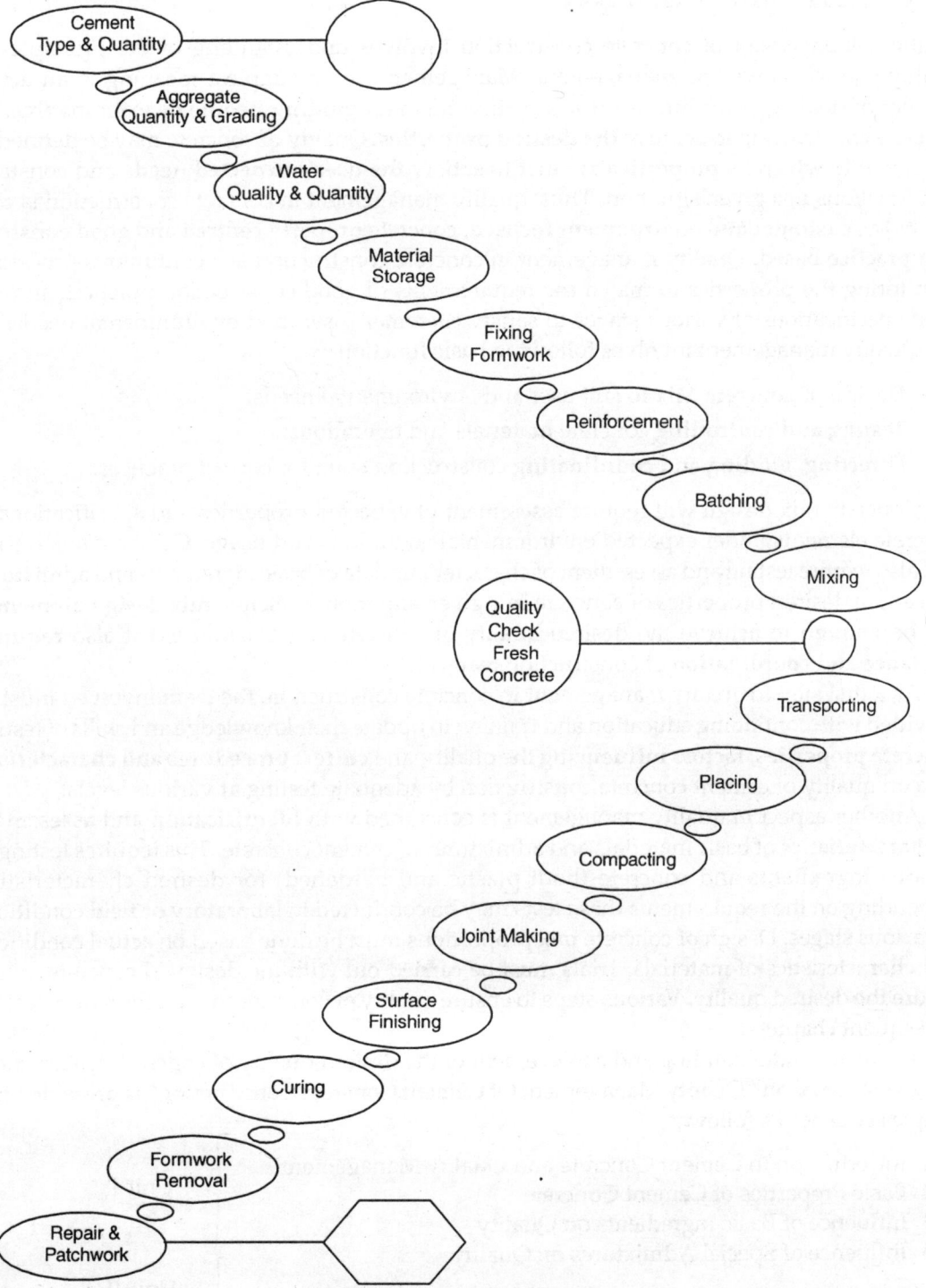

Fig. 1.1. Concrete Quality Chain

1.5 QUALITY MANAGEMENT

Quality Management of concrete construction involves understanding of the principles of quality control as well as management. Management, in its simplest meaning is an act or manner of dealing **planning, controlling, directing** and **guiding people** or team involved in concrete construction to **achieve the desired properties.** Quality of concrete may be defined as the extent to which its **properties are met to satisfy the user's expected needs** and construction functions in a given situation. Thus, quality management in concrete construction is considered as **customer and environment focused,** concrete **property centred** and **good construction practice** based. Quality management in concrete construction is a continuous process of monitoring the properties to match the requirements of good construction practices and desired specifications at various stages to satisfy customer (user) and environmental needs.

Quality management involves following basic functions:

- **Design of concrete Mix** to suit user and environmental needs;
- **Testing and controlling** concrete materials and operations;
- **Directing, guiding and coordinating** construction **team** for correct practices.

Concrete mix design will require assessment of expected properties and specifications of concrete elements under expected environmental conditions and usage. Concrete mix design will also require testing and assessment of characteristic data of basic ingredients and admixtures for certain desired properties of concrete in a given situation. Concrete **mix design alone may not be enough** to achieve the desired quality of concrete construction but it also **requires guidance and coordination** of construction team.

As a first step to quality management in concrete construction, the team involved must be provided with continuing education and training to update their knowledge and skills of testing **concrete properties, factors influencing** the quality, and **correct procedures and characteristic data** on quality of cement concrete construction by adequate testing at various levels.

Another aspect in quality management is concerned with **identification and assessment of characteristics** of basic materials and admixtures of cement concrete. This requires testing of various ingredients and concrete (both plastic and hardened) for desired characteristics. Depending on the requirements these tests may be conducted in laboratory or field conditions at various stages. Design of concrete mix proportions must be done based on actual conditions and characteristics of materials. Trials must be carried out with the designed concrete mix to ensure the desired quality. Various steps to ensure quality of concrete constructions are dealt in subsequent chapters.

To ensure understanding and achievement of the desired quality of concrete construction, the text material on "Quality Management Of Cement Concrete Construction" is presented in a simple sequence as follows:

1. Introduction to Cement Concrete and Quality Management
2. Basic Properties of Cement Concrete
3. Influence of Basic Ingredients on Quality
4. Influence of Special Admixtures on Quality

} Unit-I

5. Influence of Concreting Operations on Quality

} Unit-II

6 Basic Principles of Cement Concrete Mix Design
7 Management and Statistical Quality Control Procedures
} Unit-III

8 Cement Concrete Mix Design Procedures
9 Concrete Mix Quality Trials
10 Quality Assurance Under Special Conditions of Placement
11 Repair, Maintenance and Protection of Concrete Elements
12 Tests for Quality Assurance of Cement Concrete
} Unit-IV

For optimum learning, the reader must study the text in the same **sequence** as developed on the **basis of scientific principles of learning.** The learner must also carryout practice tasks listed at the end of each chapter. Working professionals/engineers must also conduct practical tests for quality assurance as explained briefly in chapter 12. These tests can be conducted in the laboratory or at work-site as necessary.

Quality management of cement concrete construction will go a long way in reducing the wastages and failures and also enhancing the service life of concrete structures. Concrete construction forms a major component of every national development project and hence quality assurance in concrete construction will facilitate and accelerate development through savings and enhancement of service life of structures.

1.6 SUMMARY

Considering modern era of globalization and highly competitive world market, quality plays key role in every field. **Quality plays** even more **critical role in construction** industry where **wrong doings cannot be reversed at later stage.** Modern construction industry all over the world hinges around "cement concrete construction" and hence quality management of cement concrete construction plays key role in every country's development.

Cement concrete construction got established since the **advent of portland cement in 1824** and **Duff Abram's law of W/C ratio in 1919.** With further development of many new materials and scientific principles of technology and statistical quality, the **quality management of cement concrete construction** has acquired **central stage** in construction industry.

Quality of concrete is determined by its weakest link. Concrete fails due to: **over stressing** and **structural cracks, improper construction and expansion joints,** reinforcement **corrosion, poor workmanship** and **inadequate cover, improper formwork, poor quality materials, inadequate compaction** and **curing.**

Quality management involves **design of concrete mix** for a **specific situation and purpose, testing materials** and **controlling concrete operations,** guiding and coordinating construction team for correct practices. Quality assurance of cement concrete requires thorough understanding of **principles of concrete technology and management.** Continuous education and training of construction engineers and engineering students is essential for quality management of cement concrete construction.

PRACTICE QUESTIONS

1.1　Define cement concrete in not more than 50 words.

1.2　Describe the basis of present development of cement concrete construction in about 100 words.

1.3　Describe the importance of cement concrete construction in development projects in about 100 words.

1.4　List factors leading to failure of concrete.

1.5　Explain the concept of concrete chain in relation to quality.

1.6　Explain the importance of quality management in construction.

CHAPTER 2

Basic Properties of Cement Concrete

LEARNING OBJECTIVES

The learner **understands** the basic **properties of cement concrete** and will be able to:

- **List main properties** of cement concrete both in **plastic and hardened states;**
- **Explain importance** of **workability** in quality management of cement concrete construction;
- **Define** the terms of **workability, segregation,** and **bleeding** in concrete;
- **Explain** the steps of **controlling bleeding** in cement concrete;
- **List five methods** of **measuring workability** with limitation of each;
- **Explain slump test** for measuring workability;
- **Explain compacting factor** test for cement concrete;
- **Write short notes** on **Vee-Bee** degrees and flow in cement concrete;
- **List type of strengths** in cement concrete;
- **Explain inter-relationship** of compressive strength with flexural and bond strengths of cement concrete;
- **Explain durability** of cement concrete;
- **Explain resistance** of cement concrete **to chemical attack;** .
- **Explain importance of impermeability** of cement concrete on its service life and quality;
- **Explain importance of dimensional changes** in cement concrete;
- **Explain** the concept of **elastic modulus** in cement concrete;
- **Differentiate** between **shrinkage and creep** in cement concrete;
- **Describe unit weight** of cement concrete;
- **Explain the importance of expansion** in cement concrete.

2.1 INTRODUCTION

Every construction project manager has endeavour to achieve the **best quality** construction at the **most economical cost.** Cement concrete construction forms a major part in most of the construction projects. In their endeavour to achieve quality and economy, all engineers, supervisors, contractors and technicians must understand the properties of cement concrete. Understanding of properties and behaviours exhibited by cement concrete at various stages will facilitate suitable control on quality of construction. Cement concrete has two distinct stages of construction viz. **fresh** (plastic) and **hardened** (solid). For achievement and control of quality of cement concrete construction, it is, therefore, essential to clearly understand the properties and behaviours in both stages and their inter-relationship.

2.2 PROPERTIES IN FRESH STATE

The properties during **fresh or plastic state** are very **important** as it is this stage that the control on the concrete can be exercised to obtain the desired behaviour even in the hardened concrete. The **properties cannot be modified** after hardening of concrete and necessary adjustments shall have to be done in fresh concrete only. Properties of fresh concrete during **plastic state** are thus very **important for the control of its ultimate quality** in hardened state. These properties in fresh state are:

- Good **Workability**;
- Freedom from **Segregation**; and
- Freedom from **Bleeding** and **Harshness.**

2.2.1 Good Workability

The **strength** of concrete of given mix proportions in its hardened state is seriously affected by the degree of its **compaction during plastic state.** It is, therefore, important for fresh concrete to have suitable consistency for **full compaction.** Due to proper consistency of fresh concrete, it is possible to transport, place, compact and finish concrete easily without causing segregation. This fresh concrete mix which can be transported, placed and **compacted easily without segregation** is said to be **workable concrete mix.** As found by research, the strength of hardened concrete directly depends on its density. To obtain **optimum density**, it is essential to **compact concrete fully** to drive away all entrapped air from fresh concrete. For this full compaction to take place easily, the concrete mix should be of such consistency that all particles can easily move closer with the available external effort to the remotest corner of formwork. Thus in simple terms, **workability is defined as the ease with which fresh concrete can be placed and compacted fully without segregation.**

The **workability** as merely ease of placement and resistance to segregation is too simple and will result in different workabilities for the same concrete mix depending on the type of structure, placing conditions and method of compaction. Thus, the workability should be considered as an important **physical property of fresh concrete** alone without reference to method of compaction and particular type of construction. For achieving compaction either by hand ramming or by vibration, the process essentially consists of the **elimination of entrapped air** from the concrete mix until it has reached as close a configuration as is possible in the given mix. In this process, the work is done to overcome the friction between the individual particles in the concrete mix (called **internal friction**) and also between the concrete and the surface of the mould or reinforcement (called **external surface friction**). Some work is also done (wasted) in vibrating the mould or already consolidated portion of concrete. Since only the **internal friction is an intrinsic characteristic** of the mix, workability can be scientifically defined as the **amount of useful internal work necessary to produce full compaction.**

It is not only important to study workability in achieving full compaction of fresh concrete but also very vital as far as finished hardened concrete is concerned. The concrete must have a workability such that **compaction to maximum density** is possible with reasonable amount of work under given conditions. From relationships between the degree of compaction (as measured by density ratio) and the resulting strength (as measured by strength ratio), the importance of compaction for finished concrete is quite evident. The degree of compaction is expressed as **density ratio**, i.e. a ratio of the **actual density** of given concrete to the **density of** the same mix if **fully compacted.** Similarly, the strength ratio is considered as the ratio of the strength of the concrete as actually (partially) compacted to the strength of the same mix when fully compacted. From the relationship, it is evident that even a small reduction in density due to presence of voids results into very large reduction in the strength. **Five percent of voids (density ratio 0.95)** can result into **lowering of strength by as much as 30%** (strength ratio about 0.70) and even **2 percent voids** can result in drop of strength of more than **10 percent.** Voids in concrete are either bubbles of entrapped air or empty spaces left after evaporation of excess water. The volume of these voids depend on water-cement ratio of the concrete mix. The air bubbles are governed by the grading of the fine particles in the mix and are more easily expelled from a wetter mix than from a dry one. It follows, therefore, that for any given method of compaction there may be a **optimum water content** of the mix at which the sum of the volumes of air bubbles and water space will be a minimum. At this optimum water content the highest density ratio of the concrete would be obtained. Optimum water may vary for different methods of compaction.

2.2.2 Freedom from Segregation

In general terms it is implied that workable concrete should **not segregate** i.e. it should be **cohesive.** However, strictly speaking the absence of a tendency to segregate is not included in the definition of a workable concrete mix. For full compaction, it is essential that there is no appreciable segregation. **Segregation** can be defined as **separation of the constituents** (specially coarse aggregate) of a heterogeneous mixture so as to make their distribution non-uniform. In cement concrete, the difference in the size of the particles and specific gravity of the ingredients are the primary causes of segregation. The aggregate used in concrete has coarser and finer particles in various proportions. When the proportion of **coarse particles** is comparatively **large than fine particles,** the coarse particles tend to segregate from the rest of the mortar. Apart from improper proportioning, the incorrect handling of mixed concrete during discharge, transportation, and placement as well as **overcompaction** may also cause the segregation. Segregation can be checked by suitable grading and by careful handling.

The segregation occurs in two ways. In the first, the coarser particles tend to separate out by travelling farther along the slope or by setting more than the finer particles. The second form of segregation occurs, particularly in wet mixes, by separation of grout (cement plus water) from the mix. For very wet and lean mixes, second type of segregation may occur. By proper grading of aggregates, the segregation can be controlled to a great extent. Another way of controlling segregation in concrete is by proper handling and placing concrete. The concrete should be placed as near its final position as possible to minimize the danger of segregation during motion of concrete ingredients through considerable height of chutes. For laying concrete through considerable heights or chutes, specially with changing direction, cohesive concrete mixes should be used for avoiding segregation. Adoption of proper procedures in handling of concrete shall help in reduction of segregation in concrete mix.

The separation of mortar from the coarse aggregate keeps the void unfilled and results into honey combing which causes decrease in density and ultimately reduction in strength of concrete. **Segregation can be prevented** by:

(i) Ensuring a certain **minimum proportion of finer** material in concrete mix;
(ii) **Properly grading** the aggregate;
(iii) **Controlling water** content in concrete mix; and
(iv) Adopting **correct handling** procedures.

Air entraining in concrete reduces the danger of segregation. Segregation is difficult to measure quantitatively, but is easily detected by visual inspection at site. A good information of cohesion of the mix is obtained by the flow test. Tendency of segregation in the mix can be observed by visual observation of coarse aggregate distribution in the concrete mix.

2.2.3 Freedom from Bleeding and Harshness

Bleeding is a form of segregation in which some of the water in the mix tends to rise to the surface of the freshly placed concrete. This is caused by the inability of the solid particles of the mix to **hold all of the mixing water** during their downward settlement in the process of compaction. This happens when the quantity of water in the mix is more than is necessary for the cement paste to lubricate the aggregate particles. **Over compaction** also causes the coarser materials to settle down forcing the water to rise up and appear at the surface along with some cement. **Bleeding** could also take place due to the coarsely ground cement or due to less fines in fine aggregate. The separation of water from cement and sand or separation of cement paste from the mortar allowing the water or cement paste to appear at the surface is called **bleeding**.

As a result of bleeding the top surface of concrete layer becomes very wet. If subsequent concrete layer is laid on this excessive wet layer of concrete, the surface water of bleeding is trapped in the concrete which results into **weak, porous** and **non-durable** concrete.

Finishing of top surface alongwith mixing of bleeding water causes a very **weak wearing surface.** This can be avoided by delaying finishing operation till evaporation of bleeding water. If evaporation from the surface is faster than the bleeding rate, it may cause plastic **shrinkage cracks.**

Sometimes, the rising water gets trapped on the underside of coarse aggregate particles or reinforcement, thus creating zones of poor bond. The water so entrapped leaves behind capillaries, resulting into **increased permeability.** Appreciable bleeding also causes increased danger of frost damage specially in thin slabs.

Bleeding need not necessarily be harmful if it is undisturbed and the surface water evaporates reducing the effective water/cement ratio of the mix. On the other hand, if the rising water carries with it a considerable amount of the finer cement particles, **laitance** will be formed at the surface. Such a **laitance** at the top surface of a slab will result into **porous and dusty surface.** If such a layer of laitance is formed between the consecutive layers of concrete, it will result in inadequate bond between the layers of earlier and freshly laid concrete. For this reason **laitance** should always be **removed** by brushing and washing the old surface.

The tendency of bleeding depends mainly on the properties of cement. Fineness of cement reduces the bleeding. Cements having high alkali content, and a high proportion of C_3A or calcium chloride shall have lesser tendency of bleeding. **Rich mixes are less prone to bleeding** compared to lean mixes. Pozzolana and flyash addition to concrete reduces its bleeding tendency. Air entrainment also reduces bleeding so that finishing takes place faster.

It may be summarised that bleeding can be prevented by:

  i. **Controlling the water** content;
 ii. Using **finer grading of fine** aggregate;
iii. Using **finely ground cement**;
 iv. Using **cements with higher alkali** and C_3A content;
  v. **Controlling compaction**;
 vi. **Delaying finishing** operation; and
vii. Using **rich cohesive mixes.**

Harshness can be defined as the resistance to surface finishing of fresh concrete. Harshness is caused when the mix proportions contain too less fines along with too less mixing water. The quantity of fines is not enough to keep the mix cohesive and the quantity of water is not enough to lubricate or wet the surface of the aggregate particles to provide the desired workability. The workability of the concrete is very poor and there is no mortar or cement slurry available at the surface for finishing operation. Thus for the concrete mix to have good surface finishing qualities, it should be **cohesive, workable** and contain **just sufficient quantity of mixing water.** The fresh concrete mix should provide enough creamy mortar or paste at the surface during floating and finishing operation without causing bleeding or segregation. Harshness is generally observed in lean and dry mixes.

2.3 MEASUREMENT OF WORKABILITY

There is no unique test which can directly measure the workability. However, there are numerous methods of determining certain physical quantities which try to correlate workability to some extent. None of these methods is fully satisfactory although these provide useful guidance regarding variation in workability within certain limits. Since the **workability of fresh concrete plays important role in controlling quality of hardened concrete**, its measurement and control is of great significance.

The phenomenon of workability is associated with four concepts:

  i. **Ease of flow;**
 ii. **Cohesiveness** - Movement without tendency to segregate;
iii. **Prevention of bleeding;** and
 iv. **Prevention of harshness.**

The concrete mix is said to be workable if it flows easily without segregation, or causing any bleeding and harshness. Various methods of measurement of workability takes into account some of these concepts. Following are the methods of measuring workability:

  i. **Slump Test**

    ii. **Compacting Factor Test**
   iii. **Vee-Bee Test**
    iv. **Flow Test**
     v. **Remoulding Test**
    vi. **Ball Penetration Test** (Kelly Ball Test)

In India the first four methods are used, out of which first two are the most common.

2.3.1 Slump Test

This test is used extensively at work sites all over the world. Although the slump test for workability of concrete is easy to carry out, but it does not measure the workability of concrete

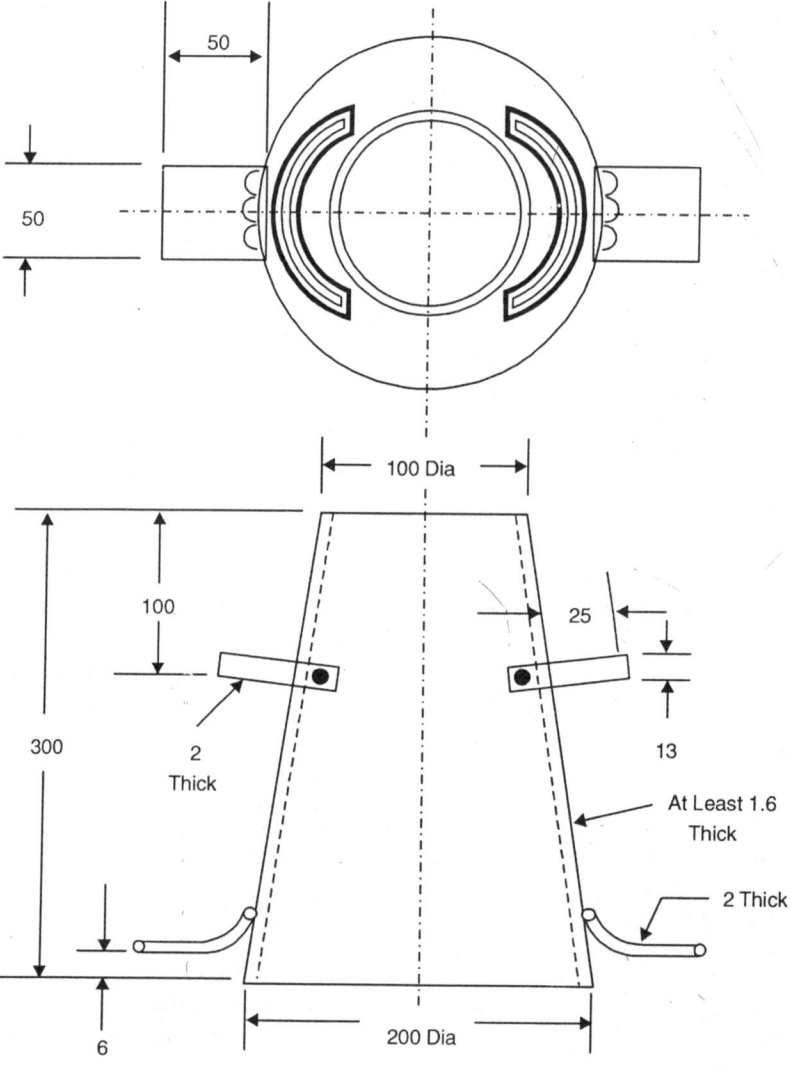

Fig. 2.1. Typical Mould for Slump Test (All Dimensions are in mm)

directly. This method is quite useful in detecting variations in water content and uniformity of a mix of given proportions.

The slump test is specified in IS:1199-1959 (Fig. 2.1). The slump test apparatus consists of a truncated cone having 100 mm top diameter, 200 mm bottom diameter, and 300 mm height. The cone is filled in four layers, each approximately one-quarter of the height of the mould. Each layer shall be tamped uniformly with 25 strokes of bullet pointed 16 mm diameter and 600 mm long tamping rod. Top surface of moulded concrete cone is struck off with trowel and finished quickly. The mould is then lifted gently and vertically. The laterally unsupported concrete settles under its own weight and hence the top surface subsides vertically. The vertical subsistence from the original top surface is called **slump** and can be measured with the help of scale in millimetre. In other words, **the decrease in the height of the highest top surface of the slumped concrete, is called slump.** For reducing the influence of the mould friction, the inside surface of the mould must be moistened or oiled before the beginning of every observation.

The concrete may slump as **true slump, shear slump,** or **collapse slump** (Fig. 2.2). Any slump specimen which collapses or shears of laterally gives incorrect result and if this occurs, the test shall be repeated with another sample from the same mix to confirm the tendency of the mix. If the observation repeats, the fact is recorded in the test results.

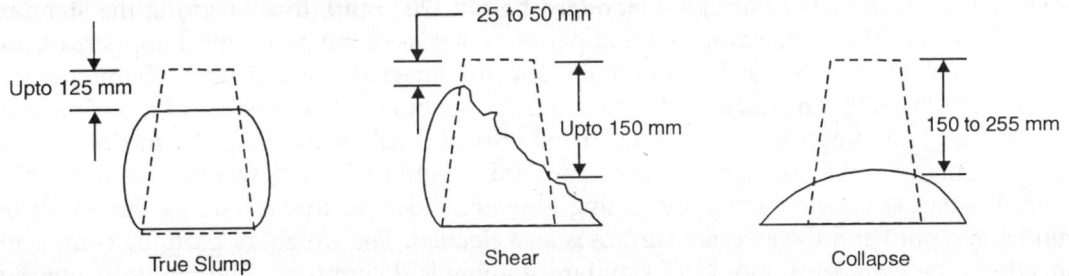

Fig. 2.2. Slump: True, Shear and Collapse

The test of cohesiveness and workability of the mix can also be obtained after completion of slump measurement. The steel plate base, of the moulded slump concrete cone after the slump test, is tapped gently with the tamping rod. The tendency of concrete cone to slump further can be observed. The concrete having appreciable slump may gradually slump further if it is cohesive and well-proportioned and may fall apart if it is poorly proportioned.

Concrete mixes of dry and stiff consistency have generally zero slump, and hence in such mixes the variation of workability cannot be detected in mixes of variable workability by using slump test. While in rich mixes of medium consistency, the slump test is quite sensitive to variation in workability and can be used for control of workability of the mix. In case of lean mixes with tendency of harshness, the slump may be true, collapse or shear pattern and indicate a vast variation even in the mix of the same workability. In some cases, the concrete mixes of different workabilities may show similar slumps. This makes the slump test unreliable in such cases. Thus there is **no unique relationship of slump and workability.**

Despite of these limitations, the slump test is the **most commonly adopted** test on the construction site to check variation in the quality and quantity of materials fed into the mixer. An increase in slump may indicate the increase in moisture content of aggregate or change in grading of aggregate (such as reduction in sand content or coarse grading). Thus sudden rise or fall in slump values, provide immediate warning to mixer operator to remedy the situation.

The procedure of conducting these tests may be studied in chapter 12 of this textbook.

2.3.2 Compacting Factor Test

Since the slump test has no direct relation with the internal energy required to achieve full compaction of concrete, other methods are developed to relate the external work applied to the internal work required for full compaction. The most reliable test developed so far uses the inverse approach i.e. the degree of compaction achieved by a standard amount of work is determined. The **compacting factor** is based on the principle that the work applied includes the **work against the internal friction** and surface friction. The work done against the surface friction can be reduced to a minimum by proper lubrication of the internal surface of the mould although, this surface friction also depends on the workability of the mix. The compacting factor test is a very useful measurement of workability (Fig. 2.3) to detect the variation specially in dry mixes.

The **degree of compaction**, called the **compacting factor**, is measured by the **density ratio** i.e. **the ratio of the density actually achieved in the partially compacted concrete for standard amount of work to the density of the same concrete fully compacted.** In the test, the concrete is allowed to fall through a **standard height (203 mm)**, thus ensuring the standard amount of work. The compacting factor apparatus consists of the two conical hoppers (**A** and **B**) mounted above a cylindrical mould (**C**) **of 152 mm** internal diameter and 305 mm internal height (IS: 1199-1959). The distance between the bottom of lower hopper B and top of cylinder C is 203 mm. The compaction of concrete in cylindrical mould is produced by the destruction of kinetic energy of the falling concrete of standard weight from the lower hopper B to Cylinder C. The excess concrete is cut by sliding two steel floats or trowels across the top of the cylindrical mould and the external surface is also cleaned. The weight of partially compacted concrete in the cylindrical mould of standard volume is determined. The mould is emptied and then **refilled in four layers** of about 75 mm with the same concrete to achieve full compaction by heavy ramming or suitable vibration. Weight of this fully compacted concrete in the same standard cylindrical mould is determined.

The **ratio of weight of partially compacted concrete** by standard work and **weight of fully compacted concrete** of the same standard volume in the cylindrical mould represents the **density ratio** or **degree of compaction** and hence called **compacting factor.**

The compacting factor test is sufficiently sensitive to enable to determine the small differences in workability arising on account of the initial process in hydration of cement. Each test, therefore, should be carried out at a constant time interval after the completion of mixing to obtain comparable results. Generally the concrete from top hopper should be released after 2 minutes after completion of mixing of concrete.

The dry mixes will have a low degree of compaction and hence low value of compacting factor. **Low value of compacting factor indicates low degree of workability** and low slump too. The wet mixes have a high value of compacting factor and consequently represent high degree of workability and high slump.

The assumption that all mixes with the same compacting factor require the same amount of useful work is not always true. Similarly, the assumption that the wasted work represents a constant proportion of the total work done regardless of the characteristics of the mix is also not correct. Even with all these assumptions, the compacting factor test provides a good measure of workability.

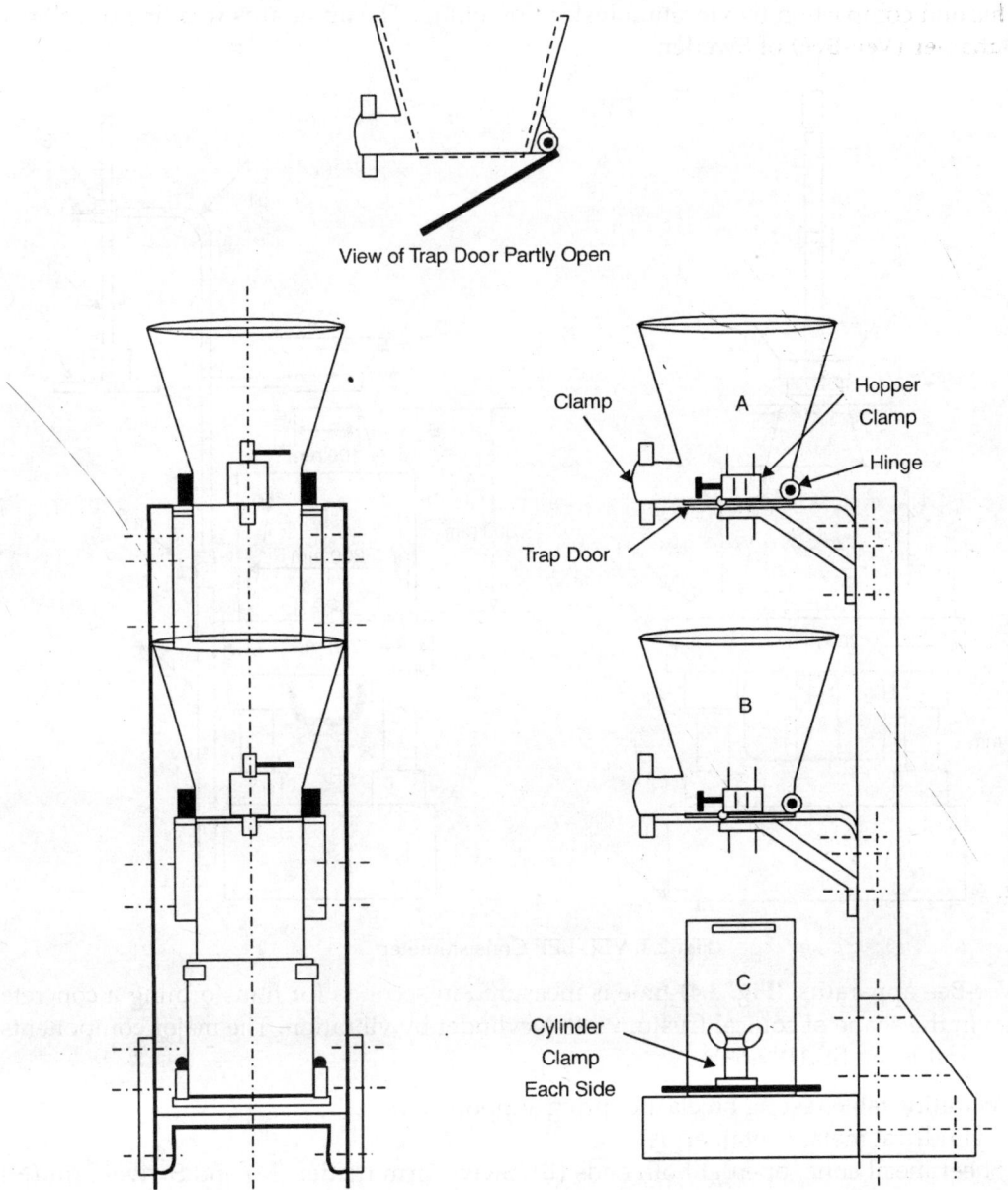

View of Trap Door Partly Open

Fig. 2.3. Compacting Factor Apparatus

An automatic compacting factor test apparatus has been developed in which the cylinder is supported by a spring balance and can be calibrated for a given mix to read workability directly. The apparatus can also be calibrated to indicate the excess or deficiency of water in litres per batch or per cubic metre of concrete.

2.3.3 Vee-Bee Test

This test is developed using the principle of remoulding test with certain modifications in the apparatus and compaction by vibration instead of jolting. The apparatus was first developed by **V. Bahamer (Vee-Bee)** of Sweden.

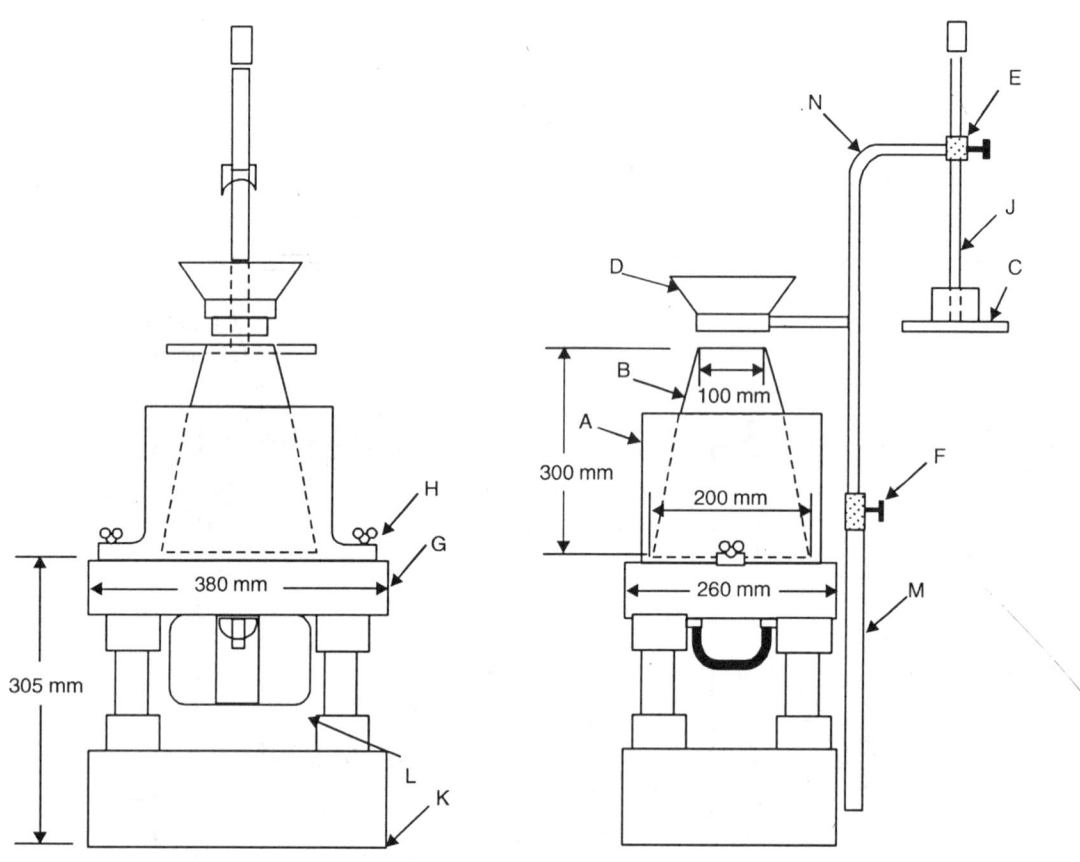

Fig. 2.4. VEE-BEE Consistometer

In Vee-Bee apparatus, **(Fig. 2.4)** time is measured in seconds for transforming a concrete specimen in the shape of conical frustum into a cylinder by vibration. The major components of the apparatus are (IS:1199-59):

a) Vibrating table resting on elastic spring supports (G)
b) Cylindrical metal container (A)
c) Sheet metal cone, open at both ends (B), Swivel arm holder (M) and Swivel arm (N) with funnel (D), glass disc (C) and guide-sleeve (E), and
d) Prime mover (electric motor)

Vibrating table (G) is 380 mm long and 260 mm wide and is supported on the rubber shock absorbers at a height of about 305 mm above floor level. The table is mounted on the base and is vibrated with a electrically operated vibrometer working on 220 volts three phase, 50 cycles alternating current. Sheet metal cone (B) is placed inside the metal cylindrical container (A) which is fixed on to the vibrating table with two holding nuts. The sheet metal cone (B) is 300 mm high and its diameter is 200 mm at the bottom and 100 mm at the top. The glass disc rider (C) is screwed at the bottom of the graduated rod (J). The rod scale (J) is used to measure the slump of the concrete cone and the volume of concrete after vibration into the cylindrical container. A standard rod of 20 mm diameter and 500 mm length is used to fill the slump cone by tamping in 4 layers.

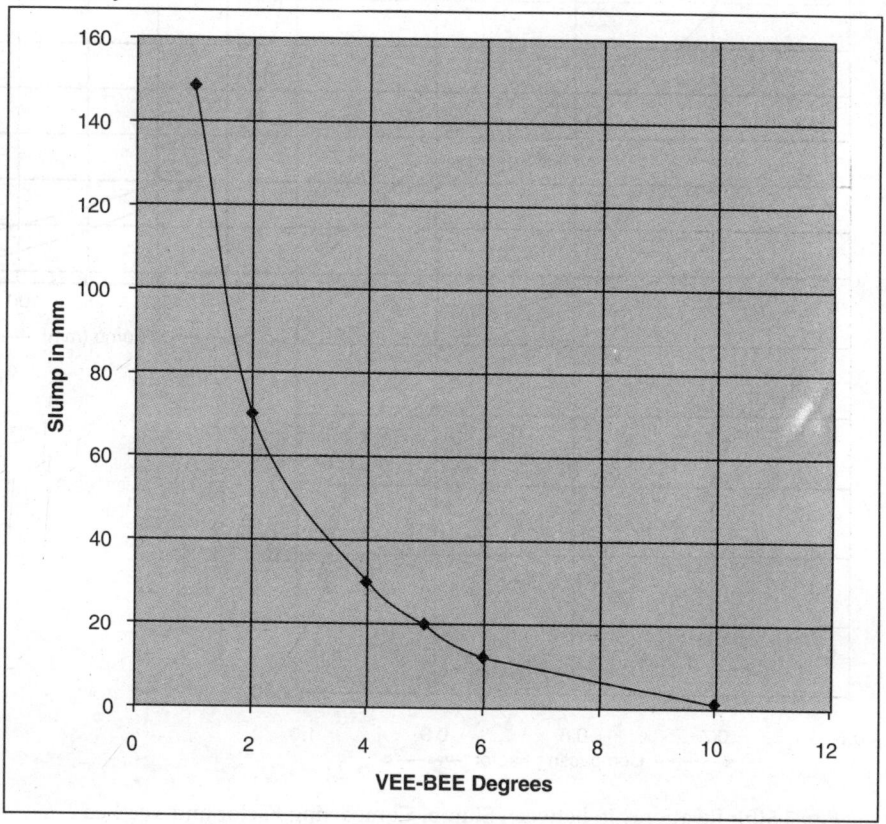

Fig. 2.5(a). Relationship between Slump in mm and Vee-Bee Degree

The glass disc rider is placed on the top of the slump cone in the cylindrical pot and the position of the disc rider adjusted and noted before filling the cone. The slump cone placed in the cylindrical container, is filled in four layers tamping with the standard iron rod (25 times each layer). The cone shall be lifted vertically and the slump noted on the graduated rod by bringing the glass disc rider in contact with the top of the concrete cone. The electric vibrator shall be switched on alongwith starting of the stop watch. The vibration shall be continued until the whole concrete surface makes full contact with the glass disc rider (representing completion of transformation and compaction of cylindrical mould from conical concrete mould).

The time taken for this operation is noted and recorded in seconds. This time in seconds is expressed as Vee-Bee degrees. The curve in Fig. 2.5 shows the relationship in slump and Vee-Bee degrees in seconds (SP23-1982). The consistency scale and Vee-Bee degree are also given in the Table 2.1.

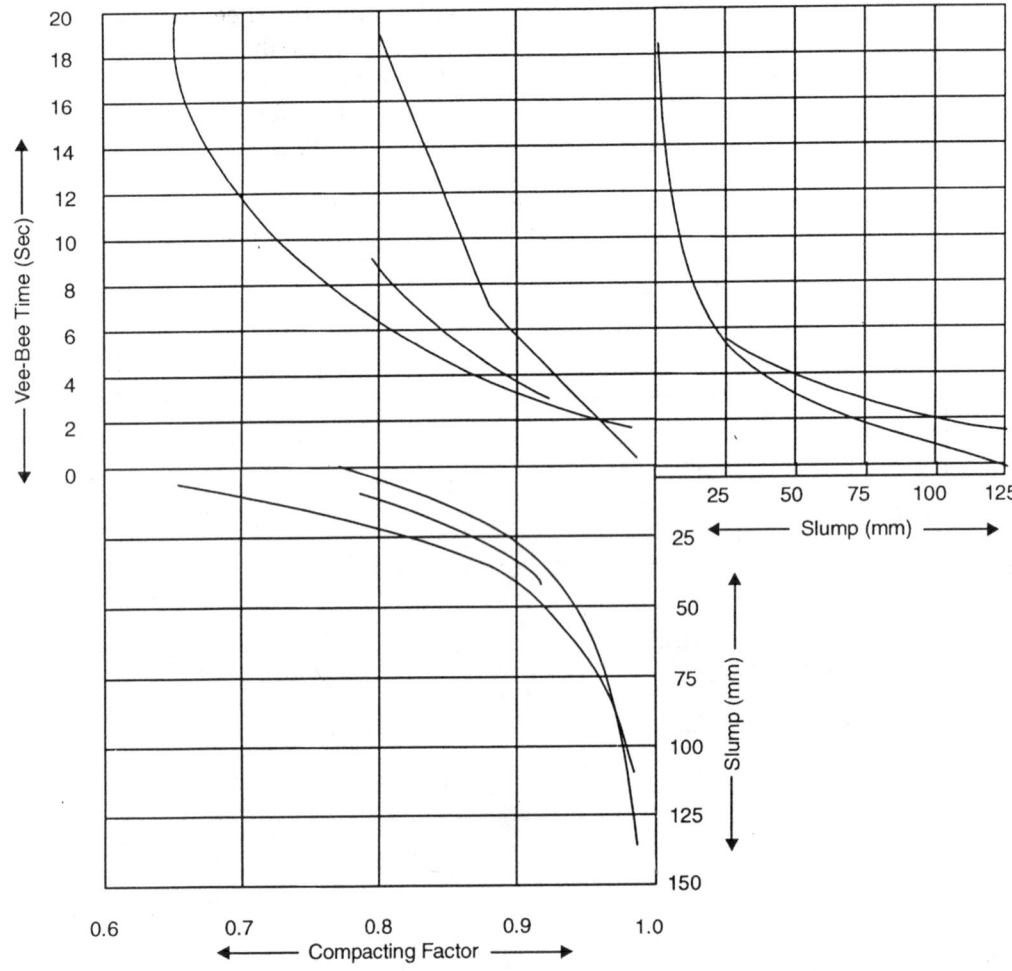

Fig. 2.5(b). Relationship between Slump, Compacting Factor and Vee-Bee Time
for Concrete of Different Aggregate-Cement Ratios

The visual observation of glass rider for fixing the end point of compaction is some times the source of error. This may be over-come by fixing automatic device for recording time and the movement of the plate.

Compaction is achieved by using a vibrating table with an eccentric weight rotating at 3000 revolutions per minute and a maximum acceleration of 3 g to 4 g. It is assumed that the input of energy required for compaction is a measure of workability of the mix, and this is expressed in terms of time in seconds, i.e. the time required to remould the concrete cone specimen to cylinder. Some times a correction, for the change in volume of concrete from V_0 before to V_1 after vibration may be applied, by multiplying the time by V_0 / V_1.

Vee-Bee test is a good laboratory test, specially for very dry mixes. This is better than compacting factor test in which there is possibility of error due to tendency of some dry mixes to stick in the hoppers. The Vee-Bee test represents more closer treatment of concrete in relation to the method of placing actual concrete at work site.

Table 2.1 Consistency and Vee-Bee Degrees (Ref. IS: 1199-1959)

Consistency	Vee-Bee Degrees (Seconds)	Characteristics
Moist-Earth	40 to 20	Particles of C.A. in concrete are adhesive, but concrete does not clot. Risk of segregation.
Very Dry	20 to 10	Concrete has the consistency of very stiff porridge, forms a stiff mound when dumped and barely tends to shake or roll itself to form an almost horizontal surface when conveyed for a long time.
Dry	10 to 5	Concrete has the consistency of stiff porridge, forms a mound when dumped, and shakes or rolls itself to form a horizontal surface when conveyed for a long time.
Plastic	5 to 3	Concrete can be shaped into a ball between the palms of the hands, and adheres to the skin.
Semi Fluid	3 to 1	Concrete cannot be rolled into a ball between the palms of hands, but spreads out slowly and without affecting the cohesion of the constituents so that no segregation occurs.
Fluid	1 to below 1	Concrete spreads out rapidly and segregation takes place.

2.3.4 Flow Test

This laboratory test gives indication of the consistency of concrete and its tendency to segregation by measuring the spread of concrete mould subjected to jolting. The flow test is of great value for segregation and it also provides good assessment of consistency of stiff, rich, and cohesive mixes. The test is useful to determine the fluidity of concrete, where the nominal size of aggregate does not exceed 38 mm. The test is covered in IS: 1199-1959. The apparatus consists essentially of a smooth brass top table of 762 mm diameter and shall be mounted on and bolted to a concrete base having a height of 400 to 500 mm and mass not less than 140 Kg. The table shall be so mounted that it can be jolted by a drop of 12.5 mm. The mould shall be made of a smooth metal casting in the form of the frustum of a cone with 250 mm bottom diameter, 170 mm top diameter, and 120 mm height. Top and bottom shall be open and at right angles to the axis of cone (Fig. 2.6).

The mould is placed at the centre of the table top and filled with concrete in two layers and each layer rodded by a standard tamping rod (16 mm diameter and 600 mm length with bullet pointed end) giving 25 strokes each. After compacting the top layer by rodding, the top surface of the concrete is struck off with trowel and all excess material removed and outside surface of the table cleaned. The mould is removed by lifting gently and vertically. The table is given 15 jolts in about 15 seconds by a wheel operating an actuating eccentric cam of 12.5 mm drop. The diameter of the spread of concrete shall be measured at 6 to 8 points and average diameter of spread noted.

The flow of concrete is defined as the percentage increase in the average diameter of the spread concrete (D mm) over the original diameter of the base (250 mm), i.e.

Flow = (Spread diameter D in mm – 250 mm) × 100/250 mm.

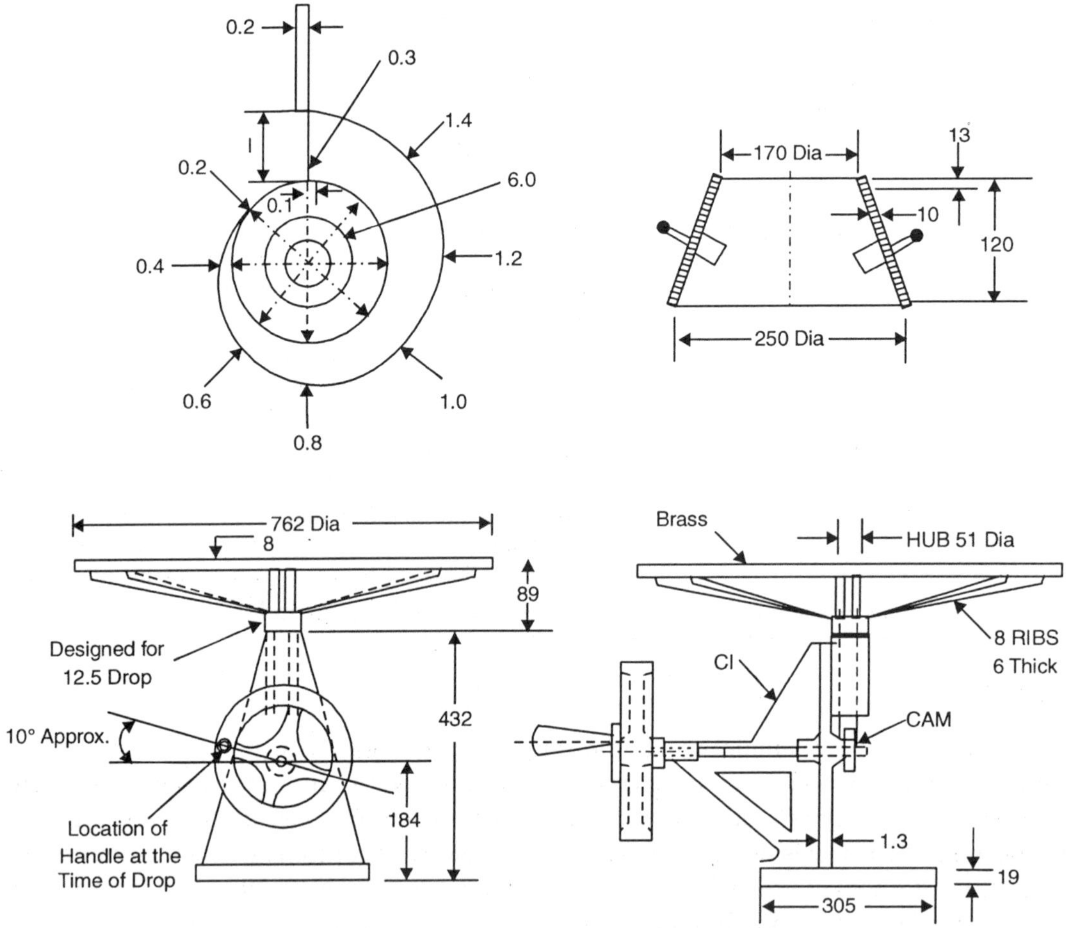

Fig. 2.6. Flow Table Apparatus (All Dimensions are in mm)

Values from 0 to 150 percent can be obtained. The jolting applied during the test induces the tendency of segregation, and if the mix is not cohesive the larger particles of aggregate will separate out and move toward the edge of the table. Another form of segregation may occur in sloppy mix where in cement paste tends to run away from the centre of table leaving the coarse material behind.

It may be noted that the flow-test does not measure the workability, as concrete having the same flow may differ considerably in their workability and also concrete with the same workability may differ in flow pattern. Flow test is specially useful in assessing the **tendency of segregation.**

2.4   PROPERTIES IN HARDENED STATE

Concrete in hardened state has following important properties:

- Strength
- Durability
- Impermeability
- Dimensional Changes
- Unit Weight

2.4.1 Strength of Concrete

The strength of concrete is defined as resistance to its failure against a system of loading. The strength of concrete is measured in various ways depending on loading pattern such as:

i. Compressive Strength
ii. Flexure Strength
iii. Bond Strength
iv. Resistance to Abrasion

All these strengths are influenced by certain specific factors and are also interdependent to some extent. The **cube crushing strength** is considered most important strength for identification of the concrete mix grade and quality.

i. Compressive Strength

It is the resistance of the concrete to **crushing.** Cement concrete has substantial compressive strength and forms a very important property for structural concrete. As per Indian Standards the crushing strength is measured as axial **load per unit area** at failure **on cubes of 150 mm** size at the **age of 28 days** and is specified in **N/mm^2.** This strength is used to specify the grade of concrete (such as M10, M15, M20, etc).

Using portland cements, concrete can be manufactured with compressive strength ranging from **10 N/mm^2** to **40 N/mm^2.** By using specially selected cement, admixtures, and other materials, concrete of compressive strength **60 N/mm^2** or even more can be obtained. Sometimes very lean concrete needed for mass concreting and base of foundations may have compressive strengths as low as 4 N/mm^2. In building construction concrete having compressive strength of **15 N/mm^2** (M15 grade) or **20N/mm^2** (M20 grade) is normally specified. For roads and bridges, structural concrete of compressive strengths ranging from 20 to 40 N/mm^2 (M20 to M40 grades) is generally specified. Higher grades of concrete (M30 to M60) are specially suitable for pre-stressed structures while concrete grades of M20 to M25 are commonly used for RCC structures.

Compressive strength of concrete is important because other strengths such as flexure, bond and abrasion resistance improve with increase in its compressive strength and it is also comparatively easier to measure compressive strength of concrete. From the quality control point of view the compressive strength of concrete is accepted as the main criterion for the acceptance of the quality of structural concrete. The ultimate failure under the action of a uniaxial compression is either a tensile failure of cement crystals or of bond in a direction perpendicular to the applied load, or a collapse caused by the development of inclined shear planes.

The compressive strength of cement concrete mainly depends on:

-   the **type, quality** and **quantity** of cement;
-   the **type, size, shape, strength** and **grading** of aggregates;
-   the **water-cement** ratio;
-   the degree of **workability** and **compaction**;
-   the **type, quality** and **age** of **curing**;
    -   the **shape** and **size** of specimen; and
    -   **rate of loading** and condition of the specimen at the time of loading.

ii. Flexural Strength

The resistance of concrete offered to tension under flexural loading (bending) is called its flexural strength. Cement concrete is comparatively **weak in tension.** Generally the flexural tensile strength of cement concrete is about **one-eighth** to **one-tenth** of its compressive strength. It is difficult to measure direct tensile strength and hence flexural tensile strength is commonly measured. The flexural tensile strength is slightly higher than the direct tensile strength because in the latter case there is no surrounding adjacent under-stressed material layer to stop the fracture. In fact the strength of the concrete, is to be considered as the strength of its weakest element. Hence the flexural strength is considered as the resistance to cracking under flexural loading. This flexural stress in concrete at the time of cracking is also called **modulus of rupture.** It is observed that there exists a relationship between tensile stress at cracking in flexure and compressive stress at cracking in direct compression specimen. This ratio is generally equal to Poisson's ratio of concrete and may vary from 0.11 to 0.21.

For optimum utilization of cement concrete in RCC structures, the **concrete is designed to resist most of compressive forces,** while **steel is designed to resist most of tensile force** under flexural loading. For simplicity, **the tensile resistance of cement concrete is neglected,** being small. Tensile strength of concrete becomes important in the design of concrete roads, airfield pavements, and concrete pipes.

Flexural tensile strength of concrete is affected greatly by the shape and texture of the aggregates. Other factors which generally influence the compressive strength of concrete also affect the flexural strength. The concrete with angular crushed aggregate offer more flexural strength in comparison to rounded smooth gravel due to better mechanical bonds in the former. The **ratio of tensile strength to its compressive strength decreases with age** of concrete and also with increase in its grade. According to IS: 456-2000, the flexural strength (f_{cr}) in **N/mm²** can be approximately determined from the relation to its characteristic compressive strength (f_{ck}) in **N/mm²** given as

$$f_{cr} = 0.7 \sqrt{f_{ck}} \text{ N/mm}^2 \qquad \text{(Ref: § 6.2.2 of IS 456-2000)}$$

iii. Bond Strength

The structural concrete is used, along with steel in reinforced cement concrete elements. The strength of bond between the two materials plays an important role for the strength of RCC elements. Bond arises primarily from friction and adhesion between concrete and steel, and may also be affected by the shrinkage of concrete relative to steel. The property of **adhesion (grip) between steel and concrete is called Bond.** Bond also involves the mechanical properties of the concrete.

In general terms, bond is related to the quality of the concrete, and the compressive strength of concrete. The bond strength is approximately proportional to compressive strength upto certain grade of concrete. For higher grades of concrete, the proportionate increase in bond strength becomes progressively smaller in comparison to increase in compressive strength. It may be stated that **better the compressive strength of concrete, better will be the bond** strength. The property of bond strength of concrete is very useful in the design of reinforced cement concrete and other structures.

iv. Resistance to Abrasion

The resistance of concrete to abrasion can be determined in several ways by simulating a mode of abrasion which occurs in practice. The simulation of the real condition of wear is not easy and it is also difficult to establish the relation of test results and the actual resistance of concrete to the type of wear. In all tests the **loss of weight of the specimen due to wear is** considered as a **measure of abrasion.**

In one of the method of the steel ball abrasion test, a load is applied to rotating head separated from the specimen by steel balls. The eroded material is removed by flowing water. Another method of dressing wheel test uses a drill press modifies to apply a load to rotating dressing wheels in contact with the specimen. Silicon carbide is used as an abrasive material. The dressing wheel and the steel ball tests estimate to relate the resistance of concrete to wheeled and heavy foot traffic in actual situation.

The **resistance to abrasion** of concrete is generally found to be **proportional to its compressive strength** and it is normally assumed that the concrete with high compressive strength has better resistance to abrasion. The resistance to abrasion is also influenced by the **type and shape of aggregate** in addition to the factors which are responsible for affecting the compressive strength of concrete. Softer aggregate results in greater wear in concrete in case of steel ball, and dressing wheel tests. While harder aggregate tends to splinter in the shot-blast type of test and causes a greater loss of weight representing higher wear.

Actual concrete which bleeds very little has a stronger surface layer and therefore, offers more resistance to abrasion. For concrete to develop high resistance to abrasion, proper moist curing is essential. The resistance to abrasion also improves with the use of medium, coarse and hard aggregate in concrete.

2.4.2 Durability

Durability of the concrete is defined as the **resistance to deterioration and disintegration against weathering** agencies in the environment and the conditions for which it has been designed to withstand over a period of time. The absence of durability may be as a result of either the external environment of concrete or by internal causes within the concrete itself. The external causes can be **physical, chemical,** or **mechanical** and may be created by **weathering,** occurrence of extreme temperatures, abrasion, electrolytic action and attack by natural or industrial liquids and gases. The internal causes are the **alkali-aggregate reaction, volumetric changes** and **permeability** of the concrete. Deterioration of concrete takes place, generally, due to more than one causes and it is difficult to isolate and assign one single specific cause. Permeability of concrete makes it prone to attack by various agencies:

Durability comprises of:

- Resistance to weathering;

- Resistance to chemical attack;
- Resistance to corrosion of steel; and
- Resistance to high temperature;

i. Resistance to weathering

This mainly refers to the effect of hot and cold weather on concrete. As the temperature of saturated hardened concrete falls below freezing point, the water held in the capillary pores freezes to **ice crystals and expands in volume.** This creates expulsion of some pore water through creation of **pore water passages and cracks.** With increase in temperature above freezing point, these pore water crystals melt. Refreezing of pore water takes place with temperature falling below freezing point again. These alternate cycles of freezing and thawing exerts radial pressures on capillary pores and results in disruption and disintegration of concrete. If the pressure exerted by freezing of pore water exceeds the tensile strength of concrete, then it causes severe damage by **cracking and disintegration** of concrete. These damages are more severe in case of exposed surfaces and where salts are used for de-icing the concrete surface such as concrete roads in cold climates. These pores exist in concrete due to **presence of excessive mixing water** which has not reacted fully with cement particles to form hardened concrete.

Although the resistance of concrete to frost attack depends on its major properties (such as strength, extensibility, and creep), the factors responsible are **degree of saturation** and type of **pore structure** of cement paste. Concrete with high **water-cement ratios** are generally **more porous** and therefore, offers less resistance to frost attack. Concrete with **low water-cement ratios** are generally **more dense and strong** if compacted properly. Such concrete will have lesser pores, and therefore, offers better resistance to frost attack. The **resistance** to frost attack **increases with the age** of concrete due to increase in strength and reduction in capillary pores due to hydration of cement.

The conditions of exposure of concrete are also important. The greater the exposure of concrete with the moisture, greater will be the attack of frost. Road slabs, kerbs, retaining wall, bridge decking, railing, piers and external concrete walls are completely exposed to weather and frost attack, and therefore, these must be constructed to have better resistance to weather and higher strengths. For exposure conditions of severe frost attack, the concrete with air entraining agents should be used. **Air entraining agent develop very large number of minute air bubbles** (approximately 0.5 mm size) which act as barrier and absorber for pore water pressures and thereby reducing disintegration and disruption of concrete structure. Air entraining is done by use of animal and vegetable fats, natural wood resins, alkali salts of organic compounds (such as Darex).

The effect of hot weather on hardened concrete is not generally severe as compared to cold weather. Effect of hot weather is quite undesirable on fresh concrete. During very hot weather the pore water near concrete surface may partially be converted into vapour and evaporate leaving behind porous surface. This porous surface may allow ingress of external air and moisture, and thus cause damage to steel reinforcement near the surface.

The important ways to **improve the weathering resistance** of concrete are summarized as:

- **Low water-cement ratio** for obtaining high strength and dense concrete;
- **Proper compaction** for obtaining dense concrete;
- **Appropriate grading** of aggregate for maximum density;

- Continuous **moist curing** over a longer period; and
- Suitable **air entraining** in concrete for high workability and better resistance to frost attack.

ii. Resistance to Chemical Attack

Concrete made with ordinary cement has low resistance to chemical attack. The common sources of chemical attack are: leaching out of cement, and the action of sulphates, seawater, and natural acidic waters. The resistance of concrete to chemical attack varies with the type of cement used. The resistance of concrete increases in the following order with use of different cements:

- Rapid hardening and ordinary portland cement;
- Portland Blast Furnace Slag Cement, and Low Heat Portland Cement;
- Supersulphated Cement; and
- High Alumna Cement

Solid salts generally do not attack concrete, but sulphate salts of magnesium, sodium and calcium, when present in solution react with hardened cement paste of concrete. Some clayey soils contain magnesium and calcium sulphates, and hence the ground water in such a place is a sulphate solution. The sulphate reacts with calcium hydroxide $Ca(OH)_2$, and calcium aluminate hydrate (C_3A) present in the cement concrete. The products of the reactions, gypsum and calcium sulphoaluminate, have a much greater volume than the compounds which are replaced. These reactions with sulphates lead to expansion and volumetric instability and hence disruption of concrete.

Some of these reactions such as sodium sulphate, magnesium sulphate are given below:

$$Na_2SO_4.10\ H_2O + Ca(OH)_2\ =\ CaSO_4.2H_2O + 2NaOH + 8H_2O$$
$$(gypsum)$$

$$3(Na_2SO_4.10\ H_2O) + 2(3CaO.Al_2O_3.12\ H_2O)\ =\ 6NaOH + 2Al(OH)_3 + 17H_2O$$
$$+ 3CaO.\ Al_2O_3.\ 3CaSO_4.\ 31H_2O$$
$$(Calcium\ sulpho\ aluminate)$$

Calcium sulphate attacks only calcium aluminate hydrate and forms calcium sulphoaluminate hydrate. Magnesium sulphate attacks calcium silicate hydrate as well as calcium hydroxide and calcium aluminate hydrate. These reactions are given as under:

$$3(MgSO_4.\ 7H_2O) + 3CaO.\ SiO_2 = 3(CaSO_4.\ 2\ H_2O) + 3Mg(OH)_2 + SiO_2 + 12H_2O$$

$$3(CaSO_4.\ 2H_2O) + (3CaO.\ Al_2O_3.\ 12\ H_2O) + 18H_2O\ =\ (3CaO.\ Al_2O_3.\ 3CaSO_4.\ 36H_2O)$$
$$(Calcium\ sulpho\ aluminate)$$

$$3(MgSO_4.\ 7H_2O) + 2(3CaO.\ Al_2O_3.\ 12\ H_2O)\ =\ 3CaO.\ Al_2O_3.\ 3CaSO_4.\ 31H_2O + 3Mg(OH)_2$$
$$+ 2Al(OH)_3 + 8H_2O$$

The reaction of magnesium sulphate is more severe than other sulphates. The rate of sulphate attack depends on the concentration of the solution. The concentration of the sulphates is expressed as the number of parts by weight of SO_3 per million (PPM). 1000 PPM of sulphates is considered moderately severe and 2000 PPM is very severe, especially if magnesium sulphate is the main constituent.

Concrete exposed to the pressure of sulphate bearing water on one side, the attack will be maximum. Similarly alternate drying and saturation with sulphate bearing water also leads to faster deterioration of cement concrete. When the concrete is buried in the ground and there is no flow of sulphate bearing water, the condition shall be less severe.

Concrete attacked by sulphate has a characteristic whitish appearance on the surface. The damage starts at edges and corners and progresses by cracking and spalling which reduces concrete to a friable or even soft state.

The resistance of cement concrete to chemical attack can be improved by using cements having a low proportion of tri-calcium aluminate (C_3A). Cements containing less than 7 percent of C_3A can be considered to have fairly good resistance to sulphate attack. Proportion of C_3A can be reduced by suitably adjusting the raw material oxides during the manufacturing process to obtain sulphate resisting cement. Improved resistance to sulphate attack can also be obtained by the addition of, or partial replacement of cement by pozzolanas (such as flyash, slag and surkhi). Sulphate resisting cements are blast furnace slag cement, low heat portland cement, Portland pozzolanic cement, sulphate resisting portland cement, and supersulphated cement.

The resistance of concrete to sulphate attack also depends on its impermeability quality. Concrete must be dense and rich. High cement content and low water-cement provides better resistance to sulphate attack and hence higher durability.

iii. Resistance to Corrosion of Steel

Good quality concrete provides good protective coating around steel reinforcement. The **alkalinity** of cement concrete leads to the formation of a thin invisible **protective oxide film** on steel reinforcement. The calcium hydroxide liberated during the hydration of cement reacts with CO_2 from the atmosphere to form calcium carbonate. This process of carbonation increases the shrinkage on drying and tends to promote the development of cracks. The **carbonation also reduces the alkalinity** which reduces the effectiveness of the protective film. Soluble chlorides such as NaCl or $CaCl_2$ reduce the effectiveness of the alkaline protective film on steel.

To achieve good protection against corrosion of steel, sufficient thickness of cover of good quality (dense and impermeable) concrete should be provided. The **concrete cover** should not allow **ingress of moisture and air** which are necessary for corrosion of steel to occur. In porous concrete, the reinforcement is likely to corrode quickly. On **corrosion the volume of steel increases** and this causes concrete cover to crack and further increase the process of corrosion by providing easy ingress of moisture and air.

Sands containing salts if used in concrete, absorb moisture from air and cause efflorescence and weaken the concrete cover leading to ingress of moisture to steel reinforcement. Use of sea water or water containing **sulphates and chlorides** beyond certain permissible limits in cement concrete **causes volumetric instability** and unsound porous concrete cover to steel. This results into corrosion of steel. **Chlorides accelerate corrosion** of steel to some extent.

The inadequacy of reinforcement may also cause cracks in concrete cover, thus allowing the moisture to reach steel for initiating corrosion. The **permeability** of concrete is the most important factor which affects the process of **corrosion of steel** reinforcement due to ingress of water. All factors which reduces the permeability of concrete are also responsible to improve the resistance of concrete to corrosion of steel. Thus the concrete should be prepared with **low water-cement ratio** and **well compacted** to achieve high density and high strength to provide better resistance to corrosion of steel.

iv. Resistance to High Temperature and Fire

Fire introduces high temperature gradient in concrete, and as a result the hot surface layers tend to expand and separate from the cooler interior layers. This encourages the formation of cracks near joints in poorly compacted parts of concrete, or in the planes of reinforcing bars. Once the reinforcement is exposed, it further accelerates the action of heat by conduction.

The effect on the strength of hardened concrete is small and irregular upto temperatures of 250°C. The concrete strength decreases with higher temperatures beyond 300°C. In some cases the concrete strength around 800°C remains only 20 to 25 % of its original strength. Leaner mixes suffer relatively lower loss of strength compared to rich mixes. Flexural strength is more severely affected than compressive strength of concrete. Concrete containing aggregate with no silica, suffer lower loss of strength. Low conductivity of concrete improves its fire resistance and hence **light weight concrete** with low thermal conductivity, offers **better fire resistance** as compared to ordinary dense concrete.

The concrete changes in colour at various temperatures and hence from the colour of concrete, maximum temperatures during fire and the residual strength can be estimated to assess the damage done by fire. Cement concrete becomes **pink or red during temperature range from 300 to 600°C** while it turns **grey or buff beyond 600°C**. The concrete whose colour has changed to pink or red due to fire is doubtful, while concrete that has changed beyond the grey stage shall be friable and porous. Such behaviour of concrete is important in atomic reactors, chimneys, and furnaces.

2.4.3 Impermeability

The **resistance** of the concrete to **passage of moisture** or flow of water through its body (pore spaces) is called **impermeability.** For good compaction, good workability is necessary and hence certain quantity of water more than that essential for complete chemical combination with cement is used. This excess mixing water occupies space and evaporates later and on drying leaves behind pores in concrete. In addition to these, there are air voids also. All such pores allow water flow through them and makes concrete permeable. The movement of water through the body of concrete can be caused not only by a pressure head of water but also by **humidity differential** on the two sides of concrete or by **osmotic effects.** Both the cement paste and the aggregate contain pores. In addition, the concrete as a whole contains voids due to incomplete compaction or bleeding. Since the paste envelops the aggregate particles, it is the permeability of the cement paste which has maximum effect on the permeability of concrete.

The pores in cement paste comprises of **gel pores** (about 28%) and **capillary pores** (0 to 40%) depending on W/C ratio and degree of hydration. The permeability of cement concrete is not a simple function of its porosity, but it depends also on the size, distribution, and continuity of the pores. Water can flow easily through the capillary pores than through the much smaller gel pores. Therefore, the **permeability** of cement paste is controlled mainly by the **capillary pores** of the paste. The permeability of cement paste reduces with the progress of hydration due to increase in volume of hydrated gel (about 2.1 times) compared to the volume of the unhydrated cement and thus filling the original water spaces. In mature paste the permeability depends on the size, shape and concentration of gel particles and on whether or not the capillaries have become discontinuous. **Permeability is lower** with higher cement content i.e. **lower Water-Cement ratio.** It is generally observed that there is steep reduction in permeability below a Water-Cement ratio of 0.60. This is due to creation of discontinuity in capillary pores.

The properties of cement, especially its fineness, also effects the permeability initially at an early age due to degree of hydration, but ultimate permeability may not differ. In general, it may be stated that higher the strength of concrete, lesser will be the permeability except when the shrinkage of cement causes rupture in the gel between the capillary pores and thus open new passage for water.

Impermeability forms a very important factor for the **durability** of any concrete. The impermeability or the resistance to the penetration of water increases the resistance to weathering (frost attack), and resistance to chemical attack. The resistance to ingress of water and air through concrete cover to steel, protects steel from corrosion. Thus for durability of exposed structures, water retaining and hydraulic structures, the property of impermeability in concrete plays very vital role. Thus for **better impermeability** of concrete, it is necessary to have cement concrete with:

- **Low water-cement ratio** for minimum capillary pores;
- **Sound, dense and well graded aggregate** for minimum voids and pore spaces;
- **Proper compaction** for minimum voids and maximum density;
- Adequate **continuous moist curing** at low temperature to ensure proper hydration of cement for minimum capillary pores and shrinkage cracks.

The permeability of concrete can be measured by a simple test. In this test the rate of water flow through a given surface area and given thickness is measured under a given water head. The permeability is expressed as **coefficient of permeability** (K). K is measured in mm/sec.

2.4.4 Dimensional Changes

The cement concrete elements also undergo change in dimensions under various conditions of loading, temperature changes, and cement hydration. Various dimensional changes in concrete members occur due to:

- Elasticity
- Shrinkage
- Creep, and
- Thermal Changes

i. Elasticity

When concrete specimen is loaded it undergoes deformation. On removal of load, the deformation disappears to some extent. This portion of deformation which disappears on unloading is known as elastic deformation. This property of concrete by virtue of which **deformation occurs on loading and disappears on unloading** is called elasticity of concrete. Concrete does not exhibit the perfect elasticity property with higher and sustained loading. Concrete may be considered **elastic within certain limits only.** The **elasticity** is measured by modulus of elasticity and is defined as the change of stress with respect to elastic strain (unit deformation). It can be calculated as:

$$\text{Modulus of Elasticity} = \text{Unit Stress}/\text{Unit Strain}$$

It represents the resistance of concrete to deformation.

From a stress-strain diagram of concrete specimen loaded in compression, it is observed that within working stress (initial portion) the stress-strain curve is straight line to some extent (Fig. 2.7). For higher stresses, the stress-strain relationship becomes curved. The modulus of elasticity is generally applicable to this **initial straight portion of stress-strain diagram.** In case, there is no straight portion, the modulus of elasticity (tangent modulus) may be considered as **slope of tangent to the Stress-Strain Curve at the origin.** Tangent modulus of concrete may also be determined as slope of tangent to stress-strain curve at any other point of stress but such a value shall be useful only for those loads which are very near the loads at which the modulus has been measured.

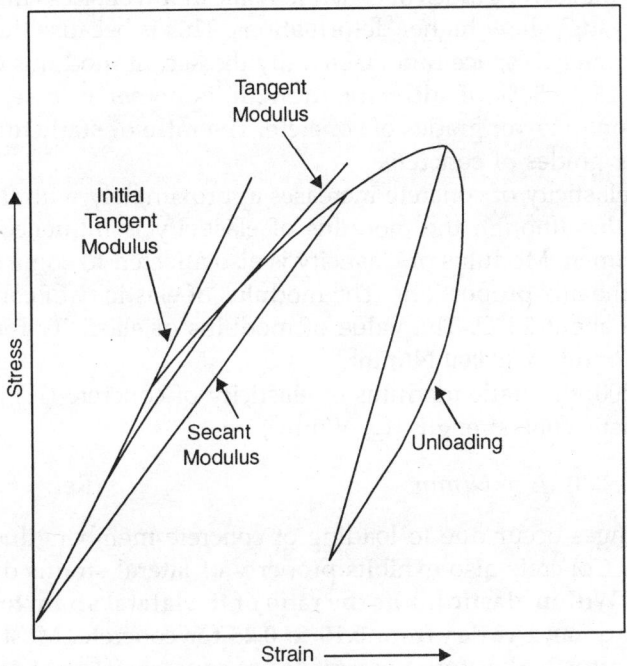

Fig. 2.7. Typical Stress-Strain Curve for Concrete

The deformation in concrete increases with time (i.e. with sustained loading) and the stress-strain relation does not remain straight with respect to total strain but remains approximately straight with respect to **initial instantaneous strain** (considered as elastic). The increase in strain with time under sustained loading is due to **property of creep** in concrete. Another method of measuring elastic modulus in concrete is by way of "**Secant Modulus**". The secant modulus is represented by the slope of a line drawn from the origin to any point on the curve at a certain standard stress or strain.

Many concretes exhibit a slight curvilinear relation of stress-strain even at working loads. The tangent modulus cannot be determined accurately because the stress-strain curve and tangent are drawn only by eye judgment. The secant modulus is the most practical and commonly used because there are no uncertainties in actual deformation. The secant modulus may be measured at stresses ranging from **15 to 50% of its ultimate strength.** The secant modulus decreases with an increase in stress level and hence the **stress level at which the modulus has been determined, should always be stated.** Secant modulus is determined from the experimental stress-strain relation. The secant of stress-strain curve on unloading is often parallel to the initial tangent to the loading curve. The **secant modulus** is also known as **static modulus.**

Repeated loading and unloading reduces the creep in subsequent loadings and the stress-strain curve becomes straight after 3rd or 4th loading. In **dynamic modulus** of concrete, the specimen is subjected to a **very small stress** and the modulus so obtained is similar to tangent modulus at zero stress, and therefore higher than any Secant Modulus. The dynamic modulus of elasticity takes into account mainly the elastic strain only.

The concrete exhibits a peculiar phenomenon that the maximum strain at failure in compression is higher for the lower grades of concrete while at lower stress than ultimate strength, the higher concrete grades show higher deformations. This is because the strength of cement paste is governed by the gel/space ratio. Generally the secant modulus of elasticity, as **measured at a stress of 15 to 50% of ultimate strength**, is higher in case of higher grades of concrete in comparison to lower grades of concrete. The **ratio of static to dynamic modulii** is also higher for higher grades of concrete.

The modulus of elasticity of concrete increases approximately with the **square root of its characteristic strength**, although the modulus of elasticity is influenced by the saturation condition of the specimen. Modulus of elasticity is also affected to some extent by the nature of aggregate and by the mix proportions. The modulus of elasticity of concrete is not affected by temperature upto about 23°C. The value of modulus of elasticity for concrete generally varies from **50,000 N/mm²** to **30,000 N/mm².**

As per IS: 456-2000, the static modulus of elasticity of concrete (E_c) is approximately related to its characteristic cube strength (f_{ck} **N/mm²**) as:

$$E_c = 5000 \sqrt{f_{ck}} \ \text{N/mm}^2 \qquad \text{(Ref: § 6.2.3.1 of IS 456-2000)}$$

Dimensional changes occur due to loading of concrete members due to the property of elasticity of concrete. Concrete also exhibits property of lateral strains due to application of longitudinal stresses. **Within elastic limits the ratio of the lateral strain to longitudinal strain is called Poisson's ratio and varies from 0.10 to 0.25** for concrete. Most common values lie between **0.15 to 0.20.** It is higher for lower grades of concrete in most cases.

Generally the factors which **increase the modulus of elasticity** for concrete are summarized as:

- **Low water-cement** ratio
- **Richness** of mixes
- **Longer curing** periods
- Quality, type and **grading of aggregate**

ii. Shrinkage

The volume of concrete specimen also changes during the hydration process in cement (i.e. reaction of cement with water). The evaporation and change in water content of the mix also causes change in volume of the concrete. The chemical combination of cement and water results in reduction of volume causing shrinkage in concrete. Some times cement reacts with certain aggregate (reactive aggregate) and causes large expansion. The concrete in general, **shrinks on drying and expands on wetting.**

Cement, after hydration, consists of crystalline material plus calcium silicate gel resulting from the combination of cement and water. The quantity of the gel increases with the age of hydration and is greater for higher water-cement ratios and for finer cements. The quantity of gel also depends on the chemical composition of the cement, because **hydrated dicalcium silicate is mostly** gel, while hydrated **tricalcium silicate** is more than half gel. For most commonly used water-cement ratios in concrete, **the gel occupies larger volume than crystalline portions.** The gel is finely porous and undergoes larger volume changes on wetting and drying and thus the quantity of **calcium silicate gel determines the extent of shrinkage** upon drying of hydrated cement. Water is held in the pores of the gel and the concrete is subjected to some moisture volume changes.

While in the plastic state, **fresh concrete undergoes an appreciable reduction in volume.** This occurs due to settlement of the solids and bleeding of free water to the top where it may be lost by evaporation. Most of this settlement and bleeding occurs within an hour or so after placement of concrete. This total volume change may be **about 1%** and is not of great importance, as the concrete is still in a plastic or semi plastic state.

The **products of hydration** have a **lesser volume** than the total volume of the original components (i.e. **cement plus water**), and hence the **cement gel shrinks** as the hydration continues. After the cement paste has set, the solids form a structure having system of pores containing free water. The total volume of hardened solid mass expands, due to formation of new solids and expansion of gel although the volume of cement plus water decreases. The free water available in the pores and loosely held water in the gel is utilized by continuation of hydration of cement. The loss of water from the gel causes the **gel structure to contract**, resulting in decreased volume of the concrete mass. With the two contradictory factors of expansion and contraction in volume of concrete, the contraction becomes more dominant at later stage of hydration. Thus concrete exhibits the property of **shrinkage** (i.e. reduction in volume with hydration of cement).

Various factors which influence the behaviour of a particular concrete are:

- Cement **composition and fineness** (affecting the nature, rate and products of hydration);
- Quantity of **mixing water and water-cement ratio** (affecting the rate of hydration, porosity of paste, and free pore water);
- Type, nature and **grading of aggregate** (affecting the rate of hydration, porosity of the paste and free pore water);
- **Temperature** and moisture conditions (influencing the rate and quality of hydration);
- Absorptiveness of forms (affecting the moisture movement during the plastic stage);
- **Age of concrete and curing** method (affecting the extent and rate of hydration);
- Use of **admixtures** (affecting the rate, and products of hydration); and
- Quantity and distribution of reinforcement (affecting the volumetric strains and resistance to deformations).

It is observed that the shrinkage is mainly affected by physical structure of the cement gel. **Higher** the aggregate-cement ratio, **less is the shrinkage** of concrete. With lower aggregate-cement ratio (i.e. for richer mixes) the concrete exhibits about 50% higher shrinkage compared to lean mixes after about 4 months keeping the same water-cement ratio. Aggregate containing clay particles increases shrinkage. Similarly, higher the water-cement ratio, greater is the shrinkage. Although cement composition and fineness of cement has sufficient effect on shrinkage of neat cement paste but there is minor effect in case of shrinkage of concrete. Total shrinkage of concrete made with aluminous cement is not much different from ordinary portland cement concrete except that it is much faster in case of aluminous cement concrete.

The addition of calcium chloride increases shrinkage by 10 to 15% due to faster gel formation. It is generally observed that the concrete made with portland pozzolana cement exhibits about **20% lesser shrinkage** after 4 months than the concrete made with ordinary portland cement. The concrete prepared **with low heat portland cement** exhibits 20% more shrinkage than ordinary portland cement concrete due to slow strength gain.

Prolonged moist curing delays the shrinkage and also reduces the shrinkage cracking because of increased strength of concrete. In fact the shrinkage of neat cement paste increase with the hydration of all the cement grains but the strength also increases. The age of curing does not affect shrinkage too much.

The final magnitude of shrinkage is largely independent of the rate of drying. Very rapid rate of drying does not allow the relief stress by creep and hence may lead to initial cracking. The relative humidity of the medium surrounding the concrete greatly affects the magnitude of shrinkage. Shrinkage increases with lesser relative humidity during curing.

The shrinkage is determined with the help of frame fitted with a micrometer gauge or extensometer or strain gauges using concrete bars. Drying of surface occurs more than the inside and hence causes differential shrinkage which shall be dependent on the size and shape of the specimen, being a function of the surface/volume ratio.

If the concrete is placed in water (or at a higher humidity) after drying in air at a relatively lower humidity, it will swell and part of the shrinkage (0.30 to 0.60 of the drying shrinkage) gets reversed. If drying is accompanied by carbonation, the cement paste becomes insensitive to moisture movement and the residual shrinkage is more. It has been observed that CO_2 present in atmosphere reacts, in the presence of moisture, with hydrated cement compounds. $Ca(OH)_2$ changes to calcium carbonate $(CaCO_3)$ and other cement compounds are also decomposed. Carbonation is accompanied by shrinkage. Carbonation shrinkage is maximum at 50% relative humidity. **Alternate wetting and drying** in air containing CO_2, carbonation shrinkage increases progressively and also results in **increase of irreversible shrinkage** and crazing on the surface of concrete. **Carbonation** of concrete also **results in increased strength** and reduced permeability.

The shrinkage varies with mix proportions but in general, **total shrinkage** may be assumed as **0.0003** for most of the concrete. It may be noted that half of the total shrinkage occurs during the first month and that about **three quarters of the total shrinkage** takes place in the **first six months.**

The shrinkage before the concrete sets can be minimized by prevention of water absorption by form work and by proper prevention of evaporation. **Shrinkage during setting** is mainly because of **cement gel,** and can be minimized by:

- Keeping quantity of **cement to a minimum;**
- Keeping water-cement ratio and **water-content minimum;**
- Maintaining **adequate humidity** around concrete during curing and keeping minimum exposed surface;
- Using a **coarse grading of sand,** and avoiding over sanding and excessive silt;
- Using **non-porous aggregate** capable of resisting compression and change of volume;
- Providing suitable and well **distributed reinforcement** which may reduce shrinkage cracking by 50%;

- Curing under **low temperatures and maximum humidity** and avoiding sudden change of humidity;
- Placing **suitable** contraction and expansion **joints** in large slabs, roofs, roads, and wall.; and
- Using admixtures which reduce the water requirement but do not affect the strength.

The property of shrinkage is important in structures which cannot expand and contract freely. It is also **important** in structures with large exposed surfaces from **durability, impermeability,** and **strength** point of view.

iii. Creep

The relation of stress and strain is dependent on time. The strain continues to increase with time for the same stress. This **increase in strain under sustained loading** is called **creep.** Creep may be several times as large as the elastic strain obtained on loading. Creep in concrete is due to yielding of concrete which may be caused by viscous flow of the cement-water paste, closure of internal voids, crystalline flow in aggregates and seepage of colloidal (absorbed) water from the gel formed during hydration of the cement. Seepage of colloidal water from the gel is the major cause for the creep. The rate of expulsion of the colloidal water is a function of the applied compressive stress, and the friction in the capillary channels. The greater the stress, the higher the rate of moisture expulsion and deformation. The phenomenon is associated with that of drying shrinkage.

Under normal conditions of loading, the instantaneous strain recorded depends on the rate of loading and the strains include part of creep in addition to the elastic strains. The modulus of elasticity increases with age, but generally its value is measured initially only. The creep is therefore, for convenience considered as an increase in strain above the initial elastic strain. Creep occurs similarly under both compressive and tensile stresses in concrete.

When the loaded concrete specimen is also drying, it undergoes shrinkage as well as creep strains which are both additive. Creep is calculated as the difference between the total time deformation of the loaded specimen and the shrinkage of similar unloaded specimen stored under the similar conditions through the same period.

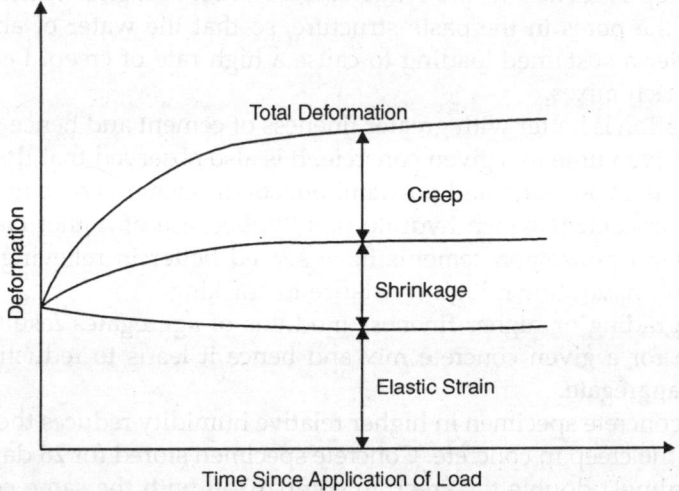

Fig. 2.8(a). Time dependent deformation in concrete subjected to a sustained load

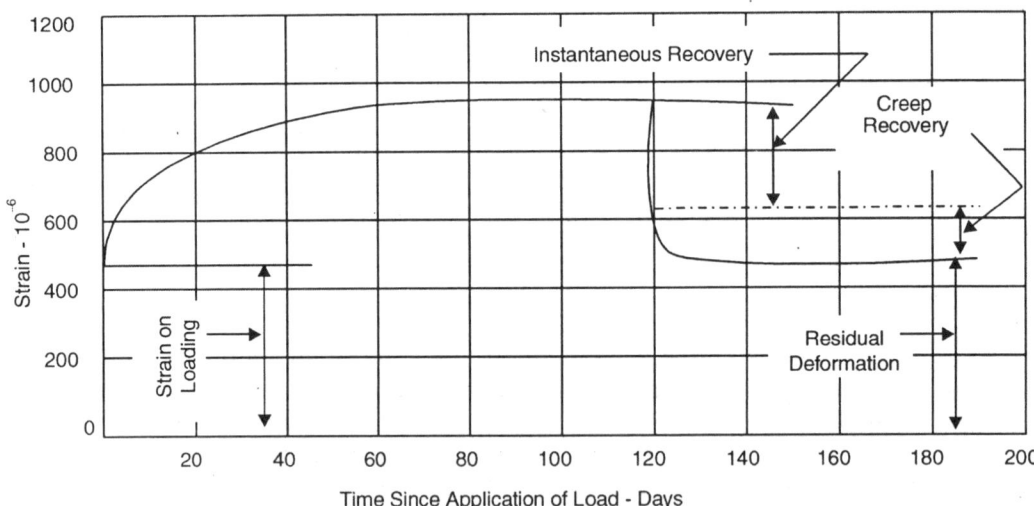

Fig. 2.8(b). Creep and recovery of a mortar specimen stored in air at a relative humidity of 95%, subjected to a stress of approx. 15 N/mm² and then released

The rate of creep is relatively rapid at early ages after loading and then decreases gradually, until after a few years it becomes negligible. About **one-fourth** of the ultimate **creep** occurs within **first month**, and **one-half** occurs within the **first year**. On releasing of the sustained load, the strain decreases immediately equal to elastic strain of the given age, generally less than the initial elastic strain. This instantaneous recovery is followed by a gradual decrease in strain, called creep recovery. The creep recovery occurs much rapidly. The creep recovery is not complete, and creep is not a fully reversible phenomenon (Fig. 2.8 (a) and (b)).

It is observed that the greater the degree of hydration of cement at the time of loading, the lower is the rate and total quantity of creep. Since degree of hydration reflects strength of a given concrete, it can be said that the **creep varies inversely as the strength** for a given concrete. Hence creep is related to the water-cement ratio. A higher water-cement ratio increases the size of the pores in the paste structure, so that the water of absorption may be expelled easily under a sustained loading to cause a high rate of creep. **Lean mixes exhibit greater creep** than rich mixes.

Degree of hydration is better with greater fineness of cement and hence there is reduction in creep strain at a given time in a given concrete. It is also observed that the **slow-hardening cement such as low heat portland** and **portland pozzolana cement creep more than ordinary** and **rapid hardening cement** which hydrate rapidly. Because of higher creep, the low heat portland and portland pozzolana cements have served better in relieving stresses in large dams on cooling and have shown high resistance to cracking.

Using coarser grading or higher fineness modulus of aggregates results in reduction of water-cement ratio for a given concrete mix and hence it leads to reduction in creep with increase in size of aggregate.

The storage of concrete specimen in higher relative humidity reduces the flow of moisture and hence reduces the creep in concrete. Concrete specimen stored for 28 days in 70% relative humidity exhibits almost double the creep in comparison with the same concrete specimen stored under water for the same period when subjected to a stress of about 5 N/mm². In general the creep is higher for lower relative humidity. At an advance age of loading the ambient relative humidity has little influence on the creep. Alternate wetting and drying results into increased magnitude of creep.

For specimens of larger size, the creep reduces because of more resistance to seepage of expelled pore water in longer pore passages and also due to effects of shrinkage.

Upon release of a sustained load, there is an immediate elastic recovery of length followed by a further creep recovery which continues for a period of several days. The magnitude of this creep recovery depends on the period of the sustained load and pervious creep. **Longer the loading period, lesser is the creep recovery.** Recovery of creep reaches its maximum value faster than the maximum creep attainment period.

Creep in concrete is sometimes beneficial as it reduces the tendency of cracking due to shrinkage in restrained structures by relieving stress. **Creep is disadvantageous** in case of **prestressed concrete** structures because creep reduce tension in steel wires.

iv. Thermal Changes - Coefficient of Expansion

Temperature change in concrete causes change in dimensions of concrete. Cement concrete has positive coefficient of thermal expansion but its value depends on the composition of the mix and on its hygral state at the time of the temperature change. Two main constituents of concrete i.e. cement paste and aggregate have different thermal coefficients, and hence the coefficient for concrete is a resultant of the two values.

The coefficient of thermal expansion of cement paste varies between about 11×10^{-6} and 13×10^{-6} per°C and is generally higher than the coefficient of aggregate 5×10^{-6} to 13×10^{-6} per°C). The coefficient of thermal expansion of concrete depends on the quantity of aggregate and the coefficient of thermal expansion of aggregate itself. Thus if the coefficients of thermal expansion of the coarse aggregate and of the cement paste differ too much, it will cause differential movement and break bond between aggregate and paste. Due to differential coefficients the concrete mass may be subjected to a severe thermal stress when the concrete is exposed to high temperatures.

The coefficient of thermal expansion of paste depends on the saturation condition. For the paste which is **completely dry or fully saturated**, the coefficient of **thermal expansion is less** than when the paste is partially saturated. These coefficients are maximum at relative humidity of 50 to 70%. The **coefficient also decreases** with the **age of hydration** due to increase in crystalline material in the hardened paste. These variations in the coefficient of thermal expansion are not found in steam cured pastes. Some times the increase in temperature is associated with drying shrinkage also and hence results in lesser expansion.

Thermal changes in concrete are due to two reasons:

- Internal (chemical combination of water and cement)
- External (due to change of temperature in environment).

The internal heat generated by chemical reaction in mass concrete dissipates quickly at the surface in comparison to the interior and thus creates temperature gradient which may cause thermal cracks. The rise in temperature causes expansion of concrete and result into thermal stress and cracks if restrained. These cracks become more prominent if the coefficient of expansion of aggregate and cement paste differs too much.

The thermal coefficient of expansion of concrete is the change in unit length of concrete member due to unit degree of change in temperature. For plain cement concrete using normal aggregate the coefficient of thermal expansion generally varies between 7×10^{-6} to 11×10^{-6} per degree Celsius. IS:456-1978 proposes the following values:

a) Plain cement concrete : 10.6×10^{-6} per°C
b) Reinforced cement concrete : 11.7×10^{-6} per°C

As already explained the coefficient of thermal expansion is mainly influenced by the quantity and nature of aggregate, condition of saturation, and age of concrete. Aggregates derived from gravel and quartzite have higher coefficients of thermal expansion as compared to lime stone, granite and sand stone.

Thermal expansions are important in concrete structures which are restrained. With the rise in temperature, the restrained concrete is subjected to compressive stresses. Similarly on fall of temperature, the restrained concrete is subjected to tensile stresses and may cause cracking if the temperature change is high. Thus to avoid development of severe tensile stresses and cracking, the concrete should be placed comparatively at low temperatures and low heat or portland-pozzolana cements should be used especially in mass concrete. Suitable expansion and contraction joints should be provided to allow for free movement of concrete due to temperature changes. The concrete with higher coefficient of thermal expansion are less resistant to temperature changes than concrete with a lower coefficient.

Thermal Conductivity

Thermal conductivity is important in mass concrete, and also for insulating properties of concrete walls and floors. Structural concrete made of normal aggregates conduct heat easily in comparison to light weight concrete, and other non-structural concretes. Conductivity, is low with use of light weight aggregates, and in case of air entrained concrete. To avoid heat losses in air conditioned rooms the walls, floors, and roofs should be made of concrete having low conductivity of heat. Concrete made with pumice and cinder has low thermal conductivity. Saturation of concrete also improves conductivity compared to dry concrete containing air voids. Since conductivity of water is about half of the cement paste, conductivity of hardened concrete is much better with low water-cement ratio.

Thermal conductivity is usually calculated from the diffusivity which represents the rate at which temperature changes within a mass.

Specific Heat

Specific heat represents the heat capacity of concrete. Specific heat of concrete is the quantity of heat required to heat a unit mass (kg) of concrete through a unit temperature (°C). Specific heat increases with increase in moisture content of the concrete.

2.4.5 Unit Weight

The weight of concrete becomes important from various quality control point of view during preparation of fresh concrete and functional requirements during hardened concrete.

The unit weight of concrete depends mainly on unit weight, quantity, and size of aggregate in the mix. Unit weight of concrete is also influenced by air entrainment. For ordinary concrete, nominal natural aggregate is used. For light weight concrete, light weight aggregate is used. For high density concrete, special high weight aggregate such as iron ore, magnetite, steel, and lead are used. Unit weight increases with use of bigger size aggregates. For plain cement concrete, the unit weight is normally assumed as 2400 kgf/m^3 (24 KN/m^3).

For reinforced cement concrete, the unit weight is considered as 2500 kgf/m^3 ($\approx$25 KN/m^3). Unit weight of reinforced cement concrete increases with increase in percentage of steel reinforcement. .

Unit weight of concrete is also influenced by degree of compaction at the time of placing. Higher air content or air voids in concrete results into lower unit weight of fresh concrete. The ·unit weight of fresh concrete is useful for determination of yield of concrete.

2.5 SUMMARY

Cement concrete is the most common and important material of construction. Quality and economy in any construction project depends largely on cement concrete construction. For managing and controlling quality and economy it is necessary to understand the properties and behaviours of cement concrete both during **fresh** and **hardened** states.

Cement concrete in its fresh state is required to exhibit property of good **workability**, freedom from **segregation**, and freedom from **bleeding** and **harshness. Workability is defined as the ease with which fresh concrete can be placed and compacted fully without segregation.** Segregation is defined as separation of the constituents (specially coarse aggregate) to make concrete composition non-uniform. The separation of water from cement and sand or separation of cement paste from the mortar allowing the water or cement paste to appear at the surface is called **bleeding.**

Bleeding in concrete can be prevented by using finer cement and finer grading, and controlling water content, compaction and finishing. Workability and segregation can be controlled by proper grading of aggregate and proportion of water. Overall phenomenon of workability includes bleeding and segregation controls. Workability can be measured by **slump, compacting factor, flow, remoulding, Vee-Bee degrees,** and ball penetration characteristics of concrete. Slump is the most common site test for control of workability. Slump test does not provide correct measure of workability specially for very rich and very lean mixes. Compacting factor and Vee-Bee degrees represent workability fairly well for various concrete mixes except for very fluid concrete.

Hardened concrete exhibits properties of **strength, durability, impermeability, dimensional changes,** and unit weight. Various type of strengths are mostly governed by its basic compressive (crushing) strength which in turn depends on cement, W/C ratio, aggregate grading and size, and concreting operations (compaction, curing, etc.). Flexural strength $f_{cr} = 0.7\sqrt{f_{ck}}$ N/mm^2. Durability of concrete depends on impermeability, resistance to weathering corrosion and chemical attack. Impermeability is affected by W/C ratio, adequate compaction and curing and dense grading of aggregate. Dimensional changes occur due to elasticity, shrinkage, creep and thermal variations. Creep occurs due to sustained loading and affects prestressed concrete adversely. Appropriate joints are constructed to avoid excessive stresses in restrained structures. Unit weight of concrete represents its **mass per unit volume** (m^3) and depends on the sp. Gravity of aggregates.

The concrete in various engineering structures had to withstand different types of forces and exhibit appropriate behaviour. In concrete structures failure of concrete occurs due to:

Lack of **durability**
Lack of **strength**
Lack of **impermeability**
Lack of resistance to **dimensional changes**

Concrete structures fail due to **inadequate cover** leading to **corrosion** of reinforcement, **over stressing** of weak and/or porous concrete, cracking as a result of bad construction and expansion joints, and deterioration due to use of inappropriate concrete ingredients. For avoiding failure, the concrete of appropriate characteristics should be developed according to the functional requirements of the structure, its exposure conditions and strength requirements.

All structures are required to resist forces and loads, and therefore, it is necessary for structural concrete to exhibit certain desired strength. Some concrete structures are exposed to atmospheric and weathering forces of wind, rain, water, chemicals, temperature, fire and requires the concrete to be durable against such conditions.

The quality and characteristics of concrete mix should therefore, be designed optimally to exhibit suitable properties and functions in given conditions of exposure.

PRACTICE QUESTIONS

2.1   Define each in not more than 50 words : grade of concrete, workability, strength, bleeding, harshness, and impermeability.

2.2   List main properties of cement concrete in (a) plastic state (b) hardened state

2.3   Explain the importance of workability in quality control of cement concrete.

2.4   Differentiate between :
  - bleeding and segregation,
  - strength and durability
  - creep and shrinkage
  - coefficient of expansion and elastic modulus

2.5   Explain the steps of controlling bleeding in cement concrete.

2.6   Explain each of the method for workability measurement in not more than 100 words
  - Slump test
  - Compacting factor test
  - Vee-Bee test
  - Flow test

2.7   Explain various conditions of fresh cement concrete mix in relation to Vee-Vee degrees

2.8   Explain strength characteristics of cement concrete.

2.9   Explain Impermeability and durability for cement concrete

2.10 Explain mechanism of chemical attack on cement concrete.

2.11 Explain importance of properties in plastic stage to control concrete properties in hardened state.

2.12 Differentiate between:
  - Elasticity and durability   ·
  - Impermeability and conductivity of heat
  - Chemical attack and corrosion
  - Compressive and flexure strengths

2.13 Explain role of cement concrete properties in management of its quality in not more than 100 words.

2.14 List five most critical properties of cement concrete essentially considered for controlling and managing its quality.

CHAPTER 3

Influence of Basic Ingredients on Quality of Concrete

LEARNING OBJECTIVES

After studying this chapter the learner **understands the influence of basic ingredients** on the quality of concrete and will be able to:

- **List characteristics** of cement and its influence on concrete;
- **Explain the influence** of **fineness, strength, soundness, heat of hydration** and **hardening** of cement on the quality of concrete;
- **Specify Indian standard specifications** of ordinary portland cement;
- **Explain characteristics** of aggregate;
- **Explain the effect of size, shape, strength** and **texture** of aggregate on the properties of cement concrete;
- **Explain the effect of aggregate grading** on the quality of cement concrete;
- **Explain bulking** of sand;
- **Explain fineness modulus** of aggregate;
- **Explain Alkali-aggregate** reaction;
- **Explain the influence of sand grading zone** on cement concrete property;
- **Explain the quality of water** in relation to the quality of cement concrete.
- **Explain the importance** of basic **ingredients** in achieving the quality of cement concrete.

3.1 INTRODUCTION

Normal cement concrete is produced by using basic ingredients: **cement, aggregate** (fine and coarse) and **water**. To control the quality of concrete it is essential to understand the influence of each ingredient on the property and behaviour of cement concrete. **Cement act as binding material, water reacts chemically** with cement, **aggregate as inert material** to increase the bulk (coarse aggregate as crushed rock or gravel for **volumetric stability** and sand **as fine aggregate for cohesiveness**) in the mass. For achieving the final concrete each of these basic ingredients shall be selected, measured and prepared to undergo a suitable process of concrete making. In addition to these basic materials, sometimes **special admixtures** may also be used to modify certain properties of concrete. These admixtures have been dealt separately in chapter-4. The quantity and quality of these basic ingredients influence the properties of concrete both during plastic and hardened states. Characteristics of these basic ingredients and their influence on concrete are discussed in subsequent paras.

3.2 CEMENT

Cement is most important ingredient and acts **as a binding material** (having adhesive and cohesive properties). Cement is obtained by **pulverising clinker formed by calcining** raw materials primarily comprising of **lime (CaO), silica (SiO$_2$), alumina (Al$_2$O$_3$)** and **ferric oxide (Fe$_2$O$_3$)** alongwith some minor oxides. **Joseph Aspdin,** a brick layer in England, developed **portland cement in 1824.** Further improvements lead to the present form of portland cements exhibiting a variety of properties suitable for variety of functional requirements of **strength, durability, impermeability** and other dimensional constraints. Cement when mixed with water forms a **paste which sets and hardens under water and binds the aggregate** together to produce a continuous compact mass. The characteristic behaviour of this concrete mass in a given condition depends on the **type, grade, quality** and **quantity** of cement.

The portland cements comprise of four principal compounds such as **tricalcium silicate (3CaO. S$_i$O$_2$), dicalcium silicate (2CaO. SiO$_2$), tricalcium aluminate (3CaO. Al$_2$O$_3$)** and **tetra calcium alumino ferrite (4CaO. Al$_2$O$_3$. Fe$_2$O$_3$).** The composition of these principal compounds is affected by the proportions of basic raw material oxides. The composition of these principal compounds affect the behaviour of cement in concrete. The actual proportions of these principal compounds differ from cement to cement and accordingly forms the basis of suitability under different conditions and requirements of a structure. Apart from variety of portland cements, there are other special cements which are suitable under special conditions of placement, environment and special structural requirements.

Different portland cements have different basic raw material oxide composition and is specified in Table 3.1.

Cements are selected for preparation of desired concrete required for structures placed under special conditions of loading, and environmental exposure on the basis of its characteristics. Various physical properties of cement are:

- **fineness;**
- **setting** and **hardening;**
- **strength;**
- **soundness;** and
- **heat of hydration.**

Table 3.1 Basic Raw Material Oxide Composition and Principal Compounds Formed

Oxides/Compounds	Ordinary	Rapid Hardening	Low Heat	Sulphate Resisting
Basic Oxides Lime (CaO)	63.1	64.5	60.0	64.0
Silica (SiO_2)	20.6	20.7	22.5	24.4
Alumina (Al_2O_3)	6.3	5.2	6.2	3.7
Iron oxide (Fe_2O_3)	3.60	2.9	4.6	3.0
Principal Compounds (CaO-C, SiO_2-S, Al_2O_3-A, Fe_2O_3-F)				
C_3S	40	50	25	40
C_2S	30	21	45	40
C_3A	11	9	6	5
C_4AF	11	9	14	9

3.2.1 Fineness

It may be recalled that cement is obtained by **grinding clinker mixed with gypsum.** The hydration of cement particles occurs immediately at its surface and hence it is the total surface area of cement particles that represents the material available for immediate hydration. Fineness of cement is determined by the particle size and its distribution. This is generally determined in terms of percentages of particles retained and **passing through 90-micron sieve.** This provides only indication of percentage of particles larger and smaller than 90 microns and does not provide the full detail of particle size distribution. This method is simple and can be used in the field.

Another method is developed which is based on **specific surface area (cm^2 or m^2) of cement particles per unit mass** (gram or kg). Particle size distribution of cement depends on the method and extent of grinding. During recent past air permeability (Refer Fig. 3.1) method of determining **specific surface area** was developed by **Lea and Nurse** and **Blaine** and is adopted by most of the countries as a standard method. It is based on the relation between the **flow of a fluid** (air) through a granular **cement bed** and the surface area of the particles comprising the bed. The surface area per unit mass of cement bed material is related to the permeability of the bed. The surface area per unit mass depends on the porosity (volume of pores in total volume of bed) of a given cement bed. Pressure difference is created and a stream of dry air is passed through the bed of cement at a constant velocity and the resulting pressure drop is measured by a manometer. Specific surface area is calculated by a suitable formula which provides a good guidance of relative fineness of the sample. Greater specific surface area represents greater finess.

Fineness of various cements is specified as:

Ordinary portland cement (OPC)	225 m^2/Kg	(2250 cm^2/g)
Rapid hardening portland cement (RHPC)	325 m^2/Kg	(3250 cm^2/g)
Portland blast furnace slag cement (PBFSC)	225 m^2/Kg	(2250 cm^2/g)
Low heat portland cement (LHPC)	320 m^2/Kg	(3200 cm^2/g)
Portland pozzolana cement (PPC)	300 m^2/Kg	(3000 cm^2/g)

Since the reaction of water and cement particles starts at the surface of cement particles, the total surface area of cement plays an important role in rate of hydration and strength development. Thus **finer the cement, rapid will be the rate of hydration** and **gain of strength.**

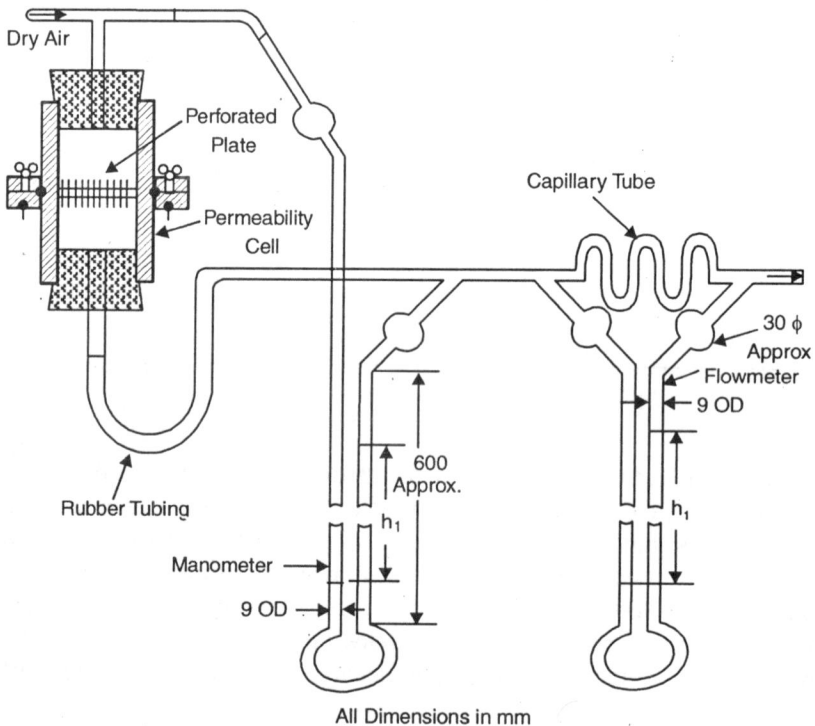

Fig. 3.1(a). Permeability Apparatus with Manometer and Flowmeter

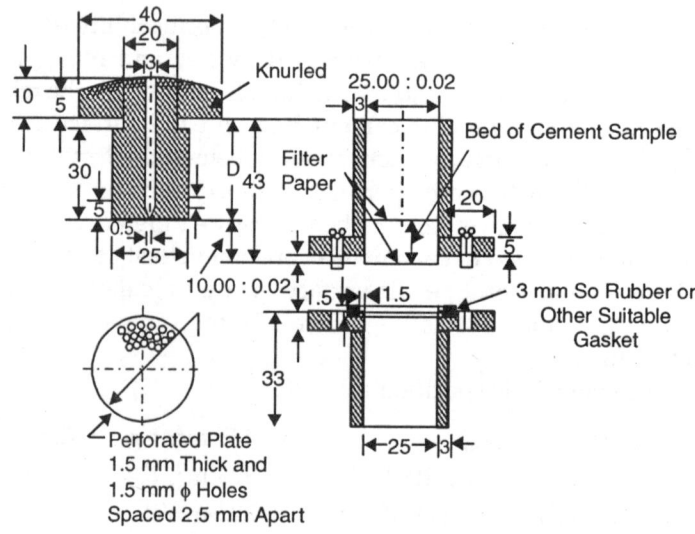

All Dimensions in mm

Fig. 3.1(b). Details of Permeability Cell

With rapid rate of hydration, the heat evolution will also be rapid. The cost of grinding will also be higher for higher fineness. Further **deterioration** of cement due to exposure to atmosphere during storage shall be **rapid in case of finer cements**. The **shrinkage** of cement paste also **increases with increase in fineness.** Finer cements have **lesser tendency to bleed.** With finer cements **higher percentage of gypsum is needed** to balance the effect of C_3A to retard setting. Quantity of water required for a paste of **standard consistency is higher for finer cements.** Workability of concrete mix improves slightly with increase in fineness of cement. **Fineness** is therefore an important property **of cement for controlling the quality** of concrete.

3.2.2 Setting and Hardening

When water is added to cement to form paste, the reaction starts and the paste loses plasticity gradually with time and finally becomes stiff. During this reaction C_3A **tends to react quickly** to produce a jelly like compound which starts solidifying. Due to addition of **gypsum** during manufacture of cement, the **reaction of C_3A delays** and C_3S starts setting first. If C_3A is allowed to set first, a **porous calcium aluminate hydrate forms** and the other compounds hydrate within this porous framework and affects the strength characteristics of cement paste adversely. The term **setting is used to describe the stiffening (loss of plasticity)** of the cement paste capable of withstanding an arbitrarily defined pressure. During this, the fluid state of the cement paste changes to rigid or solid state. Cement paste **acquires very little strength during setting** and should be neglected. With further **continuance of hydration of C_3S and C_2S, the cement paste gains strength** which is **referred as hardening.** The first part of hydration which causes **loss of plasticity or fluidity is termed as setting** while later part of hydration which causes gain of strength is termed as hardening.

Setting is described at two stages arbitrarily chosen as **initial setting** (beginning) and **final setting (ending)**, and are measured as times from the moment of adding water to cement. It is difficult to measure these exactly and hence arbitrary method is employed to locate these stages. It is important that the setting should neither start very early nor takes too long. The concreting operations of mixing, transporting, and placing should be completed before the concrete mix loses plasticity and the initial set begins. Various standards, therefore, specify a minimum initial setting time for cement. Once the initial setting has commenced, it is desirable that it should harden or gain rigidity as rapidly as possible, so as to remove side shuttering. Standards of various countries, therefore, specify the maximum time limit for final setting and minimum time for initial setting.

The water content of a paste has a marked effect upon the time of set as well as on other properties of cement paste. For acceptance tests of cement, the water content is regulated by bringing the cement paste to a standard condition of plasticity, called **normal** consistency. Generally, the water required for normal consistency varies from 25 to 32% by weight of cement in most of portland cements. The quantity of water required to make a paste of normal consistency is determined on the basis of penetration of a plunger of 10 mm diameter and a standard weight, in a neat cement paste test block in Vicat apparatus. The percentage of water by weight of cement at which the standard plunger is capable of penetrating the cement paste test block upto **5 mm to 7 mm from the base** of the mould is called the percent of water required for the **paste of normal consistency.**

The Indian standard specifies two arbitrary points which relate to the setting of cement. The initial setting time is defined as the period elapsing between the time when water is added to cement and the time at which **a needle of 1 mm square section fails to pierce the cement paste test block to a depth of about 5 to 7 mm from the bottom** of the mould. A period of **30 minutes** is the minimum initial setting time specified for ordinary and rapid hardening portland cements. Final setting time is defined as the period elapsing between the time when water is added to cement and the time at which **the needle of 1 mm square section (with 5 mm diameter attachment) makes an impression on the cement paste test block, but the attachment fails to make an impression** on the test block (Refer Fig. 3.2). A period of **600 minutes** is the maximum time specified for the final set for various portland cements. Setting times are controlled by using appropriate percentage of gypsum during grinding of clinker (2 to 3%).

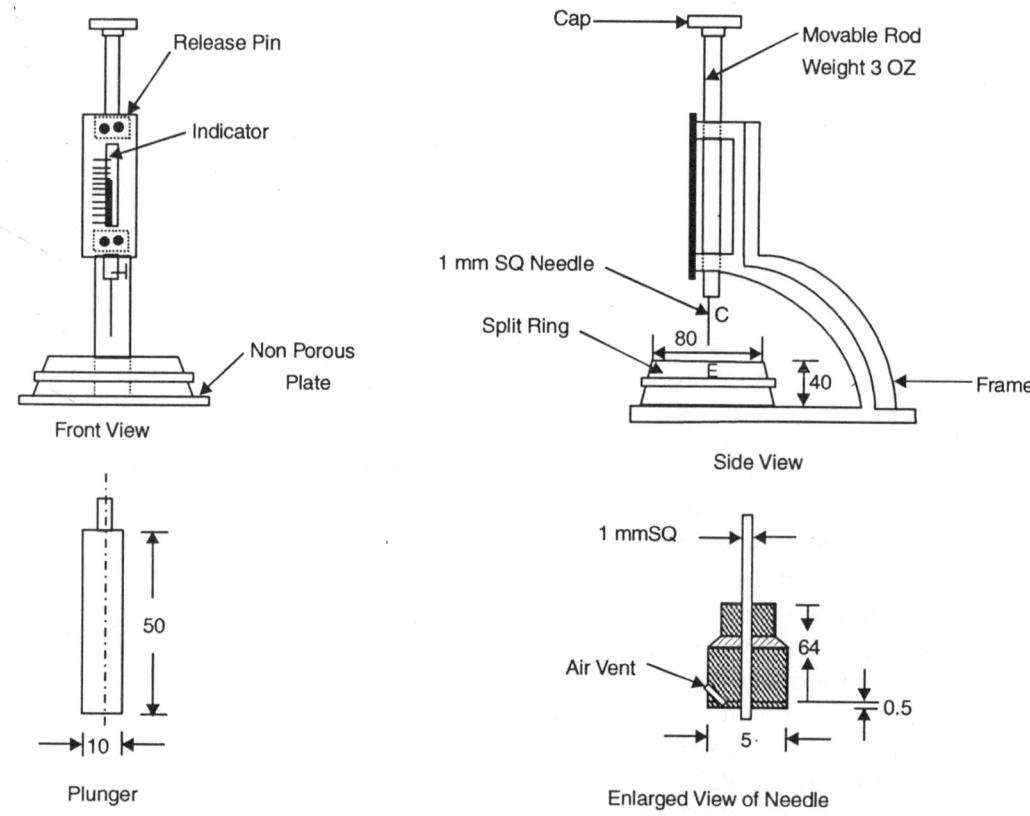

Fig. 3.2. Vicat's Mould

The time of setting of cement depends on **chemical composition, fineness, water content** of paste, and **storage temperature of paste. Fineness reduces the time of setting. Higher quantity of water delays setting.** High temperature of environment of cement paste reduces the setting time.

3.2.3 Strength

The resistance of cement paste to deformation under loads and pressures is called its strength. This gain of strength of cement paste is due to formation of products of hydration. According to H. Lechatelier the strength is due to **high cohesive and adhesive properties of interlaced elongated crystals** produced during hardening of cement paste. According to W Michaelies the strength is due to gelatinous mass of **calcium silicate hydrate** formed during hardening of cement paste. The gel mass gradually hardens and exhibits property of cohesion due to loss of water by external drying and hydration of inner core of cement grains. The gel contains some crystals of **calcium silicate hydrate** including calcium hydroxide. The strength of the gel is due to cohesive bonds created by **physical attraction and chemical bonds.** The cement gel consists of microscopic pores containing gel water and free water. Gel also contains chemically combined water (non evaporable) which reflects the degree of hydration.

The quality of gel and its rate of hydration depends mainly on the compound composition of cement gel and other surrounding conditions. It is found from research that primarily calcium silicate hydrates (C_3S and C_2S) are responsible for the strength of cement paste. C_3S is responsible for an **early strength** (first 4 weeks) while C_2S **influences later strength.**

Cement has the maximum strength in compression but has very little strength in tension (both direct and flexural). Thus it is the compressive strength of hardened cement which forms important consideration for the properties of hardened cement concrete. Since moulding of neat cement paste specimen is difficult, tests are conducted on **standard cement-sand mortar cubes** for determining cement strength. The strength of mortar or concrete depends on the cohesion of cement paste, adhesion of cement paste to the aggregate particles (bond), and on the strength of aggregate particles itself. Test shall, therefore, be conducted with mortars using **standard sand** (as aggregate) and **eliminating the possibility of failure due to poor bond or poor strength of aggregate.** According to Indian Standards mortar cubes are prepared by taking **one part of cement** and **three parts of standard Ennore sand** by adding suitable quantity of water on the basis of the normal consistency of neat cement paste. The mortar cubes are compacted, cured and tested under standard conditions. Standard sand is used to eliminate the effect of variation in properties of sand in determining the compressive strength of cement sample.

The strength of hardened cement concrete is the property required for various structural purposes. The strength tests are, therefore, important to specify the cements. The **compressive strength of cement serves as a guide** and check on the quality of cement and helps to classify the cement as ordinary, rapid hardening, low heat, or pozzolanic portland cement. Cements can also be classified by its grade of strength as **33, 43** or **53 grades**. Grade represent 28 days compressive strength. The compressive strengths are determined at the ages of 1 day, 3 days, 7 days or 28 days depending on the type of cement. The tensile strength test is not considered important for cement and hence it is not specified in standard specifications.

3.2.4 Soundness

Cement paste after it has set, should not undergo a large change in its volume specially when the concrete structure is restrained for movement. These changes in volume induce stresses due to restraint in structures. The volumetric expansion of cement paste after setting causes **cracks, disruption,** and **disintegration of mass** and is called **unsoundness** of cement. These volumetric changes are caused by slow hydration or reaction of free lime, magnesia, and calcium sulphate with water. The volume of products of these reactions increases in relation to the original volume of compounds.

During the burning of raw materials in the kiln, if raw materials contain more lime and magnesia than that which can combine with acidic oxides, these remain in free state. These hard burnt lime and magnesia (specially periclase) slake (hydrate) very slowly and the **volume** of compounds formed after slaking is **very large as compared to the original volume** and hence results in large scale cracks. Thus the cements which exhibit such expansions are called unsound.

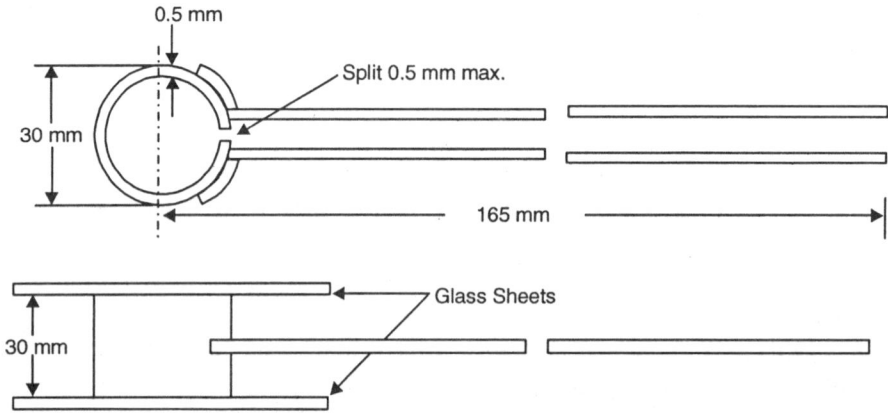

Fig. 3.3. Lechatelier Mould

Lime added to cement does not produce this delayed expansion because it hydrates rapidly before the cement paste has set. Free **lime and magnesia present in clinker get intercrystalized** with other compounds and cannot be fully exposed to water for its hydration before the cement paste has set. Presence of such **crystalline lime and crystalline magnesium oxide is harmful.** Calcium sulphate present in excess of that required for retarding setting action also causes expansion due to formation of calcium sulphoaluminate at a later stage. These **delayed reactions and volumetric expansions** may occur after quite long periods, sometimes, extending to many months or years.

Fine grinding of raw materials brings them into more intimate contact when burned, so that there is less chance of free lime existing in clinker. Thorough burning of raw materials further reduces the quantity of free lime. Fine grinding of the clinker tends to expose the fine particles of free lime for hydration quickly before hardening of cement gel, thus rendering the cement sound.

The method of detecting unsound cements is by **Lechatelier** apparatus (Fig. 3.3) or autoclave test. In case of Lechatelier test, the specimen of cement paste is prepared and cured by boiling in water. The expansion of specimen is measured in the apparatus and expressed in mm. Generally this expansion does not exceed 10 mm for sound cements. In autoclave (steam boiler) test, cement paste bars are prepared and cured at high pressure and temperature **after 24 hours of casting for 3 hours.** The lengths of these bar specimen are compared with the original gauge lengths and the expansion is expressed as percentage. The cements exhibiting not more than 0.50% expansions are considered within sound limits.

3.2.5 Heat of hydration

The chemical **reaction of water and cement** is known as **hydration.** This chemical reaction of hydration of cement is **accompanied by liberation of heat.** This heat liberated during setting and hardening of cement is called heat of hydration. The rate and the quantity of heat of hydration depend on the compound composition of cement and surrounding environment of cement mass. Concrete has a low thermal conductivity and hence in a large concrete mass such as dams, the heat generated in the interior during setting and hardening is not readily dissipated. This causes immediately expansion of concrete mass due to heat of hydration and later contraction on cooling. This expansion and contraction results into cracks in the concrete mass due to restraints in the structure.

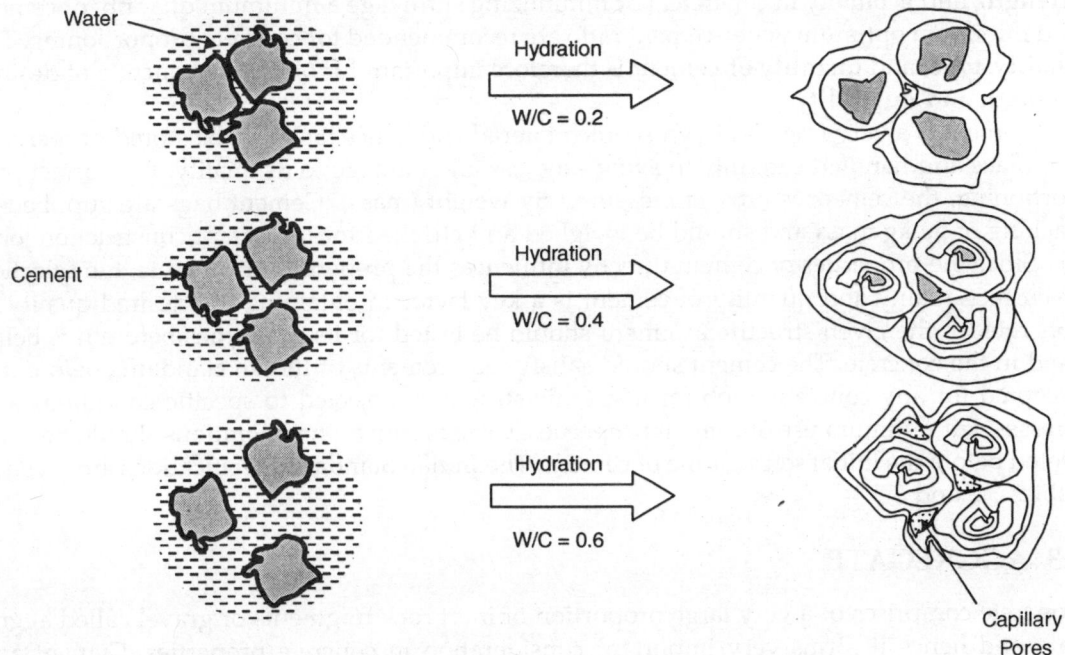

Fig. 3.4. Schematic Representation of Hydration of Cement With Insufficient (W/C = 0.2), Sufficient (W/C = 0.4) And Excess (W/C = 0.6) Water

In construction of massive concrete structures, the temperature of concrete is controlled by using a system of embedded interior cooling pipes and by rate of placement of concrete from the known value of heat of hydration of cement. The amount of heat generation can also be reduced by modifying the cement composition. Tricalcium aluminate (C_3A) and tricalcium silicate (C_3S) are mainly responsible for **rapid generation of heat** and hence by limiting the proportions of these compounds in cement, the heat generation rate and quantity can easily be controlled. During very cold weathers, the heat of hydration of cement can be utilized advantageously to prevent the capillary pore moisture from freezing. Thus **rapid hardening portland cement is more suitable in cold weather** concreting.

The heat of hydration in case of ordinary portland cement normally varies from 85 to 100 calories per gram, while that of low heat portland cement varies from 60 to 70 calories per gram.

3.2.6 Influence of cement on concrete properties

There is increasing use of cement concrete as a construction material for structures serving a variety of functions and subjected to various exposure conditions. These concrete structures in course of time may develop cracks, spalls, or disintegrate due to the action of severe weathering or aggressive chemically charged waters or soils or gases.

Cement reacts with water and forms plastic paste which hardens to impart strength to concrete. The reaction of cement is also responsible for shrinkage. For achieving concrete of desired strength and durability, a suitable type, quality and quantity of cement has to be selected. The selection of **type of cement** is governed mainly by the nature of **exposure of concrete**, while the **quantity of cement** is mainly determined on the basis of **workability, strength,** and **economy** in concrete. For minimizing shrinkage a minimum quantity of cement and minimum optimum water-cement ratio are recommended for concrete proportioning. The **quality, type,** and **quantity of cement** is therefore important for obtaining concrete of desired **strength and durability.**

Cement is a very fine and hygroscopic material and hence it should be stored properly in dry place and handled carefully to avoid any loss and damage to its quality. For correct proportioning, the cement is always measured by weight (mass). Cement bags are supplied in packing of 50 kg mass and should be weighed and checked for important construction jobs.

Since the properties of cement directly influences the properties of concrete, the selection of correct quality and quantity of cement is a key factor in obtaining the desired quality of concrete for the given structure. Cement should be tested for its properties before it is being used in the concrete. The cement should satisfy requirements of Indian standards before it is accepted for any concreting job for a specific structure subjected to specific conditions and stresses. For optimum use of cement properties, various concreting operations should be completed before the initial setting time of cement. The Indian Standard specifications are given in Tables 3.2 and 3.3.

3.3 AGGREGATE

Concrete comprises of a very large proportion of inert rock fragments or gravel called aggregate and hence it forms very important consideration in concrete properties. Cement and water forms paste which combines with aggregate by developing mechanical bond to form hard mass called concrete. Although, earlier, the aggregate was considered as inert material dispersed in the cement paste, but in fact it is not truly inert and its physical, thermal, and sometimes also chemical properties influence the performance of cement concrete. The **aggregate** imparts greater **volumetric stability** to concrete mass by offering greater resistance to deformations caused by shrinkage of cement paste. The aggregate limit the strength of cement concrete as weak aggregate cannot produce strong concrete. The properties of aggregate greatly affect the durability and structural performance of cement concrete. The aggregate is very cheap compared to cement and hence for economical production of cement concrete, maximum (optimum) quantity of aggregate and minimum (optimum) quantity of cement should be used. The selection of **type, size, quantity,** and **grading** of aggregate is done on the basis of **workability requirements** in fresh state, **strength and durability** in hardened state, and **economy** in concrete production.

A suitable grading is done from fine to coarse particles to achieve maximum density of concrete mass. Grading of aggregate particles also affects the workability of fresh plastic concrete and should be selected suitably for optimum **workability, strength,** and **economy** in concrete mass.

Table 3.2 Specifications for Physical Properties of Cements (as per Indian Standard)

Properties	Type of Cement and Relevant IS Code Requirements							
	OPC	RHPC	LHPC	HSPC	PPC	PSC	SSC	HAC
	(i)	(ii)	(iii)	(iv)	(v)	(vi)	(vii)	(viii)
	IS: 269-1989	8041-1990	12600-1989	8112-1989	1489-1991 (Pt. I)	455-1989	6909-1990	6452-1989
1. Fineness								
a) Residue (90mic) not to exceed %	10	5	—	—	—	—	—	—
b) Sp.surface(m²/Kg) not less than	225	325	320	350	300	225	400	225
2. Setting Time (min)								
a) Initial not less than	30	30	60	30	30	30	30	30
b) Final not more than	600	600	600	600	600	600	600	600
3. Strength (N/mm²)								
a) At 1 day (24±0.5 hrs)	—	16	—	—	—	—	—	30
b) At 3 days (72±1 hr)	16	27	10	23	16	16	15	35
c) At 7 days (168±2 hrs)	22	—	16	33	22	22	22	—
d) At 28 days$(672±4hrs)	33	—	35	43	33	33	30	—
4. Soundness								
a. Lechatelier expansion not more than (mm)	10	10	10	10	10	10	5 modified	5
b. Auto-Clave Expansion not more than (%)	0.8	—	0.8	0.8	0.8	—	—	—
5. Max. Heat of Hydration (calories/g)								
a) At 7 days not more than	—	—	65	—	—	—	—	—
b) At 28 days not more than	—	—	75	—	—	—	—	—

(Contd.)

Table 3.2 Specifications for Physical Properties of Cements (as per Indian Standard)

Properties	Type of Cement and Relevant IS Code Requirements							
	OPC	RHPC	LHPC	HSPC	PPC	PSC	SSC	HAC
	(i)	(ii)	(iii)	(iv)	(v)	(vi)	(vii)	(viii)
	IS: 269-1989	8041-1990	12600-1989	8112-1989	1489-1991 (Pt. I)	455-1989	6909-1990	6452-1989
6. Specific gravity (Bulk Density Kg/l)	—	—	—	—	—	—	2.8-2.9 (1.30)	—
7. Drying shrinkage (max percent)	—	—	—	—	0.15	—	—	—
8. Chemical Requirements								
a) Loss on ignition max. %	5.0	5.0	5.0	4.0	—	4	Insoluble Residue (max. 4.0%)	—
b) Wt. of MgO (max.%)	6.0	6.0	6.0	6.0	—	8	10.0	—
c) Sulphuric anhydrate (max.%)	2.75	2.75	2.75	2.75	—	3	6.0 Sulphide Sulphur max.	—
d) Insoluble material (max. %)	2	2	—	2	—	2.5	1.5 %	—
e) Lime Saturation factor %	0.66 – 1.02 max.	0.66 – 1.02 max.	—	—	—	—	—	—

OPC   -   Ordinary Portland Cement
RHPC -   Rapid Hardening Portland Cement
LHPC -   Low Heat Portland Cement
HSOPC - High Strength Ordinary Portland Cement

PPC   -   Portland Pozzolanic cement
PSC   -   Portland Slag Cement
SSC   -   Super Sulphated Cement
HAC   -   High Alumina Cement

For durability of concrete, the aggregate should be **resistant to weathering action**, should not react with cement compounds, and should not contain impurities which affect the strength and soundness of cement paste in the concrete. There may be some interaction, over a long period of time, between cement paste and surface of aggregate particles which may promote bond with concrete mass and may not be harmful in all the cases. For preparing concrete required to exhibit special properties, aggregates of special characteristics (viz. heavy weight, light weight) are used.

Thus for production of concrete of desired **strength, durability,** and **economy,** the aggregate should be considered for its:

- **Physical characteristics** and mix proportions useful for **strength** and **workability;**
- Properties which affect the **durability** of concrete; and
- Properties which offer **special characteristics** to concrete.

3.3.1 Classification of aggregate

Aggregate may be generally classified on the basis of its source, mineralogical composition, mode of preparation, and size. Such classification helps in identifying and selecting suitable aggregate for the desired concrete job. The acceptance of aggregates for use on the required job is done after getting specific information regarding their qualities.

According to source, aggregate may be called **natural** or **artificial** depending on whether the aggregate particles are formed by natural processes, or artificial industrial processes. Natural sands and gravel are produced by weathering and the action of running water, while crushed stones and stone sands are reduced from natural rock by crushing and screening. The rocks may be igneous, sedimentary or metamorphic and may or may not provide good aggregate. The mineral composition of these rocks are mainly siliceous or calcareous alongwith certain secondary minerals. The form in which these minerals are present, makes the aggregate responsible to exhibit certain specific (good or harmful) behaviours in concrete.

Artificial aggregates are produced for some special purposes. Light weight concrete aggregates are produced by burning clay nodules, cinders, and industrial bye products such as blast furnace slag and fly ash. For production of heavy weight concrete, steel nodules, steel balls may be used. Natural aggregate such as pumice is used for light weight concrete, while magnetite (iron-ore) may be used for heavy weight concrete. At some places natural pit run aggregates are also available and may be used for normal concrete preparations after proper washing. According to size, the aggregate is classified as **fine** aggregate (particles **passing 4.75 mm** sieve and **retained on 75 micron** sieve), and **coarse** aggregate (particles passing **80 mm** sieve and **retained on 4.75 mm** sieve). Further the aggregate may be classified as **single size** (containing mostly particles of one size) and **graded** (containing particles of various sizes).

The aggregate should satisfy the required specifications and qualities before used in preparation of cement concrete. Various characteristics of aggregate important for preparation of concrete are studied in subsequent paragraphs.

3.3.2 Aggregate Characteristics

Aggregate has many characteristics which influence the properties of fresh or hardened cement concrete directly or indirectly and forms useful data for the design of mix proportions. These characteristics are listed below:

  (i) **Size**;
 (ii) **Shape**;
(iii) Surface **texture**;
(iv) **Strength**;
 (v) Water absorption and **surface moisture**;
(vi) **Bulking** of sand;
(vii) Unit weight and **bulk density**;
(viii) **Specific gravity**;
 (ix) **Thermal properties**;
  (x) **Deleterious impurities**;
 (xi) **Soundness**;
(xii) **Alkali-aggregate reaction** and durability; and
(xiii) **Gradation** of aggregate.

3.3.3 Size of Aggregate

Size of aggregate is designated by its size of particles. Size of individual particles is defined by **standard sieve size through which particles pass if these are retained on the next lower standard sieve.** An individual particle is said to be 20 mm size, if it passes through a 20 mm sieve and is retained on the next lower standard sieve in the set (i.e. 16 mm sieve). An aggregate heap containing particles of one or many sizes is designated by the **maximum size** of the **particles present in substantial quantity.** For example, the aggregate containing particles of 40 mm, 20mm, and 16mm in sufficient quantities shall be designated as 40 mm size aggregate.

According to IS: 460-1962, the standard set of sieves commonly used in concrete technology are:

80 mm, 63mm, 50 mm, 40mm, 20mm, 16mm, 12.5 mm, 10mm, 6.3 mm, 4.75 mm, 2.36 mm, 1.18 mm, 600 micron, 300 micron, 150 micron, and 75 micron

(1 micron = 10^{-6} m = 10^{-3} mm)

Particle sizes in aggregate varies from **80 mm down to 75 micron. Particle sizes from 60 microns down to 2 microns are classified as silt,** while particle sizes **below 2 micron are classified as clay and does not form part of aggregate.** Silt and clay are harmful to concrete. For the sake of convenience, the aggregate is further classified into **coarse aggregate** and **fine aggregate** according to the **size of particles.** The aggregate particles which are **retained on 4.75 mm sieve and are below 80 mm** sieve size are grouped as **coarse aggregate.** The aggregate particles which **passes through 4.75 mm sieve and retained on 75 micron sieve** are grouped as **fine aggregate.** All-in-aggregate is the combination of both coarse and fine aggregate and the particles range from **75 micron to the maximum size.** Single size aggregate comprises of mainly particles of one and the same size. A graded aggregate comprises of particles of various sizes in different proportions.

Plums (boulders) have 160 mm and upto a reasonable size may be used in plain concrete work upto a maximum limit of 20% by volume of concrete when specially permitted in a particular job. The plum shall be distributed evenly and shall not be closer than 150 mm from the surface. This indicates that plums can be used only for thick layers of concrete more than 300 mm.

The **surface area of particles per unit volume increases** as the **size of particles reduces.** This can be explained by considering a simple cuboid of 20 mm size and reducing it to 8 cuboids of 10 mm size each **(Fig. 3.5).**

20 mm cuboid:

Volume V_o = 8000 mm^3, surface area S_o = **2400 mm^2**, while

8 cuboids of 10 mm:

Volume V_1 = 8 × 1000 mm^3 = 8000 mm^3, surface area S_1 = 8 × 6 × 100 mm^2 = **4800 mm^2**.

Because of change in surface area with size of aggregate, the quantity of cement paste required for certain desired workability also changes. Thus size of aggregate plays very important role in properties of fresh concrete spcially **workability.**

3.3.4 Particle Shape of Aggregate

Particle shape is described in terms of geometrical characteristics. The geometrical characteristics of particles are based on the method of their formation, and strength and abrasion of the parent rock. The shape is measured in terms of **roundness** or **angularity** of the edges and corners of a particle.

According to IS: 383-1970, the **particle shape** can be classified as given in Table 3.4 (Refer **Fig. 3.6**). The degree of packing of particles of one size depends on their shape. The **angularity of aggregate** can be estimated from the **percentage of voids** in a compacted sample. The **angularity number** is defined as the **percentage of voids in a standard compacted aggregate** in excess of that in the rounded gravel (i.e. 33%). **Angularity number** (as well as voids in well compacted aggregate) **increases as the angularity of the aggregate increases.** Angularity number (as well as voids in well compacted aggregate) decreases as the angularity of aggregate decreases or roundness increases. The **angularity number** practically ranges from 0 for completely round to about 11 for completely angular aggregate. **Angularity influence** the total surface area of particles per unit volume and hence the **workability** of the mix.

Table 3.4 Particle Shape

Classification	Description	Example
Rounded	Fully water-worn or completely. shaped by attrition	River or seashore gravels; Desert, seashore and wind blown sands
Irregular or partly Rounded	Naturally irregular, or partly shaped by attrition and having rounded edges	Pit sands and gravels; dug flints; cuboid rocks
Angular	Possessing well defined edges formed at the intersection of roughly planar faces	Crushed rocks of all types; Talus; screes; crushed slag
Flaky	Material of which the thickness is small relative to the width and/or length	Laminated rocks
Elongated	Material, usually angular in which the length is considerably larger than the width and thickness	—

The particles may also be elongated and flaky type. **Flaky** particles are those **whose thickness is less than 0.6 times the average sieve size** (average of sieve sizes through which

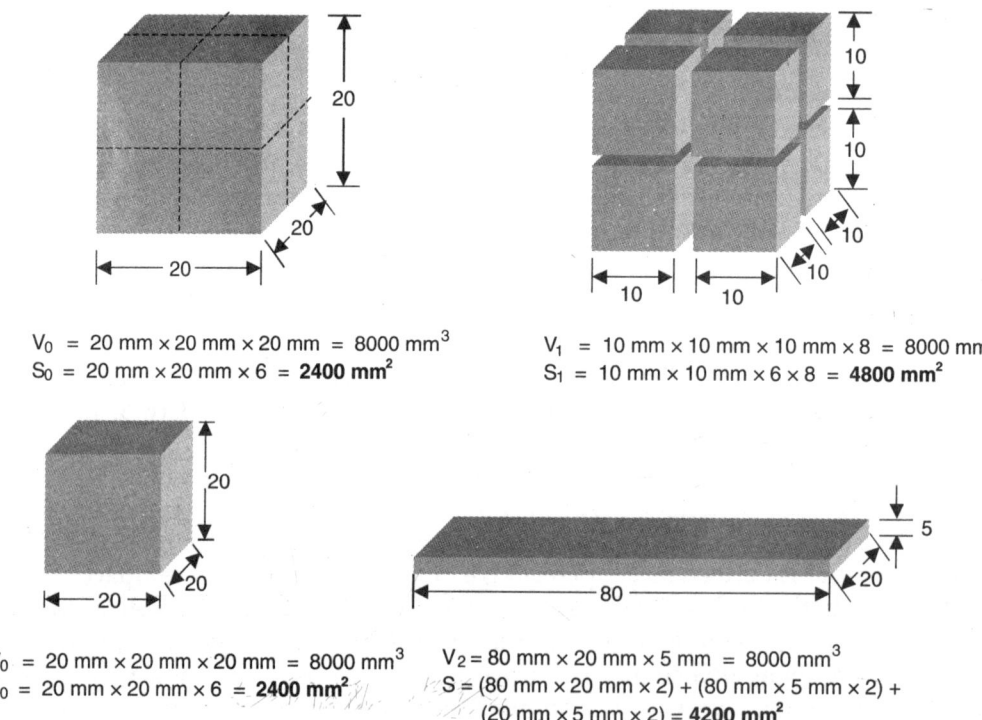

V_0 = 20 mm × 20 mm × 20 mm = 8000 mm^3
S_0 = 20 mm × 20 mm × 6 = **2400 mm^2**

V_1 = 10 mm × 10 mm × 10 mm × 8 = 8000 mm^3
S_1 = 10 mm × 10 mm × 6 × 8 = **4800 mm^2**

V_0 = 20 mm × 20 mm × 20 mm = 8000 mm^3
S_0 = 20 mm × 20 mm × 6 = **2400 mm^2**

V_2 = 80 mm × 20 mm × 5 mm = 8000 mm^3
S = (80 mm × 20 mm × 2) + (80 mm × 5 mm × 2) + (20 mm × 5 mm × 2) = **4200 mm^2**

Fig. 3.5. Influence of Particle Size and Shape on Surface Area for a Given Volume

it passes and on which it is retained). **Elongated** particles are those **whose length is more than 1.8 times** the average sieve size. **Flakiness index** is the **percent by weight of flaky particles** in the aggregate. **Elongation index** is the **percent by weight of elongated particles** in aggregate. Generally flaky and elongated particles in excess of **10 to 15 %** of coarse aggregate **affects the workability and durability** of concrete adversely due to orientation of such particles in one plane and forming **air voids** and **water pockets.** No limit of flakiness is specified in Indian standard but British Standard BS: 63 specifies a maximum limit of **35 and 40 % flakiness** for aggregates of 6.3 mm to 25 mm and 25 mm to 50 mm respectively.

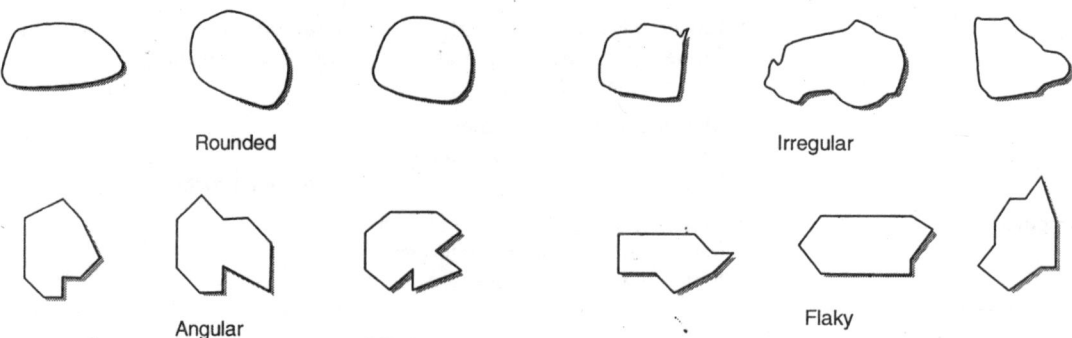

Rounded

Irregular

Angular

Flaky

Fig. 3.6. Particle Shape

3.3.5 Surface Texture

Surface texture of aggregate particles describe the **nature of the surface.** It depends on the **hardness,** the **grain size** and **pore characteristics** of the parent rock material. Table 3.5 specifies surface textures as given in IS: 383-1970 and BS: 812-1960.

Table 3.5 Surface Texture of Aggregate

Surface Texture	Characteristics	Example
Glassy	Conchoidal fracture	Black Flint, Vitreous Slag
Smooth	Water worn or smooth due to fracture of laminated or fine grained rock	Gravel, Chert, Slate, Marble, Stone, Rhyolite
Granular	Fracture showing more or less uniform rounded grains	Sand Stone, Volite
Rough	Rough fracture of fine or medium grained rock containing no easily visible crystalline constituents	Basalt, Felsite, Lime Stone
Crystalline	Containing easily visible crystalline constituents	Granite, Gabbro, Gneiss
Honeycombed and porous	With visible pores and cavities	Brick, Pumice, Foamed Slag, Clinker, Expanded Clay

The **surface texture influences** the surface area and hence the **workability and strength.** **Rougher** the surface, **greater is its area** and results in a greater adhesive force between the particles and the cement matrix thus **reducing the workability.** Due to roughness of surface better bond between cement paste and aggregate is realised. Sometimes the bond is developed due to chemical composition of aggregate particle's surface and cement paste. It is difficult to measure this bond of aggregate and cement paste directly but judged indirectly from the fractured surface of concrete. Surface texture becomes important in high strength concrete specially in air field pavements.

3.3.6 Strength

The compressive strength of concrete depends on the strength of the bulk of aggregate. It is not easy to directly measure the crushing strength of aggregate.

The strength of aggregate is measured by indirect methods so as to check its suitability in concrete. These indirect methods are:

- Compressive strength of prepared prismatic sample from the **parent rock** of aggregate;
- **Aggregate crushing** value or **aggregate impact** value; and
- **Performance** of aggregate **in concrete.**

Preparing parent rock samples are costly and difficult and hence this method is not very commonly used. Performance of aggregate in concrete is judged from the previous experience of using a particular aggregate in a given concrete or by trial use of such aggregate in desired concrete. If the aggregate under test leads to a lower compressive strength of concrete, and there are number of fractured aggregate particles, it indicates that the aggregate strength is less than the normal strength of concrete mix. Although there is no direct relationship of **aggregate crushing value**, or **impact value** and compressive strength, but **qualitatively** there is sufficient agreement in the two tests. The aggregate crushing value or impact value represent indirectly the value of compressive strength. These tests can also be carried out easily. The aggregate crushing value, impact value and abrasion value tests are described in IS: 2386 (part iv)-1963 (Fig. 3.7).

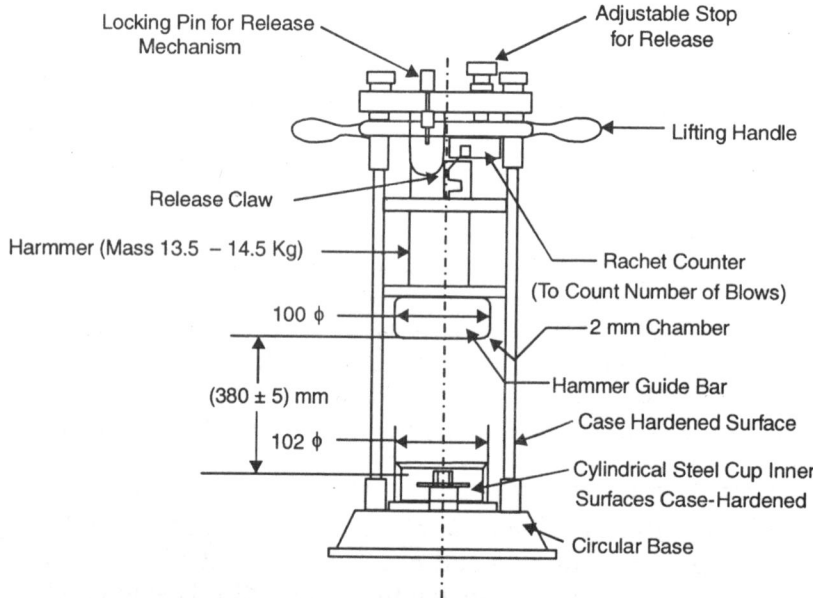

Fig. 3.7. Aggregate Impact Test Machine

To determine the aggregate impact value, a sample is obtained from the aggregate passing through **12.5 mm sieve and retained on 10mm sieve.** The same is cleaned and dried upto 100 ^{0}C, cooled, and filled in a standard cylindrical mould and tamped as specified in the code.

The sample is subjected to **15 blows of standard hammer** and then sieved through a 2.36 mm IS sieve and the material passing is weighed. The **aggregate impact value** is defined as the **fraction of material passing 2.36 mm IS sieve by mass of the total sample** (passing 12.5 mm and retained 10 mm). Similarly **aggregate crushing value** is also represented as the **fraction of material passing 2.36 mm Sieve** after subjecting the sample to a **compression of 40000 kgf (400KN)** by mass of total sample (passing 12.5mm and retained 10mm). Another test of strength of aggregate is conducted through **LOS Angeles** test of abrasion. In this test the sample of aggregate is placed in cylindrical drum alongwith a charge of **steel balls** and the drum is rotated for specified number of revolutions. The falling of aggregate and steel balls causes abrasion which is measured as **percentage by mass of material passing 2.36 mm sieve** with respect to the mass of the sample.

As a general guide, **higher crushing values**, impact values or abrasion values **represent lower compressive strength** of aggregate. IS: 383-1970 specifies that the aggregate used in concrete shall satisfy the following requirements:

	Aggregate Crushing Value	Aggregate Impact Value	Aggregate Abrasion Value
(i) Concrete in Wearing surfaces (shall not Exceed)	30 %	30 %	30 %
(ii) Concrete in other than Wearing surface (shall not Exceed)	45 %	45 %	50 %

3.3.7 Water Absorption and Surface Moisture

Aggregate particles have minute **permeable and impermeable pores** of various sizes which influence the water absorption and **permeability** characteristics of aggregate. These characteristics in turn affect the **bond** between cement paste and aggregate, **resistance of concrete to freezing and thawing,** and **chemical stability.** The pores in aggregate may exist as closed one (inside the solids) or as open to surface of the particles. Water can enter or evaporate from permeable pores. The moisture condition of aggregate can be determined according to tests prescribed in IS: 2386-1963 (part iii).

According to the moisture condition of aggregate various states are described as given below:

(i) **Bone dry** (or oven dry): In this condition the moisture from all permeable pores is completely evaporated. This state is reached if aggregate is heated in an oven at 100°C for about 24 hours.

(ii) **Air dry**: When moist aggregate is allowed to dry in air, part of the moisture from the permeable pores evaporates, and such a state of aggregate is called air dry.

(iii) **Saturated surface dry**: When the permeable pores are completely filled with moisture, but there is no free moisture on the surface of aggregate, such a state is called saturated surface dry. This is an important balanced state of moisture in aggregate for its performance in concrete.

(iv) **Damp** (or wet): In this state, aggregate have all the pores fully saturated and also contains some free moisture on the surface.

These four states of aggregate are shown diagrammatically in **Fig. 3.8.**

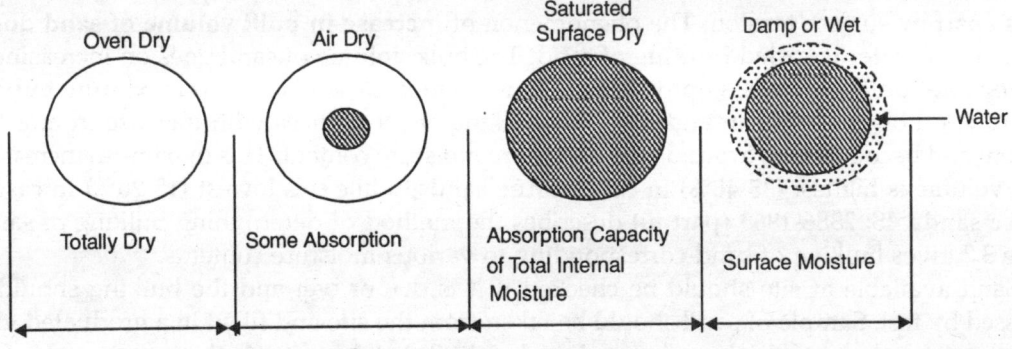

Fig. 3.8. Moisture State of Aggregate

The knowledge of moisture condition of aggregate at site is an important consideration in proportioning of concrete mix and quality control. Based on the moisture condition of aggregate the mix proportions (including quantity of mixing water) are adjusted at site so that correct workability and cement hydration are achieved as envisaged in the mix design. IS: 456-1978 specifies approximate amount of surface moisture carried by aggregate in its different conditions of wetness in Table 3.6.

Table 3.6 Approximate Surface Water Carried by Aggregate

	Condition of Wetness Percent by mass	Quantity of surface water litre/m^3
i. Very wet sand	7.5	120
ii. Moderately wet sand	5.0	80
iii. Moist sand	2.5	40
iv. Moist gravel or crushed aggregate	1.25 to 2.5	20 to 40

The aggregate shall be used in its saturated surface dry condition. The coarse aggregate can be brought to the saturated surface dry condition by sprinkling water 24 hours before using in concrete. If it is not done, necessary adjustments in the quantity of water must be made depending on the moisture condition of aggregate. The approximate moisture condition of aggregate can be determined at site by visual inspection or accurately by laboratory tests depending on the importance and quality of the concrete job. The moisture condition may be estimated as percent by mass of water in relation to aggregate mass.

Completely dried or oven dried aggregate needs certain additional quantity of water to become saturated surface dry. This quantity of water can be found by tests. Similarly water requirement or excess of surface water available can be determined and estimated for suitable adjustments in water in mix proportions. The presence of moisture in fine aggregate or sand causes certain volumetric changes which should be studied properly.

3.3.8 Bulking of Sand

When sand is **moist,** its **volume is more than its actual volume.** The increase in volume of sand is due to formation of film of water around sand particles which push these sand particles apart by surface tension. The phenomenon of **increase in bulk volume of sand due to presence of water is called bulking** of sand. The bulk volume of sand goes on increasing as its moisture content increases upto a certain limit. In most of sands there is **maximum bulking at 5 to 6%** moisture content, and then this bulking effect reduces with increase in moisture content and becomes **zero around 20% and more** moisture content. This maximum increase in bulk volume is highest **(35-40%) in case of fine sands,** while it is lowest **(15-20%) in case of coarse sands.** IS: 2386-1963 (part iii) describes the method of determining bulking of sands. Table 3.7 gives bulking of sand corresponding to various moisture contents.

Sand available at site should be checked if it is dry or wet and the bulking should be assessed by test. Sample of sand should be taken from the site and filled in a graduated glass jar upto certain height loosely and record the height (h_1). Add water in the jar till sand is fully submerged. Stir the jar and keep for some time. Due to excess water sand comes to its original volume without bulking. Measure the level of sand (h_2).

$$\boxed{\text{Bulking} = \frac{100(h_1 - h_2)}{h_2}\%}$$

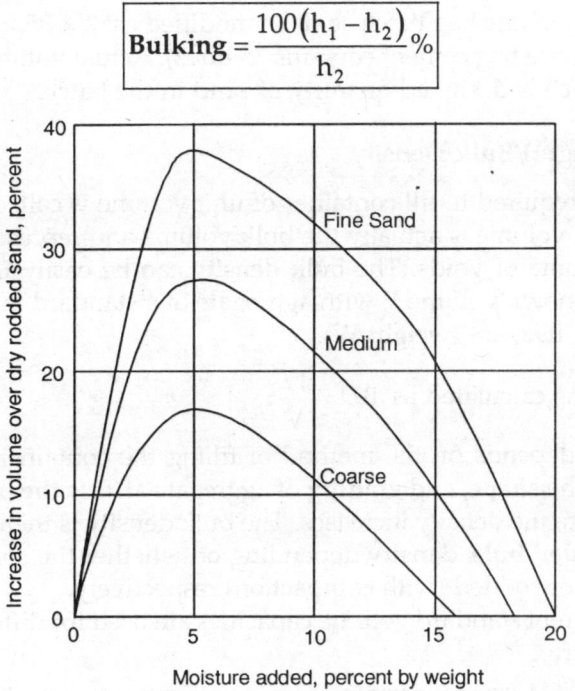

Fig. 3.9. Bulking of Sand

Table 3.7 Bulking of Sand for various Moisture Contents

| Moisture Percent (i) | Percentage Bulking | | |
	Fine Sand (ii)	Medium Sand (iii)	Coarse Sand (iv)
1	16	8	6
2	16	16	12
3	32	22	15
4	36	27	17
5	38	29	18
6	37	28	18
8	35	26	16
10	32	22	12
12	28	19	8
15	22	12	2
17	18	7	0
20	0	0	0
27	0	0	0

Volume measurement of sand must be modified to account for bulking effect in moist sand at site. Water content should also be estimated and adjusted from the designed quantity of water. For example if mix proportions by volume are 1 : 2 : 5 and sand at site has 25 % bulking, the sand measurement in one bag batch shall be modified as: $2 \times 35 \times (1 + 25/100) = 88$ litres of moist bulked sand (one bag cement contains 35 litres). Actual volume of sand shall be $88 \times 1/1.25 = 70$ litres, which is designed quantity of sand in the batch.

3.3.9 Unit Mass (Weight)/Bulk Density

The mass of material required to fill container of unit volume is called unit mass (weight) or bulk density. The unit volume is actually the bulk volume and represents the sum of volume of solids and the volume of voids. The bulk density can be easily measured by filling the containers of certain known volume V with aggregate in a standard manner levelled with the brim and determining its mass (weight-W).

Bulk density is thus calculated as $\mathbf{BD} = \dfrac{W}{V}$

The bulk density depends on the method of filling the container and how densely the aggregate is packed. The **shape**, and **grading** of aggregate **affects the packing** and with better packing by compaction, the density increases. The bulk density is therefore specified as **loose bulk density** and **rodded bulk density** depending on whether the container is filled loosely (without compaction) or rodded (with compaction) respectively.

Containers of different standard volume capacities are used for different sizes of aggregate and are given as under:

Size of Aggregate:	4.75mm and under	4.75mm to 40 mm	over 40mm
Volume of Container:	3 litres	15 litres	30 litres

Method of determining bulk density is-specified in IS: 2386-1963 (part iii).

Various factors affecting bulk density of aggregate should be considered and specified while measuring the bulk density. These factors are:

Degree of compaction	Higher compaction leads to higher bulk density
Shape of particle	With the same compaction rounded aggregate has higher bulk density than angular aggregate.
Grading of aggregate	Well-graded aggregate has lesser voids and hence higher bulk density compared to single size aggregate, provided the compacting effort is the same.

Some values of bulk densities are given in Table 3.8

Table 3.8 Bulk Densities

Material/Aggregate Type	Bulk Density (kg/litre)
Cement	1.44
River sand: Fine	1.44
Medium	1.52
Coarse	1.60
Beach or river single	1.60
Crushed stone	1.60
Stone screening	1.44
Broken granite	1.88

For batch purposes loose bulk density is used for conversion of mass (weight) of aggregate into bulk volume or vice versa. Rodded bulk density is used for detecting the uniformity of shape and grading of aggregate.

3.3.10 Specific Gravity

Aggregate generally contains pores, both permeable and impermeable, the term specific gravity has to be carefully defined with reference to these pores. Absolute specific gravity refers to the weight of the solids to the weight of an equal volume of distilled water both considered at the same temperature. In order to eliminate the effect of enclosed impermeable pores, the **material needs to be pulverised.** This is not very important in relation to concrete.

Apparent specific gravity is defined as the ratio of weight of aggregate solids to the weight of distilled water of an equal volume of solids excluding permeable pores. Specific gravity based on surface saturated condition is defined as the ratio of **weight of aggregate solids in saturated surface dry condition** to the weight of distilled water of an equal volume to that of **solids including permeable pores.**

IS: 2386-1963(part iii) specifies methods of determining the specific gravity for different sizes of aggregate.

Most of aggregates have specific gravity (SSD basis) between 2.5 and 2.90. Specific gravity of aggregate form an important parameter in design of concrete mix proportions and for determining void ratio in aggregate. Heavy weight concrete needs aggregate of high specific gravity while light weight concrete needs aggregate of low specific gravity. Values of specific gravity of some aggregates are given in Table 3.9.

Table 3.9 Specific Gravity (Saturated Surface Dry Basis)

Sr. No.	Material	Sp. Gravity
1.	Cement	3.15
2.	Traps	2.90
3.	Granite	2.80
4	Gravel	2.66
5	Sand	2.65

3.3.11 Thermal Properties

Thermal properties of aggregate affects the performance of concrete. These properties are coefficient of thermal expansion, specific heat, and conductivity. Specific heat and conductivity becomes important in mass concrete and where insulation characteristics are desired. The coefficient of thermal expansion of aggregate influences this coefficient of concrete directly. Higher coefficient of expansion of aggregate leads to higher coefficient of expansion for the concrete. Large difference in coefficient of expansion of aggregate and cement paste with high change in temperature leads to distress. The thermal coefficient of expansion for majority of aggregates lies between 5×10^{-6} per °C and 13×10^{-6} per °C while that of cement paste lies between 10×10^{-6} and 16×10^{-6} per °C. Quartz changes state at 573 °C and expands suddenly and hence should not be used for fire resistant concrete.

3.3.12 Deleterious Substances

There are three categories of deleterious substances generally present in aggregate: Impurities interfering with hydration of cement such as organic matter; Coatings preventing the development of bond between aggregate and cement paste such as silt and clays; and Individual weak and unsound particles such as shale and coal.

These deleterious substances, if present in sufficient quantities, affect the concrete strength and durability adversely. IS: 383- 1970 specifies the limits of deleterious material as given in Table 3.10.

Apart from standard tests, simple test for detection of silt impurities may be carried out with a glass measuring cylinder of 250 ml capacity. Add one percent common salt solution in the cylinder upto 50 ml mark. Add aggregate to fill the jar half full. Add some more solution, vigorously shake the contents and keep the cylinder for 3 hours for settlement of silt. Silt settles slowly and forms a distinct layer at top which can be measured. The percentage of silt should not be more than 5-6 percent, and for high silt content the aggregate should be washed thoroughly with clean water. The aggregate should not contain large quantities of weak and unsound particles.

Table 3.10 Limits of Deleterious Materials

Sr. No.	Deleterious Substance	IS Code for Test	Fine Aggregate Percent By Mass Maximum		Coarse Aggregate Percent By Mass Maximum	
			Uncrushed	Crushed	Uncrushed	Crushed
1	Coal and lignite	IS: 2386-63 (part II)	1.0	1.0	1.0	1.0
2	Clay lumps	-do-	1.0	1.0	1.0	1.0
3	Materials finer than 75 micron sieve	IS: 2386- 63 (part I)	3.0	1.50	3.0	3.0
4	Soft fragments	IS: 2386-63 (part II)	—	—	3.0	—
5	Shale	-do-	1.0	—	—	—
6	Total percentages of all deleterious materials (except mica)	—	5.0	2.0	5.0	5.0

3.3.13 Soundness

Soundness of aggregate is its **resistance to disintegration** by volumetric changes under physical forces such as heating and cooling, wetting and drying, freezing and thawing due to climatic changes. Aggregate is said to be unsound when volume changes occur due to above causes and result in disintegration of concrete. This may range from local scaling to extensive surface cracking and disintegration of concrete.

Unsoundness is exhibited by porous Cherts, some Shales, some lime stones containing expansive clay, some dolarite and some sand stones. Characteristics of aggregate mainly makes it sound or unsound.

Soundness can be determined as specified in IS: 2386-1963 (part v). This test popularly known as "sulphate test" consists of subjecting a graded and weighed sample of aggregate to alternate cycles of drying and chemical immersion. Alternate **immersion** of aggregate in saturated solution of **sodium sulphate** or **magnesium sulphate** and **oven drying** is carried out for determining the weight loss after specified cycles of immersion and drying. The test is not really simulation of conditions of exposure of freezing and thawing and provides only some guidance. The limits specified in IS: 383-1970 should be carefully used alongwith the actual performance of aggregate from the service-record. As a general guide, the code specifies that the limits of average **loss of weight after 5 cycles** shall not exceed the following:

Fine aggregate **10%** when tested with Sodium sulphate, and **15%** when tested with magnesium sulphate

Coarse aggregate **12%** when tested with sodium sulphate and **18%** when tested with magnesium sulphate.

Weigh the sample of aggregate passing certain sieve size and retained on the next standard sieve size before the test. The sample is subjected to alternate cycles of sulphate test and weight is taken after sieving through the same sieve. **Percentage loss of mass (weight)** is measured to indicate soundness.

3.3.14 Alkali-Aggregate Reaction and Durability

Durability of aggregate is its resistance to disintegration due to chemical reaction of aggregate with cement paste. This reaction takes place between **active silica or carbonate** constituents sometimes **present in aggregate** and **the alkalies in cement.** This reaction causes excessive expansion and results in cracking of concrete. Such deleterious reactions are encountered in various climatic zones and are basically due to high alkali content of cement (more than 0.6 % expressed in terms of Na_2O) and presence of active silica in aggregate.

Alkali-aggregate reaction is determined with the help of "mortar bar test" as specified in IS: 2386-1963 (part vii). The test measures the expansion developed by the cement-aggregate combination in mortar bar 100mm in length and 25 mm × 25 mm in section.

Although, the test results are quite useful, but it takes several months and necessitates crushing of coarse aggregate. If the mortar bar expansion exceeds **0.05 percent at 6 months** and **0.10% at 12 months,** the aggregate is considered as **potentially reactive.** Alternative test for determining the potential reactivity of aggregate is the chemical method using sodium hydroxide solution. Although the results are obtained in 3 days but the test results are not reliable and conclusive.

The problem of alkali-aggregate reaction can be overcome by using cements with low alkali content (less than 0.6% alkali calculated as Na_2O) or by adding finely ground suitable pozzolana to the concrete mix. Pozzolana reacts with the alkalis before they attack the reactive aggregates. Generally natural aggregate in India are not reactive except some of rock aggregates containing highly granulated quartz. Some sand stones and quartzites containing more than 5% cherts also show deleterious reactions.

3.3.15 Grading of Aggregate

Particle size distribution of aggregate is known as its gradation. The aggregate is said to be graded if it contains particles of various sizes in different proportions. The size of aggregate particles is designated by the **sieve size they pass** and is determined by sieve analysis. Set of IS sieves used for sieve analysis of aggregate are as follows:

80mm, 63mm, 50mm, 40mm, 31.5mm, 20mm, 16mm, 12.5mm, 10mm, 6.3mm, 4.75mm, (square hole perforated plate type). **3.35mm, 2.36mm, 1.18mm, 600 micron, 300 micron, 150 micron, 75 micron** (fine mesh wire cloth type).

Aggregate is sieved through the set of sieves and the material retained on each of sieves is weighed. From these weights necessary grading curves are drawn and fineness modulus is also calculated for studying grading pattern.

Depending on the sieve analysis data, the grading can be called **"continuous"** or **"gap"** grading. Aggregate is said to be continuously graded when it contains all particle size groups from the maximum particle size to the minimum size in sufficient proportions and when these proportions increase or decrease progressively. These gradings are further said to be **coarser or finer** according as they contain a higher proportions of coarser or finer particles with reference to the specified grading with which comparison is being made. Aggregate is said to be gap graded when certain sizes of particles are missing in the whole lot.

The grading may also be called irregular/poor if it has **excess or deficient** proportions of certain intermediate particle sizes and the **grading is not continuous** or gap graded.

The grading of aggregate has its influence on **voids** content, **mortar** requirement, **cohesiveness** or segregation, **harshness, bleeding,** and **strength** of concrete.

Well-graded aggregate has less volume of voids than aggregates of single size or irregular/poorly graded aggregate. This can easily be verified by taking samples of single size aggregate, poorly graded aggregate and well graded aggregate in graduated glass jars and filling these jars, with measured quantity of water upto the surface of samples. It will be found that **well graded** aggregate require **minimum quantity of water** for filling the voids. Also the percent of voids remain approximately the same for different sizes. For continuous and **coarser grading voids are less** and hence the mortar required to fill the voids shall also be less.

Coarser grading, specially if not well graded, tends to segregate and hence for proper cohesiveness there should be sufficient finer material in the mix. Continuously graded aggregate produce smooth surface, while irregular grading produce harsh mixes specially when finer material is deficient. **Coarser grading** alongwith excessive water content leads to **bleeding.** Finer grading alongwith just sufficient quantity of water shall prevent bleeding in concrete mix.

Grading influences indirectly the strength of concrete. **Coarser grading** results in lower surface area and requires lesser quantity of water for certain desired workability and hence makes concrete **economical for the desired strength.** For low water-cement ratio, coarser grading providing lesser surface area gives better workability and compaction and higher strength. **Finer grading** having more surface area requires more water or higher water-cement ratio and hence **lower strength** of concrete mix.

The method of sieve analysis is described in IS: 2386-1963 (part i). A known mass of aggregate is sieved through a set of standard IS sieves and the material retained on each sieve is weighed and tabulated as shown in example given in Table 3.11. Cumulative percentage retained and passing each sieve are calculated from the data of sieve analysis, **grading curves are drawn from cumulative percentages passing** and fineness modulus calculated from **cumulative percentages retained.**

Table 3.11 Example of Sieve Analysis - Sample 1000 gm

I.S. Sieve Size	Mass retained (gram)	Percentage mass retained	Cumulative Percentage retained	Cumulative Percentage passing	Remarks
i	ii	iii	iv	v	vi
80mm	0.0	0.0	0.0	100.0	
63mm	0.0	0.0	0.0	100.0	
40mm	0.0	0.0	0.0	100.0	
20mm	7.0	0.7	0.7	99.3	
16mm	145.0	14.5	15.2	84.8	
12.5mm	510.0	51.0	66.2	33.8	
10mm	120	12.0	78.2	21.8	
4.75mm	133	13.3	91.5	8.5	
2.36mm	51.0	5.1	96.6	3.4	
1.18mm	25.0	2.5	99.1	0.9	
600micron	02.0	0.2	99.3	0.7	
300micron	01.0	0.1	99.4	0.6	
150micron	01.0	0.1	99.5	0.5	
Finer than 150 micron	05.0	0.5	—	—	
			745.7		

Total Commulative % retained = 745.7

Fineness Modulus = 7.457 ≈ 7.46

Fineness modulus of aggregate is defined as the **sum of the cumulative percentages retained** on the standard set of sieves from smallest to the largest sieve size present **divided by 100**. The fineness modulus (**FM**) can be looked upon as a **weighted average size of sieve** on which the material is retained, the sieves being counted from the finest. For example, a **FM** of 7.46 indicates that the **average size lies between 7th and 8th sieve** i.e. between 10 mm and 12.5 mm sieves. FM for coarse aggregate is higher than fine aggregate. The **fineness modulus increases with coarser grading.** The fineness modulus can not represent the distribution and the same FM may represent a large number of particle size distributions or grading curves. The fineness modulus is therefore not a good way of describing the grading of aggregate, although it is a useful tool for checking variations in the aggregate from the same source from time to time.

Grading curves are prepared on a semilog graph representing percentage passing along ordinate (on linear scale) and the sieve size along abscissa (on **logarithmic scale**). The graph provides particle size distribution of aggregate at a glance and it can be compared with standard grading curves for improvement of given sample by mixing certain sizes in the desired proportion. Steep slope of grading curve at a particular sieve size indicates greater proportion of a particular particle size, while a flat slope indicates lesser proportion of that size. It is very difficult to have exactly the same grading as specified and hence **grading zones** are specified giving **ranges of grading.** Generally, if grading curve lies within the specified zone it is accepted. If a grading curve of a given sample lies above the specified grading, it is said to be

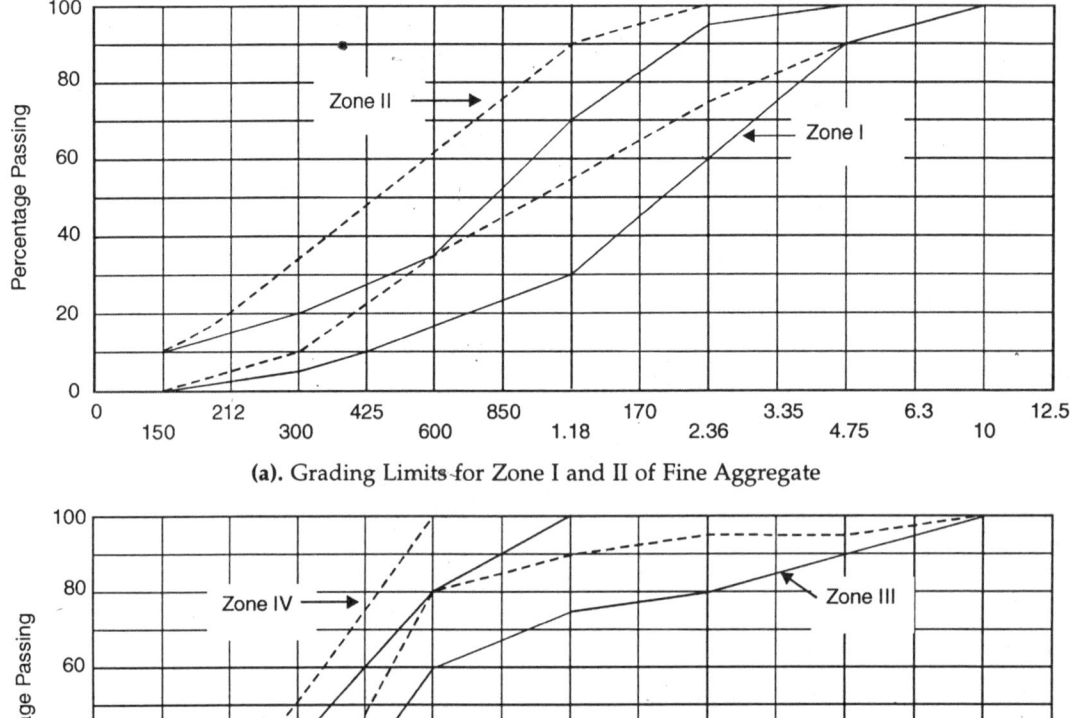

(a). Grading Limits for Zone I and II of Fine Aggregate

(b). Grading Limits for Zone III and IV of Fine Aggregate

Fig. 3.10. Grading Curves of Fine Aggregate

Table 3.12 Fine Aggregate Limits (Sand Zones)

I.S. Sieve Size	Percentage Passing				Remarks
	Grading Zone I	Grading Zone II	Grading Zone III	Grading Zone IV	
10mm	100	100	100	100	For crushed Stone
4.75mm	90-100	90-100	90-100	95-100	sands, permissible-
2.36mm	60-95	75-100	85-100	95-100	Limits on 150 micron-
1.18mm	30-70	55-90	75-100	90-100	sieve is increased to
600 micron	15-34	35-59	60-79	80-100	**20 percent** without
300 micron	5-20	8-30	12-40	15-50	**affecting** 5 percent
150 micron	0-10	0-10	0-10	0-15	allowancepermitted
					in other **sieves**

* **Note:** Zone classification is mainly based on 600 micron sieve.

finer than the specified grading. If a **grading curve** of a given sample **lies below** the specified grading, it is said to be **coarser than the specified grading.** Horizontal **flat portion** in a grading curve **indicates missing of a particular size** and represents **gap grading.**

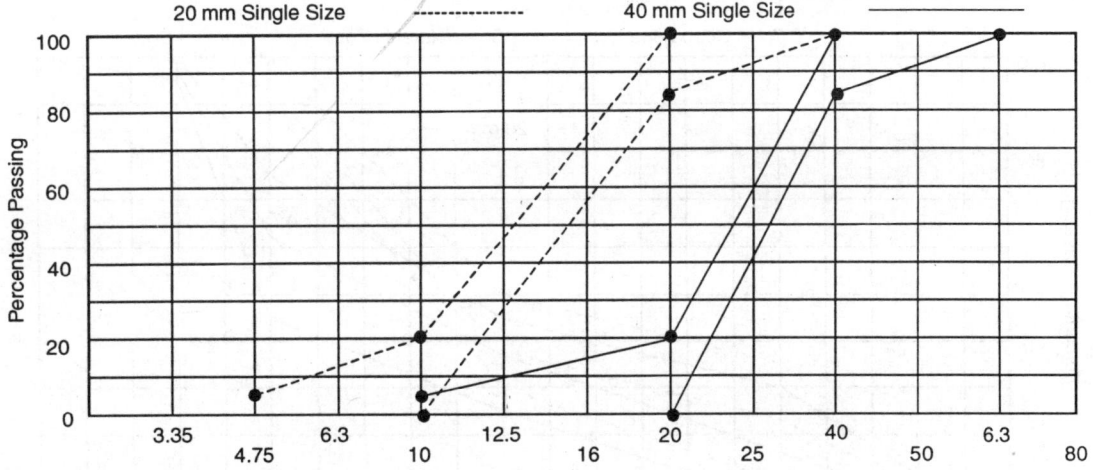

(a). Grading Limits for Coarse Aggregate 20 mm and 40 mm Single Size

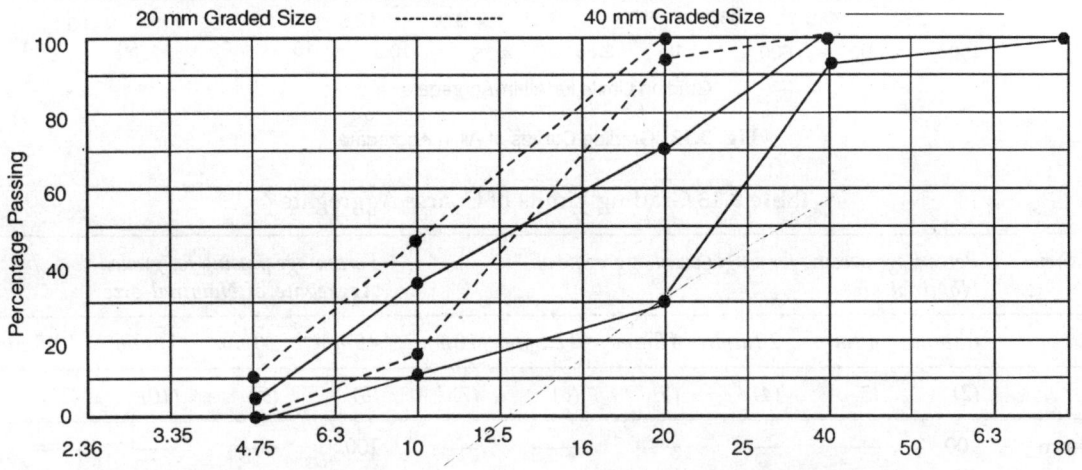

(b). Grading Limits for Coarse Aggregate 20 mm and 40 mm Graded

Fig. 3.11. Grading of Coarse Aggregate

For preparing good concrete mixes there are standard grading Zones specified in IS: 383-1970 for single size coarse aggregate, graded coarse aggregate, all in aggregate, and fine aggregate. According to grading, sands (fine aggregate) are grouped in 4 Zones: **Zone I, Zone II, Zone III** and **Zone IV** (Refer Table 3.12). **Zone I is coarsest grading,** while **Zone IV is the finest** of all and as far as possible it should not be used for important structural concrete. Most commonly available Indian sands are of **Zone II** and **Zone III.** These 4 sand grading Zones are shown in **Fig. 3.10 (a) & (b).** Grading limits of coarse aggregate of single size and graded type is shown in **Fig. 3.11,** and limits are given in **Table 3.13.** Grading limits of all in aggregate is shown in **Fig. 3.12** and given in **Table 3.14** and such a grading is obtained by combining coarse aggregate and fine aggregate in suitable proportion. Specified grading can be obtained by combining different single size or graded aggregates in suitable proportions determined by analytical or graphical method.

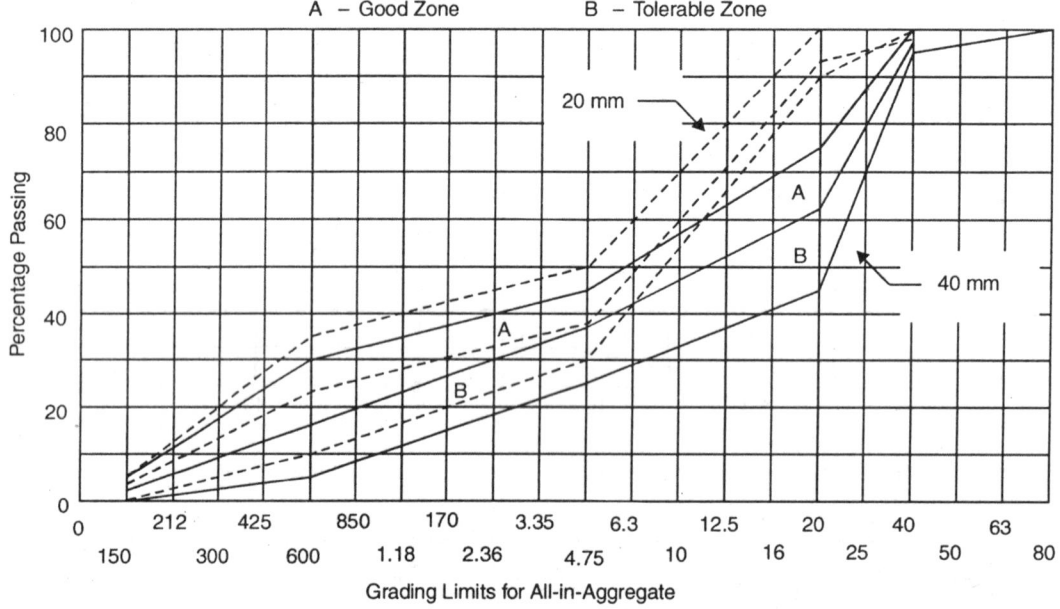

Fig. 3.12. Grading Curves of All in Aggregate

Table 3.13 Grading Limits of Coarse Aggregate

IS Sieve	Percentage passing for single-sized aggregate of Nominal size						Percentage passing for graded Aggregate of Nominal size			
	63mm	40mm	20mm	16mm	12.5mm	10mm	40mm	20mm	16mm	12.5mm
(1)	(2)	(3)	(4)	(5)	(6)	(7)	(8)	(9)	(10)	(11)
80mm	100	—	—	—	—	—	100	—	—	—
63mm	85 to 100	100	—	—	—	—	—	—	—	—
40mm	0 to 30	85 to 100	100	—	—	—	95 to 100	100	—	—
20mm	0 to 5	0 to 20	85 to 100	100	—	—	30 to 70	95 to 100	100	100
16mm	—	—	—	85 to 100	100	—	—	—	90 to 100	—
12.5mm	—	—	—	—	85 to 100	100	—	—	—	90 to 100
10mm	0 to 5	0 to 5	0 to 20	0 to 30	0 to 45	85 to 100	10 to 35	15 to 45	30 to 70	40 to 85
4.75mm	—	—	0 to 5	0 to 5	0 to 10	0 to 20	0 to 5	0 to 10	0 to 10	0 to 10
2.36mm	—	—	—	—	—	0 to 5	—	—	—	—

Table 3.14 Grading limits of all-in-aggregate

IS Sieve Designation	Percentage passing for All-In-Aggregate of	
	40mm Nominal size	20mm Nominal size
80mm	100	—
40mm	95 to 100	100
20mm	45 to 75	95 to 100
4.75mm	25 to 45	30 to 50
600 micron	8 to 30	10 to 35
150 micron	0 to 6	0 to 6

3.3.16 Effect of Aggregate on Concrete

Aggregate is generally stronger than the cement paste in concrete. The strength of concrete is the **lowest** of **strength of paste, strength of aggregate**, and **bond between cement paste** and aggregate. **Strength of cement paste** is mainly governed by **water-cement ratio** while characteristics of aggregate governs the other two strengths. In **low strength concrete** failure may occur by **failure of cement paste** while in high strength concrete failure is likely to occur by **loss of bond between** cement paste and aggregate. Bond between cement paste and aggregate depends on the **surface area of aggregate** particles which is influenced by **size, shape** and **surface texture** of particles. Thus for high strength concrete small size, angular shape and rough texture aggregate shall be more suitable due to availability of higher surface area for adhesive strength. Finer grading also provides higher surface area for higher bond strength. Aggregates of 40mm size and rounded shape may be permitted for concrete of grades M10 and M20 while aggregates of **20mm size and angular shape** are preferred for concrete of **grades M25 and above.** Angular (crushed) rough textured aggregate of **10mm size** shall be found most suitable in case of very high strength concrete (**M40, M50** grade).

The influence of aggregate **size, shape, texture,** and **grading** may be opposite in nature on workability and strength and hence these characteristics should be chosen with proper analysis and tests.

Aggregate characteristics such as specific gravity, bulk density and moisture absorption should be determined for proper design and adjustments of mix proportions for a job site. Determination of **deleterious materials, soundness** and **alkali aggregate reactivity** help in **quality control for proper durability and strength** of concrete. Thus aggregate characteristics play an important role in concrete chain for the desired quality.

3.4 WATER

Although water is an important constituent of concrete, but it does not receive due attention in preparation and quality control of concrete. Strength and other properties of concrete are developed as a result of reaction of cement and water (hydration) and thus water plays a critical role. Quality of mixing and curing water sometimes leads to distress and disintegration of concrete reducing the useful life of the concrete structure. Water use for concrete mixture should not contain substances which can have harmful effect on strength (i.e. on hydration process of cement) or durability of the concrete in service. Certain substances if present, in sufficient quantities in water, may have an injurious effect upon concrete. Water used for mixing and curing shall be clean and **free from injurious** amounts of **oils, acids, alkalis, salts, sugar, organic matter, sewage,** and **other substances** which are deleterious to concrete or steel reinforcement.

Potable water is generally considered satisfactory for mixing and curing of concrete. In case of doubt, water should be tested for its suitability. The details of various tests (physical and chemical) for water used in concrete are given in IS: 3025-1964. Permissible limits of impurities in mixing water are specified in IS: 456-1978 and is given in Table 3.15.

Table 3.15 Permissible limits for solids and concentrations of impurities (Ref. IS: 456-1978)

S.No.	Impurities (test IS: 3025-1964)	Permissible limits (max)
1.	Organic solids	200 mg/litre
2.	Inorganic solids	3000 mg/litre
3.	Sulphate solids (as SO_4)	500 mg/litre
4.	Chloride solids (as Cl)	2000 mg/litre for plain concrete work 1000mg/litre for RCC
5.	Suspended matter	2000 mg/litre
6.	To neutralise 200ml water sample using phenolphthalein as an indicator	Should not require more than 2ml of 0.1 normal NaOH
7.	To neutralise 200ml water sample using methyl orange as an indicator	Should not require more than 2ml of 0.1 normal HCl
8.	PH value	Generally not less than 6

For evaluating the effect of using a water of questionable quality, make comparative tests for time of set and soundness, and strength with water of doubtful quality and distilled water. **Compressive strength test for performance of water in concrete** shall be carried out on the 150 mm concrete cubes prepared with water proposed to be used. Average 28 days compressive strength of 3 cubes of 150mm size shall not be less than 90 percent of the average strength of 3 similar cubes prepared with distilled water. Initial setting time of test block made with proposed water and cement shall not be less than 30 minutes and not differ by ± 30 minutes from the initial setting time of control test block prepared with the same cement and distilled water.

Based on the minimum strength ratio of 85 percent, the following waters were found to be suitable for concrete making:

- Marsh water;
- Sea water (salinity not above 3.5%), for plain concrete only;
- Water with a maximum concentration of 1% SO_4;
- Alkali water with a maximum of 0.15% Na_2SO_4, NaCl;
- Pumpage water from coal and gypsum mines; and
- Waste water from slaughter houses, breweries, gas plants, paint and soap factories.

The water found unsuitable for the purpose of preparing concrete are:

- Acid waters;

- Lime soak water from tannery waste;
- Carbonated mineral water discharged from galvanizing plants;
- Water containing over 2 percent of sodium chloride or 3.5 percent of sulphates, and
- Water containing sugar or similar compounds

The lowest content of dissolved solids in these unacceptable waters was found to be over 6000 PPM except for a highly carbonated mineral water which contained 2140 PPM of total solids. Normal potable (drinking) water (except that containing sugar) is suitable and does not contain more than 2000 PPM of dissolved solids and also the specification of potable water excludes all above polluted water. SP: 23-1982 specifies tolerable limits on the basis of test results and is given in Table 3.16.

Table 3.16 Concentrations of some impurities in water considered as tolerable (Ref. SP: 23-1982)

Sr. No.	Impurity	Maximum Tolerable Concentration
1.	Sodium/Potassium Carbonates and Bicarbonates	1000 PPM (total)(if this exceeds, tests of setting time and 28 days strength conducted)
2.	Sodium Chloride	20000 PPM
3.	Sodium	10000 PPM
4.	Calcium and Magnesium Bicarbonates	400 PPM of bicarbonate ion
5.	Calcium Chloride	2 % by mass of cement in plain concrete
6.	Iron Salts	40000 PPM
7.	Sodium Iodate, Phosphate Arsenate and Borate	500 PPM
8.	Sodium Sulphide	Even 100 PPM warrants testing
9.	Hydrochloric and Sulphuric and other Common Inorganic Acids	10000 PPM
10.	Sodium Hydroxide	0.5 % by mass of cement if setting not affected
11.	Silt and Suspended Particles	2000 PPM

Wherever suitable waters are not available locally, then water processing, modification of cement quality or transporting suitable water from long distances may be resorted or loss of strength and durability may be accepted.

Except for possible discoloration, the presence of silt, oil, or salts in curing water does not appear to have harmful effects. However, water containing higher concentrations of acid or organic substances should be considered with suspicion and subjected to investigation.

3.5 SUMMARY

Cement concrete comprises of cement, fine aggregate, coarse aggregate and water as its basic ingredients. Properties and proportions of each ingredient influence the quality of the concrete during fresh as well as hardened states.

Cement acts as **binding material** and is obtained by pulverising clinker formed by calcining raw materials comprising of Lime, Silica, Alumina and Ferric Oxides. Principal compounds formed are C_3S, C_2S, C_3A and C_4AF (where C = CaO, S = SiO_2, A = Al_2O3, and F = Fe_2O_3).

Proportions of these four principal compounds influence the characteristics and type of cement. Physical properties of cement viz. **fineness, setting and hardening, strength, soundness** and **heat of hydration** directly affect the concrete properties. Fineness and strength of cement helps in classifying the cement into different grades (**33, 43, 53**). For producing desired concrete quality various characteristics of cement must be considered. Finer cements hydrate rapidly to develop higher strength and also results in better workability of concrete. Strength of cement can be evaluated by making **1:3 cement: standard sand mortar** cubes. Soundness of cement can be checked with the help of Le-chatelier apparatus. Setting time can be evaluated using Vicat's apparatus.

Aggregate (FA and CA) forms bulk of concrete mass and also provides volumetric stability to concrete. **Aggregate** properties affect **workability, strength, durability** and other special characteristics of concrete. Aggregate characteristics comprise of **size, shape, texture, strength, absorption, bulking, density, sp.gravity, thermal properties, impurities, durability** and **gradation.** Smaller the particle size of aggregate greater will be surface area per unit volume resulting in lesser workability of concrete. Aggregate crushing value, impact value representing strength is obtained as percent of sample passing 2.36 mm sieve after 15 blows of standard hammer or a pressure of 400 kN. Fineness modulus is commulative percentages of particles retained on standard set of sieves. **Fineness modulus** represents **average sieve size of aggregate** starting from the finest sieve size. Proper grading of aggregate eliminates harshness, segregation and bleeding.

Sand grading zone also affects the properties and qualities of concrete. Sand Zone I is coarsest and leads to coarser grading for production of economical and good quality concrete. Bulking of sand should be measured and adjusted for good quality concrete production. **Surface saturated dry** aggregate **forms the basis of calculations** for design of mix proportions.

Aggregate is generally stronger than cement paste and hence for producing high strength concrete, use of smaller size, angular shape and rough textured aggregate is recommended. Aggregate grading within specified grading zones can be accepted for the design mix.

Potable water is generally acceptable for the production of good quality concrete. Water should not contain deleterious impurities beyond those specified in Indian standards. Acidic water is avoided as far as possible.

PRACTICE QUESTIONS

3.1 List basic ingredients of cement concrete.

3.2 List characteristics of cement.

3.3 Describe: fineness, strength and hardening of cement.

3.4 Explain the effect of strength and heat of hydration of cement on quality of concrete.

3.5 Differentiate: strength and soundness and setting and hardening of cement.

3.6 Describe Indian standard specifications of ordinary portland cement.

3.7 List and define characteristics of aggregate.

3.8 Explain the effect of size of aggregate on the properties of fresh concrete.

3.9 Explain the effect of **shape of aggregate** on the properties of concrete.

3.10 Explain the effect of strength of aggregate on the strength of concrete.

3.11   Explain **bulking of sand** and its influence on concrete. Describe how to compensate bulking at site.

3.12   Explain the influence of alkali-aggregate reaction on the properties of concrete.

3.13   Explain the influence of **aggregate grading** on concrete properties.

3.14   Explain the influence of **sand grading zone** on concrete properties.

3.15   Define finess modulus of aggregate.

3.16   Explain the effect of quality of water on the quality of concrete.

3.17   Specify minimum standards of water required for concrete.

3.18   Explain the role of basic ingredients in achieving the desired quality of cement concrete construction.

3.19   Explain the role of site engineers for managing the quality of cement concrete construction.

CHAPTER 4

Influence of Special Admixtures on Quality of Concrete

LEARNING OBJECTIVES

The learner understands the influence of special admixtures on properties and quality of cement concrete and will be able to:

- Describe importance of admixtures in modifying properties of cement concrete;
- Define admixtures to concrete;
- State general influence of admixtures in fresh and hardened cement concrete;
- List type of admixtures used for cement concrete;
- Describe properties and influence of plasticizers on properties of concrete;
- Describe properties and influence of **Accelerators**, Retarders, and Air entraining agents on properties of cement concrete:
- Describe properties and influence of mineral admixtures on properties of concrete;
- Describe properties and influence of Water proofing compounds, Bonding admixtures, Curing compounds, Non shrinking grouting agents, and Pozzolanic admixtures on properties of cement concrete:
- Explain the criteria of selection of admixture for cement concrete or mortar.

4.1 INTRODUCTION

Cement concrete is the most widely used construction material in the world. Cement consumption in the world is around 20 billion tonnes annually (2 tonnes/capita/annum). The reasons for such widespread use of cement concrete are its easy adaptability, durability, strength, availability, easy construction and overall economy in construction. Cement concrete is the only material which can be used everywhere from buildings to big projects (viz power houses, irrigation, transportation, railways, bridges and shipyards).

Cement concrete comprises of basic materials: cement, aggregates, and water. Quality control in cement concrete construction is based on its properties both in plastic (or fresh) and hardened states. The most important property of cement concrete sought in fresh state is **"workability"** while **"strength"** and **"durability"** are important in hardened state. With the development of concrete technology, it is now possible to achieve desired properties and quality both in fresh and hardened states. Achievement of desired qualities in concrete can be made economically possible by mixing of certain additives in the cement concrete.

The strength development of concrete with age is dependent on the curing method consistent with the use of admixtures (Fig. 4.1).

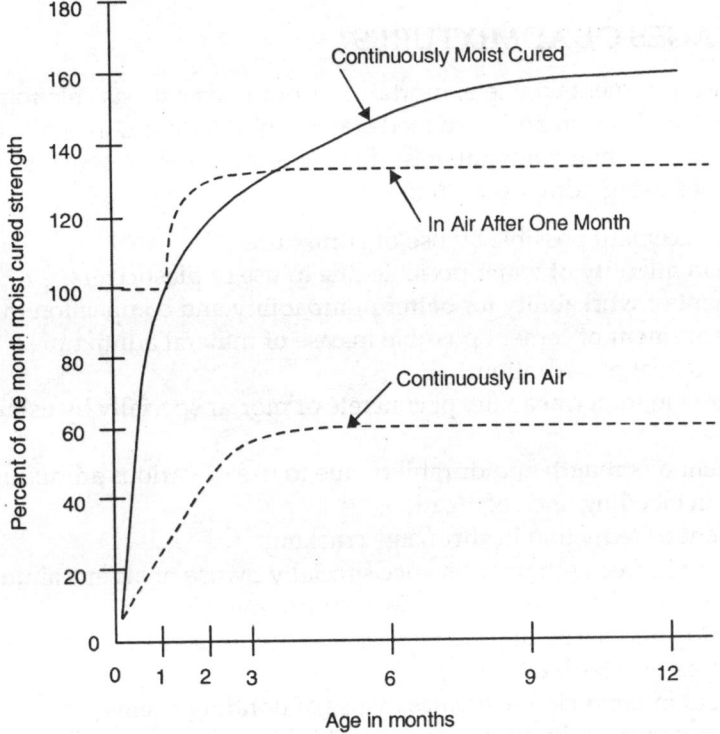

Fig. 4.1. Influence of moist curing with age on strength development of cement

An **admixture** is defined as the material added in small quantities to cement concrete, mortar or grout during mixing or before mixing to modify the properties both in fresh and hardened states. The proportion of admixtures does not generally exceed 5% by mass of the cement content in concrete, mortar or grout. Concrete admixtures are a new category of ingredients used in addition to the basic materials (cement, aggregates and water). These admixtures are added just before mixing of basic ingredients for uniform consistency.

The quality of concrete or mortar may be defined by the extent of **accomplishment of desired properties** for the specific purpose by the user at a **most optimum cost.** Several materials have been produced for addition to cement concrete or mortar to modify certain properties as required for a specific purpose and situation.

The admixtures may influence one or more than one property of cement concrete. Some of the admixtures may improve one property in positive direction while affect the other property in negative direction. The admixtures should, therefore, be used carefully only after ensuring the desired property through scientific **trial tests** and study of manufacturer's literature. Thus the proportion and the type of admixture is decided on the basis of modified property and optimum cost of concrete production.

Admixtures are normally supplied in liquid form for accurate dose measurement through automatic dispensing equipment by volume in relation to cement mass. Some admixtures are also supplied as powders.

Admixtures can be classified by the nature of their composition or the properties required to be modified.

4.2 ADVANTAGES OF ADMIXTURES

Use of admixtures in cement concrete, mortar or grout makes it possible to modify the properties of concrete both in fresh and hardened states. This facilitates in achievement of desired quality at site. There are many advantages of using admixtures in concrete, mortar or grout. Main advantages of using admixtures are:

- Reduction of cement possible by use of admixtures;
- Reduction in quantity of water possible due to use of plasticizers;
- Improvement of workability for better pumpability and compaction in general;
- Partial replacement of cement possible in case of mineral admixtures;
- Reduction in heat of hydration;
- Improvement in impermeability of concrete or mortar specially by using water proofing admixtures;
- Improvement of strength and durability due to use of various admixtures;
- Reduction in bleeding and segregation;
- Improvement or reduction in shrinkage cracking;
- Improvement in freeze-thaw resistance specially by use of air entraining admixtures;
- Improvement of resistance to aggressive environments;
- Modify setting and hardening characteristics as required in a particular situation by use of accelerators or retarders;
- Improvement in bond characteristics by use of bonding agents;
- Reduction in corrosion by using corrosion inhibitors;
- Improvement in surface resistance to abrasion and impact;
- Control and achievement of desired properties according to situations.

4.3 ADMIXTURE TYPES

Admixtures are grouped under two main categories as:

- Chemical admixtures; and
- Mineral admixtures.

Chemical admixtures are manufactured under variety of trade names to serve specific purpose of modifying properties of concrete in fresh as well as hardened sates.

4.3.1 Types of Chemical Admixtures

Chemical admixtures can be subdivided into the following types:

I. **Plasticizers** (also called **water reducers**)

   - Normal plasticizers (water reducers)
   - Super plasticizers (high range water reducers)

II. **Accelerators**
III. **Retarders**
IV. **Air Entrainers**
V. **Special Purpose Admixtures**

4.3.2 Types of Mineral Admixtures

Mineral admixtures are finely divided silicon particles added to cement concrete or mortar in large proportions. These finely divided mineral admixtures contain reactive silica for providing pozzolanic action and economical concrete. Mineral admixtures are further subdivided into: **Natural** Materials and **Bye** Product Materials.

Natural Materials	Bye Product Materials
Volcanic gases	• Fly Ash
Volcanic tuffs	• Blast Furnace Slag
Calcined clays or shells	• Condensed Silica Fumes
Diatomaceous Earth	• Rice Husk Ash

These materials can be regrouped on the basis of physical and chemical properties as:

- Cementitious
- Pozzolanic
- Pozzolanic and Cementitious
- Inert Material

4.4 PURPOSE OF ADMIXTURES

4.4.1 Chemical Admixtures

Chemical admixtures are used to modify following properties of fresh and hardened concrete:

a) Fresh State

- To increase workability;
- To accelerate or retard setting;

- To reduce settlement or create slight expansion;
- To modify the rate and capacity for bleeding;
- To reduce segregation;
- To improve pumpability; and
- To reduce slump loss.

b) Hardened State

- To reduce heat evolution during early hardening;
- To accelerate strength development at early ages;
- To increase ultimate strength;
- To increase durability;
- To decrease permeability;
- To control expansion due to alkali-aggregate reaction;
- To increase steel-concrete bond;
- To increase bond between existing and new concrete;
- To improve impact resistance and abrasion resistance;
- To inhibit corrosion of steel; and
- To produce coloured concrete or mortar;

4.4.2 Mineral Admixtures

Finely divided mineral admixtures are used for one or more of the following purposes:

- To correct the deficiency in the concrete to provide 'fines' missing in the fine aggregate so as to avoid problems of workability and finishing;
- To improve one or more qualities of the concrete, such as to increase sulphate resistance, to reduce expansion due to alkali-silica reaction, to reduce permeability, or to decrease heat generation; and
- To reduce cost of concrete-making materials, cost of concreting operations, or both.

Finely, divided mineral admixtures, particularly pozzolanas, can provide reduced cost of concrete-making materials because the optimum mix for given work usually will contain less cement, than a comparable mixture not containing finely divided pozzolanic admixtures. Apart from economy in cement concrete construction, the use of some of the bye products such as fly ash provides solution to its disposal problem. Large number of chemical and mineral admixtures are available and their characteristics and special features are discussed in next section.

4.5 CHEMICAL ADMIXTURES

4.5.1 Plasticizers (Water Reducers)

Normal plasticizers act directly on the individual particles of cement causing them to deflocculate and disperse more uniformly through the mixing water. This results in a significant increase in workability of the mix. Since water-cement ratio remains the same, the strength is not affected much. However, there is likelihood of minor improvement in strength on account of better compaction as a result of higher workability. The improved workability can be used to aid in placing due to self-leveling and easy flowing characteristics of concrete. The principal active components of plasticizers are surface-active agents comprising of long-chain molecules that are hydrophobic (nonwettable) and hydrophilic (wettable).

Such molecules tend to concentrate and form a film at the interface between cement and water and alter the physio-chemical forces acting at the interface. Further, absorption of these substances on cement particles gives rise to a negative charge thereby leading to repulsion between the particles to cause better dispersion. Due to the charge, a protective sheath of oriented water molecules is also developed which increases the mobility of fresh concrete. The water freed from the restraining influence of the flocculated system becomes available to lubricate the mix for increased workability. Thus the quantity of water can be reduced for achieving the same workability. A more uniform distribution of the dispersed cement throughout the mix also contributes to the higher strength.

These plasticizers can also be used to reduce water for maintaining the same workability and thus reducing water-cement ratio. This decrease in water-cement ratio results in increase of strength in accordance with the **Duff Abram's** principle. Thus higher strength can be achieved by keeping the cement content the same.

For keeping the workability and W/C ratio the same, water and cement content per cum can be reduced by use of plasticizer and thus economising the production cost. Cost of concrete production will reduce if the cost of the admixture added is less than the cost of cement saved.

The normal plasticizers are mainly based on lignosulphonates and/or polymeric carbohydrates. These water-reducing plasticizers are:

- Lignosulphonic Acids and their salts
- Hydroxylated carboxylic acids and their salts

These plasticizers are used in amounts of about 1% by mass of cement and are quite effective with all types of portland cements and also with high alumina cement. Water reducing admixtures may be classified as water reducing set retarding or simple water reducing. Water reducing admixtures are produced under different trade names and their optimum doses may be studied from the producer's literature.

4.5.2 Super Plasticizers (High Range Water Reducers)

These admixtures work on the same principle as normal plasticizers (water reducers) but their effect on workability or water reduction is much greater. This is partly because they can be used at much higher dosage without undesirable side effects.

Super plasticizers are mainly based on synthetic polymers such as sulphonated Naphthalene or Melamine Formaldehyde condensates. Super plasticizers can be used at higher doses and are generally expensive than normal plasticizers. Generally super plasticizers cost more than the cement saved and hence these are used to modify certain properties necessary for long-term effects. Use of super plasticizers from durability point of view is justified. The main uses of super plasticizers are:

- Very high workability to facilitate self leveling of concrete in floors;
- Very high workability to facilitate compaction through congested reinforcement;
- Low water-cement ratios for high strength concretes; and
- Low water-cement ratios to give impermeable durable concrete in adverse conditions.

Some of superplasticizers work very effectively when used alongwith mineral admixtures for cement replacement. Use of silica fume, flyash, calcined clays and blast furnace slag alongwith superplasticizer makes cement replacement quite effective and economical in achieving desired quality of concrete.

Many companies are producing liquid plasticizers under their trade names. Some of these plasticisers also known as workability agents and are: **CONPLAST, CICO, SUPAPLAST, EMCEPLAST, ZENTRAMENTSUPER,** etc. While using any plasticizer, its optimum doses and other instructions must be observed as per manufacturer's literature as well as practical test results.

4.5.3 Accelerators

Accelerators are those materials which enhance the rate of hydration to reduce the setting time and/or gain of early strength. These admixtures fall into three categories viz set accelerating, strength gain accelerating and both set and strength accelerating.

Set accelerators cause an early stiffening of the concrete mix with atleast one-hour acceleration over the standard normal mix. The most effective admixtures in this category are based on aluminates or silicates and give initial set in few minutes. Unfortunately these admixtures result in reduction of the later age strength. These admixtures are mainly useful in sprayed grout for mining and tunneling operations.

The other accelerators, which meet the 1-hour set acceleration, are chlorides. Chlorides promote corrosion of steel. Chloride based accelerators cannot be used in RCC and hence are banned. Chloride based accelerators are both set and strength accelerators. Set accelerators are mainly used in early finishing and repair of floor slabs and concrete highways. These accelerators are quite suited to repair and maintenance jobs and laying concrete in extreme cold weather conditions. Accelerators are used for plugging the leakages, quick rendering work, quick installing of railing posts and underwater concreting.

Strength accelerators give high strength at an early age upto 24 hours. This gain of an early strength may not continue at later ages. Generally inorganic salts such as chlorides, nitrites and thiocyanates form the basis of strength accelerators. Because of many side effects their use is superseded by High Range Water Reducing Agents. High range water reducing agents (superplasticizers) offer the advantage of more cost effectiveness and much higher strength gain at 1-day (200 %) along with better ultimate strength. The inorganic salts are however, more effective at very low temperatures (around 5°C) in producing very early strength (8 –10 hours).

Most common accelerator for cement concrete is calcium chloride ($CaCl_2$). Some other accelerators are sodium chloride, sodium hydroxide and sodium silicate. Some companies have started producing chloride free accelerators to offset bad effects of chlorides. There are varieties of accelerators produced under different trade names both in liquid and powder forms. Calcium chloride produces greater acceleration in hydration than sodium chloride. Maximum acceleration in hydration occurs during first few hours, whereas there is little or no effect on the 28 days strength. Research indicates that effect of calcium chloride is much more predominant in rapid hardening cement than OPC. Two percent addition of $CaCl_2$ by mass of cement is quite adequate to reduce the initial setting time of OPC by two thirds to about 1 hour and will reduce the final setting to about 2 hours. Three percent $CaCl_2$ may lead to flash set while one percent may even retard the setting.

The use of $CaCl_2$ has been found to increase the resistance of concrete to erosion and abrasion. The use of $CaCl_2$ may increase drying shrinkage and possibly creep. It may also lead to corrosion of steel and hence use of chlorides must be avoided in prestressed concrete. $CaCl_2$ should not be used in sulphate resisting concrete to avoid alkali-aggregate reaction. $CaCl_2$ should not be used in lightweight concrete placed over metal decks. $CaCl_2$ does not affect the Air Entrained concrete but it facilitates improvement in workability. Use of $Cacl_2$ enhances freeze-thaw resistance of concrete in very low temperatures.

Accelerators may also comprise of calcium lignin sulphonate, or organic catalyst such as orthohydroxyl Benzoic Acid, and Combinations of these materials depending on the environmental conditions. All instructions for use of accelerators must be followed carefully for optimum results in improving the desired properties without adverse effects on other properties.

4.5.4 Retarders (Retarding Admixtures)

Retarders are used to cause a delay in the time to retard setting and/or hardening of cement concrete. Some retarders may not confer with workability retention initially or at the time of placing. Calcium sulphate (Gypsum) is added to cement at the time of its manufacture to prevent too quick set. The addition of gypsum beyond certain proportions causes unsoundness in cement and hence requirements of retardation cannot be fully realized by gypsum especially in tropical climates. Delaying of setting may be necessary for placing concrete under high temperatures or special conditions. Retarders are necessary for grouting deep voids behind concrete arch tunnel linings, for transporting concrete over long distances, for pumping concrete for avoiding cold joints between successive lifts in mass concrete, and hot weather concreting. Retarders facilitate monolithic action and better bond in consecutive layers in continuous concreting operations.

The requirements of retarders are **workability retention, initial set, final set** and **hardening** of concrete. The concrete needs adequate workability at the point of placing for full compaction. This is achieved by using plasticizers to increase initial workability so that even after the loss of workability at the time of placing, the concrete can be compacted properly. Whenever there is likelihood of delay between two adjacent concrete pours, then the first pour still needs sufficient workability to allow fully monolithic construction. This level of workability refers to initial set requirements. Longer delays in the second pour will still bond to the first pour if the first pour has not passed its final set time due to use of retarder. Thus use of retarders result in workability retention, delay initial and final setting to **avoid cold joints.**

A special feature of most of the retarders is its ability to reduce water requirements of concrete mix and thus increase in ultimate strength in general. Retarders generally reduce the resistance of concrete to freezing and thawing but has no or insignificant effect on contraction or expansion. Water reducing effect of retarders have been studied in relation to strength and economy of concrete construction.

Some retarders also entrain air and result in increase of 3, 7 and 28 days strengths if air entertainment remains below 8 percent. Advantage of this fact could be taken to reduce the quantity of cement for the same strength or reduce W/C ratio to achieve higher strength. The water reducing aspect of retarders is very useful.

A wide variety of retarding agents are produced by different companies under their trade names. Each producer specifies optimum proportions of retarder for a specific purpose and placement conditions. Retarding admixtures are usually sugars, hydroxyl carboxylic acids, phosphates or unrefined lignosulphonates, ammonium chlorides, ferrous and ferric chlorides, calcium borates and oxychlorides, calcium tartarate, alkali bicarbonates, tannic, gallic, humic and sulphonic acids in sodium hydroxide solutions, starch, and salts of sulfuric acid. Sugar acts as a good retarder. It has been found that 0.05 to 0.10 % sugar by mass of cement has very little effect on the rate of hydration while addition of 0.20 percent sugar by mass almost stops hydration such that final set occurs after 72 hours. Skimmed milk powder has also retarding effect due to its sugar content. Use of 0.10 percent of sodium hexameta phosphate can retard initial setting to about 12 hours and final setting to about 13 hours.

The use of any retarder should be made after studying the quality requirements, placing conditions, temperatures, dosage, humidity, wind velocity, cement composition, initial workability requirements and other architectural considerations. Retarders should be selected only after proper testing and study of its specifications.

4.5.5 Air Entraining Agents

Air entraining agents are well selected and blended surfactants designed to entrain large volume of very small air bubbles into concrete or mortar. The main purposes of air entrainment is to: **Increase workability** with improved cohesiveness, enhance freeze-thaw resistance and to limit the loss of strength. The adverse effect of strength loss on account of entrained air can be compensated by use of water reducing plasticizers. Air entrained concrete becomes more cohesive, workable and easier to work. The tendency of segregation and bleeding in plastic concrete reduces considerably. Air entrainment becomes essential in adverse cold weather conditions when durability requirements are more critical compared to strength considerations. Air entrainment plays critical role in improving workability of harsh concrete with rough angular or crushed coarse aggregate and low W/C ratios. The increase in workability due to air entrainment makes it possible to reduce water and W/C ratio, thus compensating part of loss of strength due to air entrainment.

Air entraining agents can either be mixed with cement during its manufacture or added during mixing operation. **Air entrainment does not refer to large unintentional air voids present in concrete forming continuous channels.** The deliberate air entrainment consists of discontinuous channels and of much smaller size particles (less than **0.05mm** diameter). There are several billion of these minute air voids introduced by air entrainment in concrete. These bubbles provide cushioning effect against thermal fluctuations and minimize crazing and cracking in concrete.

When water in the capillary network in concrete freezes, it expands by upto 9% in volume and generates an internal crushing force to break up concrete. If the capillary network is intersected by entrained air voids then the expansion force/pressure can be relieved by flow of water particles into the air voids. 5 % air entrainment can result in good freeze-thaw resistance of concrete.

Air or gas cells can be introduced in any of 3 ways:

- Addition of Aluminium powder or zinc powder which reacts with cement to generate gas or hydrogen peroxide forming gas cells within the concrete;
- **Surface active** agents (called air entraining agents) which reduce surface tension;
- Cement dispersing agents, which are surface-active chemical compounds, which cause **electrostatic charge** to cement particles rendering them mutually repellent and thereby preventing coagulation. They do not reduce surface tension by wetting or foaming.

Aluminium powder is generally not used on construction job, as it requires strict controlled conditions for desired quality results. This can produce 60% or even more air voids (hydrogen bubbles).

Surface-active agents can further be subdivided into:

- Natural wood resins and their soaps, such as vinsol resin;
- Animal or vegetable fats and oils, such as beef tallow and olive oil;
- Wetting agents such as alkali salts of sulphonated or sulphated organic compounds. e.g. commercially produced synthetic detergent with trade name as Darex.

There are many air-entraining agents produced by different names such as Airalon, Teepol, Cheecol, Orvus, Petrosan, CicoAiren (IS9103), MC-Mischoel and AEA.

Vinsol resin is first neutralised with sodium hydroxide to convert it into soap to avoid chemical reaction with cement. These admixtures can entrain air upto 30 %. Only small quantities of these AEA are required (0.005 to 0.05 % by mass). The amount of air entrained depends on the type of cement, A/C ratio, FA/CA ratio, mixing time and type and temperature. **Five percent** air entrained can improve compacting factor by **0.07** or **slump by 45-50 mm.** Air entertainment can be done in no slump concrete. Air entrained concrete have much better workability and require much shorter period of compaction to avoid loss of air entrainment.

4.6 MINERAL ADMIXTURES

Based on their properties, the mineral admixtures can be grouped into: cementitious, inert or pozzolanic.

4.6.1 Cementitious

Cementitious materials have binding properties of their own. Cementitious materials include finely ground granulated blast furnace slag, natural cement, hydrated lime and combination of these materials. These admixtures provide additional binding properties when used alongwith main cementitious material.

4.6.2 Inert Mineral Powders

Finely divided mineral admixtures can be used with cement concrete or mortar for variety of purposes such as to correct the deficiency of aggregate grading for suitable workability, reduce permeability, decrease heat of hydration and decrease the cost by reducing cement. These mineral admixtures are ground atleast as fine as cement and usually finer than cement. Many of these mineral powders may comprise of chemically inert materials such as: limestone, quartz, bentonite, kaolin, chalk, hydrated lime, blast furnace slag, diatomaceous earth, volcanic ashes, calcined clays and fly ash. Some of these finely divided mineral admixtures may contain reactive silica, which may result in pozzolanic action.

The addition of these fine powders increase the workability by increasing the amount of paste in the concrete and hence cohesiveness. Excessive use of such finely divided powders may necessitate larger water requirement and consequently result in decrease in strength. The finely divided mineral powders are used in concrete mixes or mortars, which are deficient in fines to improve their workability to reduce the rate and amount of bleeding and possibly increase the strength indirectly. The substitution of lime for a certain portion of cement in mortars is frequently used method of increasing the workability. An excessive use of these finely ground powders may also increase the shrinkage of the concrete. Sometimes the addition of such minerals may not be economical especially in concrete due to the harmful influences on shrinkage and reduction of strength. These mineral admixtures become more useful if they fall in the pozzolanic category.

4.6.3 Pozzolanic Admixtures

A pozzolana is a finely divided siliceous material which whilst itself possessing no cementitious properties but will react in the presence of water with lime at normal temperatures to form compounds having cementitious properties. These pozzolanic materials are either natural or artificial. The action and properties of pozzolanas differ widely and their proportion and efficacy in any particular circumstances can be determined only by careful tests by considering appropriate practical situation.

The principal use of pozzolanas is to **replace a part of the cement** in making concrete or mortar to achieve improved workability, reduced bleeding and segregation, and economy in cost of construction. The improvement in the workability may not necessarily be reflected by an increase in slump or compacting factor. In fact addition of pozzolana may necessitate higher water requirements for the same slump but the concrete mix behaves more cohesive and needs lesser effort in its compaction. Other minor advantages of pozzolanas are greater **imperviousness,** better **resistance to freezing and thawing,** better **resistance to sulphate attack, reduction in alkali-aggregate reaction,** and **reduction in heat of hydration specially** in mass concrete.

Principal natural pozzolanas are: **calcined clays** and **shells,** calcined or uncalcined **diatomaceous earth, opline cherts, shales, volcanic ash,** and **pumicites.** The most active of the natural pozzolanas are diatomites, opline cherts, and some shales. Some of the natural pozzolanas need to be calcined and ground to fine powder for making them reactive. Generally the optimum amount of pozzolana as a replacement for cement may normally range between 10 to 30%.

Artificial pozzolanas are finely ground blast **furnace slag** and **fly ash** produced in thermal power plants. Fly ash or **pulverised fuel ash** (PFA) is the residue from combustion of pulverised coal and collected by mechanical or electrostatic separators from the flue gases of thermal power plants. Its properties and composition vary widely and depends on the type of fuel burnt and method of collection. Most of the pozzolanas are rich in silica and alumina and may also contain small quantities of alkalies. Specific surface area of flyash may vary from 3500 to 5000 sqcm per gm. The principal constituents are silicon dioxide (SiO_2), Aluminium Oxide (Al_2O_3), unburnt fuel as carbon, calcium oxide (CaO), and small quantities of Magnesium Oxide (MgO), and Sulphur trioxide (SO_3). The properties of pozzolanas mainly depend on the **fineness,** reactive silica and carbon contents.

The action of **pozzolanas** mainly depends on the **reaction of active silica** (amorphous form) with the free lime released during the hydration of cement. Recent studies indicate that Alumina and iron also take part in a complex reaction. It is, therefore, necessary to examine the optimum proportion and properties of cement concrete or mortar mixed with pozzolanas by laboratory testing prior to use in construction. The properties and suitability of a pozzolanic admixture can be determined by referring to code IS: 1727.

4.7 SPECIAL ADMIXTURES

There are many admixtures, which are used for producing a special purpose concrete or mortar. Some of these special purpose admixtures are:

- Gas Forming Agents/Foam Forming Agents;
- Expansion Producing Agents;
- Bonding Agents (polymer Admixtures);
- Grouting Agents;
- Pigments and Colouring Agents;
- Water Proofing and Permeability Reducers;
- Shotcrete Accelerators; and
- Curing Compounds;

4.7.1 Gas or Foam Forming Agents

Gas forming agents are used in production of cement grout and lightweight concrete. Gas forming agents are also known as foaming agents. These concrete are also called as gas concrete, cellular concrete, foamed concrete, aerated concrete or porous concrete. Gas or air bubbles are used as aggregate in gas concrete. Air bubbles are different from those formed in air-entraining agents. These agents are used to form foam during mixing in the mixing truck. The proportion of gas forming agent can control the density of concrete to as low as 500 kg/m^3. The gas or foam concrete slightly expands to fill the gaps in confined spaces in grouting ducts and machine bases.

There are variety of gas forming agents such as Aluminium Powder, Powdered Zinc, Hydrogen Peroxide, Soaps and resins, Air entraining materials, and calcium carbide (CaC_2). Hydrogen or oxygen gas is liberated by the chemical reaction between the admixtures and $Ca(OH)_2$ released from cement. Generally these admixtures are used in amounts less than 0.2 percent by mass of cement. For controlling the amount of expansion and density the optimum quantity of the admixture is determined by practical testing. When these admixtures are used in large quantities, these produce gas or cellular concrete. Appropriate use of gas forming agent can produce lightweight concrete with unit weights varying from 4 KN/m^3 to 20 KN/m^3. Special techniques are used for mixing, placing and compacting such concrete for retaining adequate air or gas uniformly in the mix. Gas forming agents are mixed first with cement. Sometimes certain quantity of preformed foam is added to freshly prepared concrete to achieve desired density.

Hydrogen peroxide breaks into water and oxygen when added to cement to produce lightweight concrete or mortar. Aluminium or zinc powder is used for production of cellular or gas concrete. Finely ground aluminium or zinc powder is mixed with the cement in the proportion of **0.10 to 0.20%** by mass of cement to form a fluid paste which is placed in respective moulds to about one-third full. Chemical reaction produces hydrogen bubbles, which fill the mould completely. Slightly excess material shall be struck off level with the top of the mould. Addition of pozzolana such as fly ash, calcined clay or blast furnace slag to this mixture reduces drying shrinkage, which helps in controlling cracking of these blocks/or moulded members. Various gas forming agents should be tested for their optimum proportion and desired effect on the quality of concrete.

4.7.2 Expansion Producing Agents

Gas forming agents cause expansion during plastic stage itself while expansion producing agents are required to produce some expansion in cement concrete during hardening process. This expansion occurs either by the expansion of the admixture itself or by its reaction with other constituents of cement concrete or mortar. The desired expansion may be obtained by using the following common admixtures:

- Granulated iron, expansive cement (Sulpho-Aluminous cements),
- Self-stressing cement, and Anhydrous Sulfo-Aluminates.

The expansion-producing agents are used for filling the cracks and confined spaces difficult to approach. These admixtures are specially suited for repair works.

4.7.3 Bonding Agents (Polymers)

The bonding agents are water emulsions or any of several organic materials that are resistant to saponification. Liquid resin latex can be added to cement or mortar grout for improving its bond strength. These admixtures are suitable for cement concrete, cement mortars, lime and gypsum mortars for increasing the bond strength between the old and new concrete. These admixtures increase the resistance of the toppings against wear and renders the mortars watertight. These materials have also been found useful in the manufacture of **cement paints**.

Commonly used bonding agents are made from natural rubber, synthetic rubber, or organic polymers. Sometimes these are available in two-pack system of epoxy resins for one time use for bonding slurry coat between old and new concrete or mortar toppings. Cement slurry with appropriate proportion of bonding agent is coated on the prepared old concrete surface prior to laying of fresh plastic concrete which may also contain bonding agent. These bonding agents are manufactured under many trade names such as: Monolithex, DP Bonding agent, Nafufil, Acrylic based bonding agent and Epoxy Resins.

4.7.4 Grouting Agents (Pumping Aids)

Grouting admixtures may exhibit the most desired property of flowability of concrete/mortar/cement slurry. These agents may be in the form of air entraining agents, workability agents, set retarders, accelerators, and waterproofers to produce the desired property of grouting material. The most important purpose of the grout is to fill the discontinuity due to holes, ducts, cracks and congested reinforcement areas. Grouting admixtures are also used for grouting machine foundation bolts and for stabilised beam supports. Grouting material with specific admixtures are used to seal oil wells, stabilize foundations, and repair of architectural features.

Grouting admixtures must ensure the following characteristics in the grouting material:

- Good flowability and pourability;
- Continuity throughout the grouted portion;
- High initial and ultimate strength;
- Excellent bonding with grouted surface;
- No shrinkage or change in volume;
- Compatibility of the grout with grouted member;
- Transfer and bearing of loads well in time; and
- Freedom from harmful chemicals (chlorides).

There are many grouting admixtures produced under different trade names such as: Emce Krete, Centricrete, CICO Pagel GP, CICO Polygrout, CICO Moushrink, Polymeric Grouting Compound, Epoxy Grout, etc.

For optimum effect of grouting admixture, the producer's instructions must be studied thoroughly and trial tests carried out before using. These admixtures are produced as fine powders, pastes, or liquids.

4.7.5 Pigments and Colouring Agents

Inorganic pigments are added to cement to produce colour in the finished concrete. All pigments must be permanent and in particular these should not be affected by the free lime in cement concrete. For better result and deep colouring effect, the pigments are ground with the cement in a ball mill. Sometimes pigments are mixed with fillers or extender such as chalk and barium sulphate, which are insoluble in water and have no harmful chemical effect. Materials, which enter into chemical reaction with cement, should be avoided. Concrete strength gets reduced when fillers or extenders are used in large quantities. For better results the pigments should be blended well with dry cement before adding mixing water. Table 4.1 gives list of pigments.

Table 4.1 Colour and Pigment Compounds

Sr. No.	Colour	Pigment Compounds
1.	Brown	Burnt Amber (Ferrous Oxide and Hydroxides, Manganese Oxides)
2.	Grey to Black	Carbon black, Magnetic Ferrous Oxide, Manganese black.
3.	Red	Red Oxide of iron
4.	Blue	Ultra Marine Blue Phathalocyanine, Barium Manganese
5.	Yellow	Natural and chemically prepared orchres, Hydroxide of iron
6.	Green	Chromium oxide and Chromium Hydroxide
7.	White	Titanium dioxide

4.7.6 Damp Proofing Agents (Permeability Reducers)

These are also known as water proofers and these block capillary pores or coat them with hydrophobic material to inhibit water transmission. Waterproof concrete are impervious and resist water absorption. By adding damp proofing agents to well designed concrete mix, the required qualities of impermeability alongwith some saving in cost can be achieved. The admixtures improve the resistance of concrete to absorption of water. These are very essential for concrete elements in contact with water or exposed to external weather conditions of rain, sun and heat.

Damp proofing agents are available in powder, paste or liquid form and consist of pore filling materials or water repellent materials. These materials are chemically active or inert. The pore filling materials are used to reduce capillary flow of moisture through concrete, which is in contact with water or moist earth. The main pore filling materials are: alkaline silicates (sodium silicate), Aluminium and zinc sulphates, and aluminium and calcium chlorides. Mineral admixtures such as flyash and other pozzolanas are also used to reduce permeability of concrete with low cement content and deficient in fines. Damp proofing materials are either chemically active or inactive. The main inactive pore filling materials are chalk, fuller's earth, and talc. These materials are usually ground to high fineness, which facilitate filling of fine pores.

Chemically active water repellent materials are soda and potash soaps to which sometimes lime, alkaline silicates or calcium chloride may be added. Chemically inactive water repellent materials are calcium soaps, resin, vegetable oils, fats, waxes, coaltar and bitumen residues. These materials fill or block pores in concrete.

Proprietary water proofers may consist of calcium, aluminium or other metallic soaps and water repellent materials. Some of the commercially produced water proofing admixtures also contains acrylic polymers.

4.7.7 Shotcrete Accelerators

These are special set accelerators mixed with concrete or mortar to cause rapid set and early strength gain. These are mainly used in mining and tunneling and repair tasks. The details of accelerators have already been discussed under 4.5.3

4.7.8 Curing Compounds

In real sense, the curing compounds cannot be considered as admixtures but these are aids for achieving desired quality under specific conditions of placement. Curing compounds are generally coated after placing the fresh concrete so as to form a **impervious membrane film** on the exposed surface. This prevents evaporation of water from the body of concrete, which is required for continuance of hydration process. These compounds are of many types as follows:

- Thinned coal tar with solvents;
- Rubber latex emulsions;
- Clear or translucent compounds (emulsified in water or volatile solvents);
- An emulsion of paraffin wax and boiled linseed oil in water and stabilised by stearic acid and triethanolamine; and
- Plastic compounds

The use of curing compounds becomes necessary for places where there is shortage of water or it is inaccessible or difficult to cure continuously and properly. Curing compounds should have no toxic solvents and should be capable of forming a **membrane film** on wet **concrete surface.**

The consistency of curing emulsions should be such that it can be sprayed easily at normal temperatures from atomising nozzles as a continuous coherent film. The film should have sufficient elasticity to form an unbroken film for atleast 7 days. It should not run when applied to a vertical surface at the specified rate. The membrane film may be pigmented. Black bituminous coatings may be white washed after 3-4 hours of application to avoid excessive rise of concrete temperature. The chemical effect of these curing compound coatings should be investigated by trial for harmful influence, if any. The curing compound coatings should be applied immediately after the free water has disappeared from the surface. The curing compounds should remain stable when stored for a period of atleast 3 months (Fig. 4.1).

Emulsions are generally cheaper than membrane compounds of solvent type due to additional cost of solvents. There are varieties of commercially produced curing compounds under different trade names such as: Emcoril liquid, Emcoril white, CICO free cure, etc.

4.8 SELECTION CRITERIA

Selection of specific admixture for specific purposes requires evaluation of primary and secondary properties from various sources. Primary properties of admixtures have been specified in previous paras. Primary properties refer to the specific purpose for which an admixture is used. For example plasticizers have primary role of **improving workability** (flowability or plasticity) while **retarders** have primary role of **reducing the rate of hydration.** After assessing the primary role to be played, the admixtures are tested for their secondary roles in respect of side effects on other properties, chemical reactivity, long term durability and cost.

Most of the admixtures have one or more secondary properties which are similar to other primary properties but of much lower effect. For example certain plasticizers selected for their primary role of increasing plasticity (workability) may also slightly affect the setting and air entrainment properties of the concrete. The secondary properties achieved by the admixture must be considered in light of the situation of placement and desired quality of concrete. The **selection of admixtures** is thus made for **primary characteristics** and positive **gain on secondary properties** required for a given situation.

All admixtures must be **tested by** preparing **trial mixes** with the same materials and conditions of placement, **prior to its use in concrete** construction. Optimum proportion of admixtures is also determined by trials to exhibit desired primary and secondary properties of the concrete.

For achieving desired quality of cement concrete, the use of admixtures plays vital role. The use and proportion of a particular admixture in a particular situation must be decided by conducting scientific trials and **study of manufacturer's literature.**

Admixtures can also play a vital role in improving durability of concrete elements placed in adverse conditions of weather, chemical and mechanical attacks. Most of the problems of durability of concrete is associated with corrosion of reinforcement and carbonation of concrete. These problems are related to the permeability of the concrete. Water reducing admixtures with or without cement replacements can be used to reduce W/C ratio. This reduction in W/C ratio leads to: reduced permeability, reduced rate of carbonation and chloride diffusion.

4.9 SUMMARY

Admixtures in small quantities can **modify the properties** of cement concrete to obtain the desired characteristics. Admixtures influence the quality of fresh as well as hardened concrete. Use of admixtures improves placing techniques and construction methods.

The use of admixtures bring about many advantages in terms of **cost, quality** and **construction techniques.** The use of admixture can facilitate achievement of desired qualities by adopting scientific mix design procedures.

Admixtures are generally grouped as **chemical and mineral.** The most common chemical admixtures are **plasticizers, accelerators, retarders, air entrainers** and other special purpose admixtures. Mineral admixtures are obtained from natural materials or bye-products in blast furnaces or super thermal power stations. Mineral admixtures are finely ground **flyash, blast** furnace slags, silica fumes, volcanic ashes, volcanic tuffs, and calcined clays. Most of these mineral admixtures exhibit pozzolanic properties and facilitate cement replacement and improvement in impermeability and other properties.

Plasticizers, also known as water reducers, improve workability of concrete and lead to better compaction and economy in concrete construction. Plasticizers comprise of lignosulphonic acids and their salts or hydroxylated carboxylic acids and their salts.

Accelerators enhance the rate of hydration. Set accelerators comprise of aluminates or silicates while hardening accelerators comprise of chlorides or nitrites.

Retarders are required to delay the setting of cement and generally consist of gypsum ($CaSO_4$), or sugars. Hydro carboxylic acids, phosphates or unrefined lignosulphonates, ammonium chloride, ferrous and ferric chlorides, calcium borates, Oxychlorides, Calcium Tartarate, Alkali Bicarbonates, Tannic, Humic and Sulphonic Acids may be used in producing retarders. Retarders facilitate better bond and monolithic construction and long distance transportation.

Air entraining agents are used to entrain large volume of very small air bubbles for better workability and resistance to freeze and thaw. Aluminium powders or vinsol resins are required for air entrainment. Sometimes vegetable and animal fats can also be used for air entrainment. There are many commercially produced air-entraining agents.

Mineral admixtures are either inert or pozzolanic in nature. Pozzolanic action is caused due to reactive silica content. Very commonly used pozzolanic admixtures are flyash, calcined clay, volcanic ashes, and blast furnace slags. These are ground very fine and facilitate in making concrete more cohesive, reducing shrinkage and replacing part of cement.

There are many admixtures available for specific purposes. The selection is based on primary purposes and secondary effects of admixtures. The use of admixtures is adopted after ensuring its characteristics. The use of admixtures facilitates controlling of quality of concrete.

PRACTICE QUESTIONS

4.1   Define admixtures.

4.2   State benefits of using admixtures in cement concrete.

4.3   State importance of admixtures in quality management of cement concrete construction.

4.4   List type of admixtures in each category.

4.5   State use of admixtures in fresh and hardened states of cement concrete.

4.6   Explain use of plasticizers in cement concrete.

4.7   Explain use of accelerators in cement concrete.

4.8   Explain use of retarders in cement concrete.

4.9   Explain the purpose and use of air entraining agents in cement concrete.

4.10 Describe type of mineral admixtures.

4.11 Explain the purpose and use of fly ash in cement concrete.

4.12 List special purpose admixtures.

4.13 Write short notes (not more than 100 words) on:
      Gas forming agents, bonding agents, damp proofing agents and curing compounds.

4.14 State selection criteria for admixtures.

UNIT II

CHAPTER 5 : Influence of Concreting Operations on Quality

Influence of Concreting Operations on Quality

LEARNING OBJECTIVES

The learner understands the influence of concreting operations on cement concrete construction and will be able to:

- **List concreting operations** affecting quality of Cement Concrete;
- Describe the influence of **formwork erection** on the quality of Cement Concrete;
- Describe the influence of **storage of materials** on the quality of Cement Concrete;
- Explain the influence of **batching** of ingredients on the quality of Cement Concrete;
- Describe the influence of **mixing** of ingredients on the quality of Cement Concrete;
- Describe the influence of **transporting** of cement concrete on its quality;
- Describe the influence of **depositing** of cement concrete on its quality;
- Describe the influence of **compacting** of cement concrete on its quality;
- Explain the influence of **joint making** and **surface finishing** on the quality of Cement Concrete Construction;
- Explain the influence of **curing** on the quality of cement concrete construction;
- Describe the influence of **formwork removal** and patchwork on the quality of cement concrete.

5.1 INTRODUCTION

Cement Concrete Construction involves concreting operations both in plastic and hardened stages. Various operations carried out in cement concrete during its plastic stage directly affect the properties, quality and serviceability of concrete construction in its final hardened stage. These different concreting operations thus play an important role in achieving good quality of cement concrete construction under variety of placement conditions. These concreting operations form important **links in concrete chain** and needs to be understood for producing the desired quality of concrete construction. Different concreting operations which has bearing on the property and quality of final concrete construction are as follows:

- **Formwork** erection
- **Storage** of ingredients
- **Batching** of ingredients
- **Mixing** of ingredients
- **Transporting** of concrete
- **Depositing** of concrete
- **Compacting** of concrete
- **Joint Making** and Finishing of concrete
- **Curing** of concrete
- **Formwork removal** and Patchwork

The description of cement concrete practices and precautions necessary in each operation to produce good quality concrete are briefly stated in subsequent paras.

5.2 FORMWORK ERECTION

Design, fabrication and erection of formwork form an integral part of cement concrete construction. Formwork erection includes the **design** considerations, **fabrication**, **alignment**, position, **size, shape** and **appearance** of finished concrete surface. The desired size, shape and appearance of concrete element depends on the use of properly designed and constructed forms. The formwork serves as a temporary support for constructed element, material, construction equipment, and workmen. The formwork should be **strong enough to resist vertical and horizontal forces** exerted during construction. It should be **braced** suitably to resist possible forces and displacements. The speed of formwork erection depends on the shapes and accuracy in formwork fabrication and their arrangements.

The main objectives of concrete formwork construction and erection are:

(a) The formwork should be **capable of supporting all dead and live loads** without excessive settlement or collapse i.e. the formwork should be safe enough to perform its function without any danger to workmen or to the concrete structure under construction;

(b) The formwork should be made accurately so that the **desired size, shape, position, alignment and surface** finish of resulting concrete are attained. The provision and erection of suitable formwork facilitates achievement of the desired quality of concrete construction;

(c) The formwork should be **economical.** The formwork construction should be efficient to save both **time and money** by its **reusability** wherever possible.

5.2.1 Design Considerations In Formwork

Design of formwork for concrete construction should be such that all the vertical (gravity) loads and lateral pressures are supported safely by it until the concrete elements harden and become capable of carrying such loads. Vertical loads on formwork consist of **weight of fresh concrete**, weight of **reinforcement, self-weight** of forms, and **live loads of workmen** and construction **equipment**. Lateral forces may be caused due to side **pressure of plastic concrete, wind pressure and vibrations** caused by construction equipment. The formwork should be designed to carry safely all these loads under most **critical situations of unsymmetrical placings,** and **impact loading.** The formwork should also maintain its **rigidity, shape, size** and **alignment** under most critical lateral loadings also. Although it is difficult to precisely calculate these loads, but the design should always be safe for the worst condition of loading likely to occur.

Apart from vertical dead loads, the plastic concrete also exerts lateral fluid pressure on the sides of formwork. The lateral fluid pressure depends on the rate of placing concrete since the bottom layer of concrete may start setting before placing top layers of fresh plastic concrete. Thus the maximum lateral pressure on formwork sides depends on the density of plastic fluid concrete at any depth "h" is $p = wh$, where **w** is density of fresh plastic concrete and **h** is depth of fresh fluid concrete. Thus lateral concrete pressure at any point goes on increasing with increase in depth of plastic (unhardened) concrete. With faster rate of placing concrete, the lateral pressures will be higher while with slower rate of placing the lateral pressures will be lesser. The hardened concrete exerts no lateral pressure on its formwork and supports its own loading. The internal force of vibrations for compaction of concrete exerts local lateral force on the formwork in proportion to the acceleration and mass of the concrete vibrated. The external vibrating device also exerts lateral pressures on the vertical faces of the formwork and must be considered for the safe design of the concrete formwork. The formwork design should be safe for the **worst combination of forces.** External vibrators transmit vibrations through the forms and hence the forms should be designed against battering under such forces. The workability of concrete and vibrating force are appropriately adjusted in relation to the formwork design and rate of placing concrete.

Temperature of the concrete and its environment during the time of placing affects the setting time of concrete. Thus the temperature of concrete will directly affect the fluidity of concrete and lateral pressure exerted by concrete on the formwork. In low temperature setting of concrete is slow and higher depths of concrete can be placed in shorter time. This aspect should also be considered when retarding admixtures are used in concrete and when speed of laying is also high. The formwork should be strong and stable against lateral pressure considering concrete as fully liquid. Proper ties should be used to maintain the **shape of forms** and **avoid its spreading and bulging.**

5.2.2 Formwork Materials

Forms can be constructed/fabricated by using different materials such as timber, plywood, steel, plastic or masonry. Metal forms do not shrink and provide precise shape, smooth surface and accurate size. Metal forms can be used repeatedly and economically in long run. Metal forms can also be used for continuous construction with slipform sliding design.

Timber is used in one or the other manner for fabricating formwork. Partially seasoned timber with moisture content more than 20 % is quite suitable for formwork. Kiln dried timber has a tendency for swelling when soaked in water. Timber boards with light joints cause bulging and distortion due to swelling on account of water absorption. On the other hand the **green timber dries and shrinks causing fins and ridges** on the concrete surface. The green timber forms should be kept wet until the time of placing to avoid drying shrinkage ripples in surface of formwork.

5.2.3 Precautions in Formwork Erection

Forms should be so constructed that these can be **used repeatedly.** Metal forms can be made more durable, strong and accurate in size and shape.

Forms should be cleaned and reconditioned before re-using. Form surfaces should be wetted and suitably **oiled prior to casting** concrete to prevent shrinkage and attain smooth surface. Oiling of forms should be done before placing reinforcement to avoid failure of bond of steel due to oil coating. Oiling prevents sticking of the concrete and facilitates removal of formwork. Form oil should not leave permanent stains on concrete and or interfere with water curing. The final surface of concrete should be obtained by preparing appropriate surface of forms instead of repairing and patchwork after removal of forms.

Light-coloured and light-bodied petroleum oil serves satisfactorily. Compounded oils of petroleum and other **oils of animal or vegetable origin, and gums and resins** are found to be **superior** to straight petroleum oils.

Some times for ornamental details in architectural concrete elements involving intricate designs, **plaster waste moulds are formed which are broken at the time of stripping off** forms. Forms are important in obtaining the **correct size, shape,** and **quality of concrete** structure and hence formwork must be fabricated and erected properly before concreting.

Defects in formwork may result in **repair cost or complete re-casting** of structural element and thus **causing delays and poor quality** of concrete construction. Formwork thus, plays an important role in achieving quality of concrete construction.

5.3 STORAGE OF INGREDIENTS

Storage of ingredients plays an important role in producing good quality concrete. Many a times storage of various concrete ingredients is not provided necessary attention. The most **critical objectives** of storage of the concrete ingredients are:

- To **protect** the ingredients **from impurities;**
- To **preserve the characteristics** of the ingredients;
- To **preserve the uniformity** of materials;
- To **reduce the wastage** of materials to a minimum.

5.3.1 Storage of Cement

Cement is a fine hygroscopic material and gets affected by water, **atmospheric humidity,** type of **packing and method** of storage. Hard lumps of cement may be formed due to atmospheric humidity which may cause difficulties in concrete mixing and loss of concrete strength. Any contact of stored cement with **moist air** and or **carbon dioxide** in atmosphere **disturbs setting properties.** Long storage of cement results in **reduction of strength** based on the period of storage. Cement can be stored in sealed air-tight drums for a longer period without losing its properties.

Stored cement should be used only after testing and ensuring its properties and specifications. Concrete made with stored cement develops much lower strengths compared to concrete made with fresh cement. Strengths of concrete using stored cement depends on the **period of storage** and is stated in Table 5.1.

Table 5.1 Percent Strength with Stored Cements

Period of storage of cement(in months)	Fresh	3	6	12	24
Percent Strength of concrete with stored cement at the age of 28 days	100	80	70	60	50

It is quite evident that proper storage of cement plays very critical role in achieving quality of concrete construction.

Appropriate practice should be adopted for storing cement at site and in a warehouse so as to **eliminate any contact of moist air with stored cement. Avoid long storage** of cement to minimize the loss of properties of concrete. Following precautions in storing cement in warehouse must be adopted to achieve quality concrete construction:

- Cement bags should **not be put directly in contact with the floor**, but separated by an additional wooden or dry brick floor base for cement bag stacks;
- Cement bags are stacked away from walls creating a corridor of about 300 mm allround the stack to **avoid contact with outer walls**;
- Bags should be arranged in close stacks to **avoid air circulation** through the stored stacks;
- **Height** of cement stacks should be limited to about **2.70 m** (15 bags) to avoid lump formation in bottom layers. The **width** of stack may also be limited to **3 m** for efficient handling.
- Avoid long storage of some batches by adopting the principle of **first in-first out** for storing the cement according to consignment and putting date tag on each stack.

At site cement is stored on **raised dry-impervious platform** and fully covered by water proof **tarpaulin** or other plastic sheets. The storage of cement at site should be restricted to 1 or 2 days consumption only. All cover sheets should have adequate overlaps of atleast 150 mm and should be fully anchored to the ground to withstand wind force.

5.3.2 Storage of Aggregate

For achieving appropriate quality of concrete, different aggregates need to be stored to **avoid impurities and variations in grading**. During storage impurities such as mud and clay may get mixed with aggregate which affects the quality of resulting concrete. Unsystematic and unorganised stacking of aggregate may cause nonuniformity in grading due to variations in proportions of different sizes. Due to improper stacking, the labour may take variable proportions in different batches and thus causing nonuniform quality. Further the stacking should **ensure uniform surface moisture** condition in aggregate. The important considerations in storage of aggregate should be to:

- Preserve the **uniformity of grading**;
- Prevent the **segregation**;

- Keep the **uniformity of surface moisture** conditions;
- Protect the aggregates from mixing with **harmful impurities.**

Aggregate should be stacked on a dry and rigid or hard base formed by dry brick flooring or GI sheets to prevent mixing of harmful materials and to maintain uniform grading. **Coarse and fine** aggregates should be **stacked separately** and wherever necessary according to size to obtain the desired grading. Segregation in aggregate while stockpile formation can be prevented by avoiding drop from heights. Aggregate stockpiles should be formed near the mixer site to avoid unnecessary movements causing segregation. Stockpiles of different sizes of **aggregate** can be maintained by providing separating partition walls.

To achieve uniform surface moisture condition, the aggregate stock piles should be large and of an average depth of 1.25 m to 1.75 m. The stockpiles should be stored for enough time to stabilize the surface moisture condition after washing and draining the stockpiles. Mixing of deleterious materials is avoided by stockpiling on rigid impervious raised grounds and washing the aggregate before use. Proper storage of aggregate helps in achieving the quality concrete construction.

5.3.3 Storage of Water

Storage of water should ensure adequate quantity and prevent mixing of any deleterious materials or impurities. The quantity of stored water should be adequate for washing aggregate, mixing and curing concrete for atleast one day's consumption. Mixing of impurities such as mud, clay, chemicals and other organic matter should be avoided by storing water in impervious tanks. Impurities in mixing water may result in disintegration and loss of concrete strength. Mixing water should be collected one day in advance so as to allow suspended impurities to settle. During storage of mixing water care should be taken to **avoid mixing of deleterious materials** like **soap, dirt, organic matter, sugarcane, leaves** and **food materials.** Inadequate quantity and mixing of impurities can affect the quality of concrete construction by causing discontinuity and disintegration respectively. Storage of good quality and adequate quantity of water is, therefore, important for quality concrete construction.

5.3.4 Storage of Admixtures

Concrete admixtures should be stored in such a way that it does **not loose its properties** due to its **contact with** atmospheric **moisture, air, water** or any **other chemical.** Different brands of admixtures should never get mixed during storage. Each admixture should be stored in water proof containers with its **labels.** These admixtures loose their properties with time and hence should not be stored for a long period. For desired effect on quality of concrete construction admixtures should be used as fresh as possible. Certain epoxy resins are mixed only at the time of using.

Correct practices in storage of concrete ingredients is, therefore, an essential requirement for accomplishing a good quality concrete construction.

5.4 BATCHING OF INGREDIENTS

Correct measurement of concrete ingredients is an essential pre-requisite for producing the desired quality of concrete in any batch. Batching is a **process of measuring concrete ingredients** in certain given proportions to prepare a batch (material lot) which can be handled conveniently in one cycle of mixing. Batching is the starting operation for actual manufacturing or producing concrete. The process of batching requires the measurement of fixed proportions of ingredients in every batch as determined in the mix design.

Batching can either be done by **weight or volume measurement** depending on the importance and extent of concrete construction. For large and important jobs batching needs to be done by weight. For small jobs batching is commonly done by volume. In weigh batching materials are measured according to their respective weights while in volume batching, materials are measured by volume. Batching of various ingredients for preparing a batch of cement concrete is explained in subsequent para.

5.4.1 Batching of Cement

Cement is generally batched by weight or in terms of 50 Kg bags on smaller jobs. In case of heavy construction jobs, cement is always measured with weighing machines for ensuring the correct proportions. Automatic weigh batching machines, **spring balances, double swing weigh machines** or platform weigh machines can be used for weighing. Volume batching of cement is **not** permitted and should be avoided as far as possible. Volume batching of cement may result in variation of cement weight in different batches due to variation in fineness, method of filling and the condition of the containers. **Weigh batching provides accurate quantity** of cement.

5.4.2 Batching of Aggregate

Batching of aggregate is done either by volume (on ordinary jobs for the production of ordinary concrete) or by weight (on important jobs for the production of controlled or design mix concrete).

Volume batching is done by using **volume containers** made in various sizes and capacities. The containers are made of 30 mm thick wooden boards with tongue and groove joints and sheet metal tops.

Necessary **bulking allowance** should be made in measurement of sand by volume batching. The bulking allowance in volume of sand is quite large as compared to moisture allowance in weight of sand due to its moisture content. The shape of container also influences the variation in volume of batch. Deeper containers with less plan area results in reduced error of volume in terms of error in height measurement. Size of containers should be so chosen that each batch requires **whole number of containers** for the convenience of correct volume batching of aggregate by labour.

Volume batching encounters following two sources of error:

- **Variation** in volume of solids in measured volume of containers **due to voids**;
- Variation in measured volume due to **variation in bulking effect** in sand particles **due to presence of moisture.**

Weigh batching of aggregate is necessary for the production of quality or controlled concrete. Weigh batching of aggregate is, therefore, preferred for all important and large size jobs. Wherever regular weigh batching is not possible, the **weights of various ingredients can be converted to volumetric proportions using bulk densities** which is checked on regular basis at site to account for any variation due to bulking or grading.

Weigh batching can be done by using:

- **Spring balance** supported on tripod
- **Physical balance** supported on tripod
- **Platform weighing** machines
- **Double swing** weigh batching machines
- **Automatic** weighing machines

Weighing device should be such that the materials can be directly emptied to wheelbarrows or other containers for transporting to the mixer. The selection of weighing device is made on the basis of the speed and extent of work. Spring balance, physical balance, platform machines and double swing weigh batcher for smaller to medium size jobs while automatic weigh batchers may be used for continuous and large size jobs. All weighing devices should be cleaned, checked and adjusted for correctness on regular basis.

5.4.3 Batching of Water

Water is required for washing aggregate, mixing as concrete ingredient and curing of set concrete. At any site adequate quantity of water should be available for various purposes. Quantity of mixing water plays most critical role in ensuring desired quality of concrete. For batching accurate quantity of mixing water special tanks fitted with special quick release valves and gauges may be used. On smaller jobs quantity of mixing water may be batched by metallic volume containers. Water is generally batched by volume gauges since volume of water can also be measured accurately and unit mass of water is normally 1.0.

Errors in batching of ingredients can result in the following approximate variations in the strength of resulting concrete at site:

BATCHING FACTORS		LIKELY VARIATION IN STRENGTH
i) Cement quantity	—	upto 50%
ii) Aggregate quantity		
a) By Volume Poor measurement	—	upto 100%
Normal measurement	—	upto 50% to 70%
Accurate measurement	—	upto 10% to 15%
b) By Mass		
Bulking in Sand	—	Compensated — upto 10%
	—	Not Compensated — upto 25%
iii) Aggregate Grading	—	upto 20%

5.4.4 General Precautions in Batching

For producing good quality concrete, following important precautions should be observed:

- For minimizing the error in proportions of materials delivered to the mixer, the **measurement of materials** during batching needs to be accurate.
- **Loss of batched materials should be avoided** during transit to the mixer.
- Weigh batching machines should be levelled, cleaned and **checked for correctness on a regular basis.**

- Volume batching **containers** should be **evenly filled** and occasionally **checked** for its correct size.
- The batching equipment or containers should be in **good working condition** and should conform to the prescribed standards.
- In volume batching of aggregate, **bulking effect** should be checked **three times a day** and compensated for minimizing variations in production of quality concrete.
- Surface moisture of wet aggregate should be estimated and compensated during batching of aggregate and water.

5.5 MIXING OF INGREDIENTS

5.5.1 Importance

Intimate mixing of ingredients is necessary to produce **homogeneous** concrete. Homogeneous concrete indicates for intimate contact between different ingredients leading to optimum hardening process and better interlocking by filling the voids between larger particles by smaller particles. Thus, intimate mixing of ingredients facilitate to produce **dense and strong** concrete. Improper mixing of ingredients leads to large variations in strength and quality of resulting concrete at site. Sometimes site engineers try to **compensate** the loss of strength due to improper mixing, by **adding extra cement** resulting in avoidable extra cost. Mixing is thus an important operation in production of quality concrete within specified economy.

5.5.2 Manual Mixing

For small and less important jobs the ingredients are mixed manually by using shovels or khasis. For manual mixing a watertight rigid platform of about 3.5 m × 2 m is formed with wooden planks or brickflooring. Alternatively a trough of 2.70m × 2.1m × 1.80m made of 2 to 3 mm thick G.I. sheet and stiffened with angle iron may be used for manual mixing. Manual mixing involves:

- Spreading the measured quantity of sand on prepared rigid platform or trough for the purpose and then spreading cement uniformly over sand and mixing these dry till uniform colour.
- **Spreading the sand cement dry mixture** evenly and spreading coarse aggregate uniformly over the dry mixture and turning the whole mixture with shovels till uniform mixture is obtained.
- **Spreading the dry mixture making a hollow in the middle** and pouring about 3/4 quantity of water gradually and mixing the ingredients taking care that **no water escapes.**
- Adding remaining quantity of water and completing the **mixing within 3 – 5 minutes of first adding water.**
- Cleaning and washing the platform or trough at the end of days work (Fig. 5.1).

Homogeneity of hand mixed concrete is doubtful and may result in variation in quality.

5.5.3 Machine Mixing

Mechanical mixing of concrete ingredients is superior and results in **better homogeneity**.

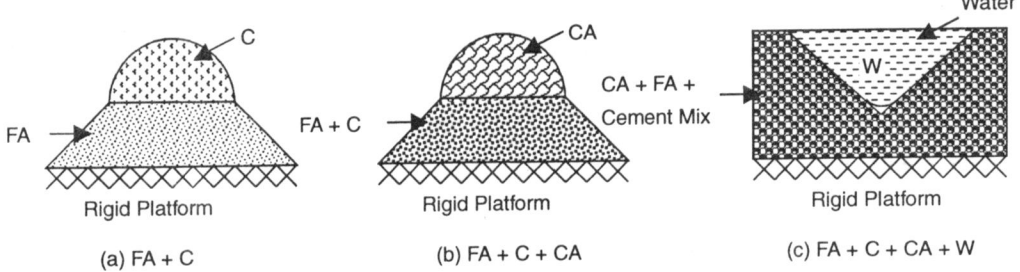

Fig. 5.1. Manual Mixing

Mechanical mixing is preferred for **speed and quality** of concrete construction. Different type of mixing machines (called mixers) can be used. Different type of concrete mixers (Fig. 5.2) are:

- Tilting (T)
- Non-tilting (NT)
- Pan
- Transit truck mixer
- Continuous paver mixer
- Other mixers

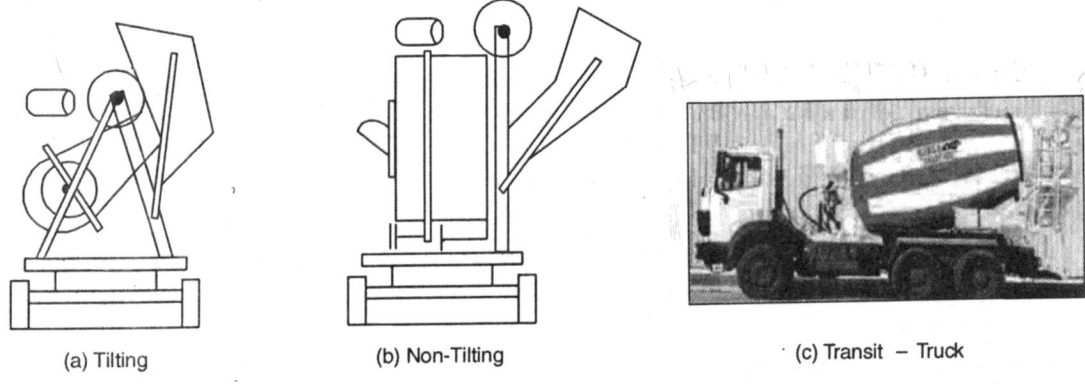

Fig. 5.2. Machine Mixing

Tilting type mixers (Fig. 5.2 (a)) are very commonly used to produce good quality of concrete on small jobs. These mixers are of **smaller capacities** and are more suited for **dry concrete mixes**. Non-tilting mixers (Fig. 5.2 (b)) are more **suited for wetter mixes** and **larger quantity** of concrete production. Pan type mixers consist of horizontally mounted pan rotating about vertical axis with eccentric paddles to create mixing action. Pan type mixers are **heavier** and are quite suitable for producing concrete of **consistent quality** specially for **precast constructions**. Transit truck mixers are suitable where construction sites are located at longer distances and **transporting is combined with mixing operation** to **save time** in delivering ready mixed concrete. Continuous paver mixers may be suitable for mixing concrete on a continuous basis specially for laying **aerodrome pavements.** There are many other type of concrete mixers used for variety of jobs.

Homogeneity, **speed** and **capacity** of concrete production are most important considerations in selection and operation of any mixer. **Installing, feeding, discharging** and **operating a mixer** should be done as per specifications for producing good homogeneous concrete. For the production of good quality concrete on a continuous basis the mixers must be **maintained in working condition** at the end of days work.

Good mixing to produce homogeneous concrete results in economy and quality of concrete construction. Manual or machine mixing method may be adopted depending on the size and the quality of concrete construction desired. The operation of mixing needs to be supervised for the quantity of mixing water and adequate mixing on a continuous basis for the production of quality concrete.

5.6 TRANSPORTATION OF CONCRETE

5.6.1 Importance

It is normally not possible to produce concrete at the place of its final laying. Concrete is, therefore, transported from the production location or mixer point to the point of final placement (which moves with the progress of work). Cement concrete looses its plasticity and starts setting in about 30 minutes. Mixing, transporting, depositing and compacting need to be completed in this short period.

Transportation of cement concrete from the mixing point to the place of laying must comply with the following main principles to accomplish good quality concrete construction:

- Transportation should be rapid so as to avoid drying and loss of workability;
- Transportation should avoid vibrations and shocks to concrete so that no segregation of concrete takes place;
- Transportation should be organised timely so as to avoid formation of unplanned construction/cold joints in any section or lift of concrete placement.
- Use of retarders may be considered in case of long distance transportation.

Appropriate timely transportation can therefore play an important role in accomplishment of good quality concrete construction.

5.6.2 Methods of Transportation

Depending on the quantum of concreting job, location of concrete depositing site, lead and lift between mixer and depositing site and availability of concrete transporting equipment, the most appropriate concrete transporting method may be adopted from the following common methods:

- Manual using pans
- Wheel barrows, handcarts
- Power barrows, powered buggies
- Dumpers, lorries, truck/transit mixers
- Elevating towers and chutes
- Cableways, crane operated skips

- Belt conveyors, mono rails
- Concrete pumps

For small jobs manual (head load) method is quite common in India. Manual method needs to be properly organised and managed to avoid delays and segregation. Cranes, cableways, belt conveyors or concrete pumps may be used advantageously for continuous and large size concrete jobs. For medium size concrete jobs other methods may be selected on the basis of availability and site situations.

5.6.3 Precautions

Whatever methods are used for transporting, the concrete should be laid and finished in its final position before it starts setting. Depending on the distances and total cycle period, suitable admixtures may be mixed in concrete to maintain its plasticity until the time of depositing and compacting. During transporting homogeneity must be maintained by avoiding vibrations and shocks. First batch must be specially prepared using cement slurry to compensate for dry surface of the tools/equipment. Care should be taken to avoid any loss of cement slurry during transportation of concrete by any method. At the end of transportation all implements must be washed clean and set in operating condition with proper lubrication of all moving parts.

Thus transportation of concrete, directly influence the accomplishment of good quality concrete construction. Selection of suitable method can also facilitate economy in concrete construction.

5.7 DEPOSITING

Depositing refers to laying concrete in its final place where this concrete matures after curing into structural element. Depositing includes preparation of subgrade surface and laying concrete in its position. Proper depositing helps in getting concrete elements of desired **shape, size** and **quality.**

Subgrade surface may comprise of:

- Natural subgrade of soil in foundations;
- Waterbound macadum or brick flooring for pavement;
- Hardened concrete base of structural slab;
- Form work for slabs, beams, columns and walls.

Subgrade surface should be **rigid, non-absorbent, trimmed to size and shape,** free from loose materials or dust, cleaned and **moistened before depositing** concrete. Reinforcements should be checked for its correctness of size, shape, length, position and appropriate **concrete cover.** Hardened concrete subgrade base surface should be roughened and provided with bonding layer of mortar before depositing fresh concrete.

While depositing it should be ensured that concrete is delivered or laid as near as possible to its final position. Concrete should be deposited in **continuous layers** instead of separate heaps or piles to avoid unnecessary movement leading to segregation. Concrete should be deposited in **required quantities** at its final location and spread without causing segregation as far as possible.

Concrete should **not be thrown from height** and suitable **chutes and skips** should be used to avoid segregation. Consistency of concrete and rate of placing concrete in deep forms should be adjusted to avoid bleeding and cracking in concrete layers.

Avoid formation of cold joints in concrete layers or lifts by suitably brushing and cleaning of previously laid concrete.

5.8 COMPACTING

5.8.1 Importance

Compaction refers to movement of **particles to come closer** to create dense mass of concrete by **removal of air voids.** The presence of air voids in concrete results in lower density which in turn affects strength of hardened concrete. It has been established by research that a reduction of **5 % in density from optimum density results in 30 to 35 % reduction in strength** of hardened concrete. **10 % of reduction in density** due to air voids may result in **50-60% reduction in strength.** Compaction of plastic concrete is therefore most vital for achieving desired quality of hardened concrete (Fig. 5.3). Optimally compacted concrete provides strong, durable and waterproof concrete elements. Optimum compaction of concrete depends on its workability. For achieving quality concrete construction the operation of compaction requires a close supervision on a continuous basis.

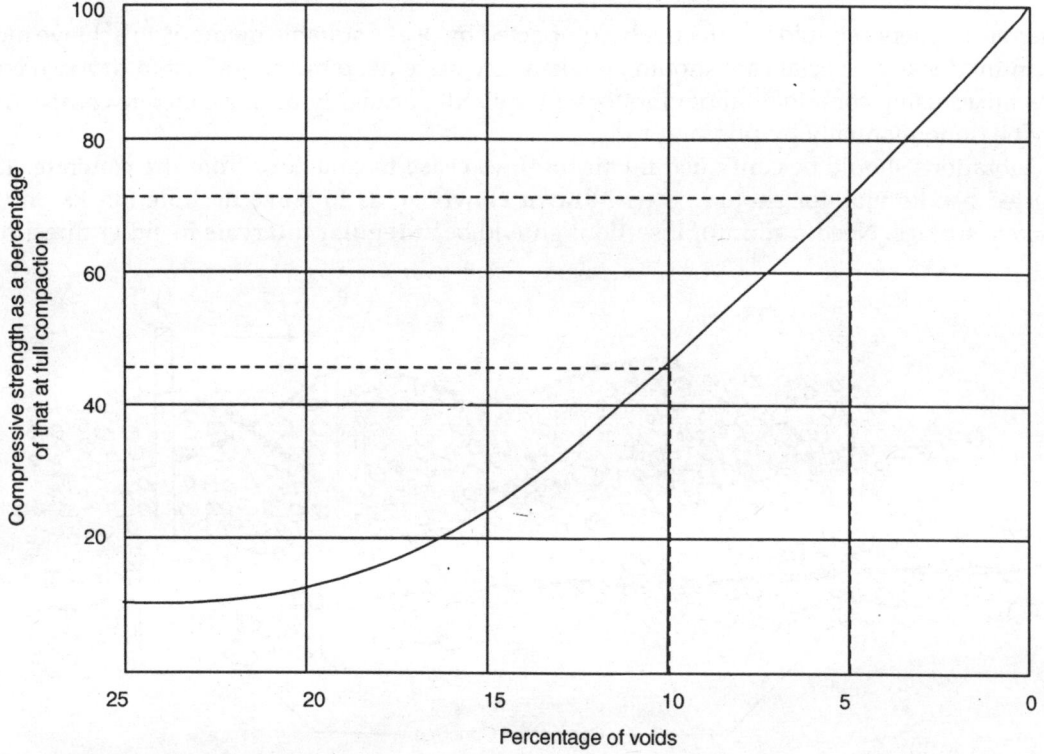

Fig. 5.3. Relationship between percentage of Voids and Compressive Strength of Concrete

5.8.2 Methods of compaction

Compaction of concrete is carried out manually or mechanically. In compaction the particles of concrete mass are moved closer to **expel entrapped air** in voids. In manual approach motion is imparted to concrete particles by punning rods or hand tampers to expel entrapped air in voids. In mechanical method concrete particles are imparted **vibrations to move closer** and **expeling air in voids.** These vibrations are caused through **needle or immersion rod**, surface or screed vibrators and shutter or form vibrator by mechanical device. Vibrations are caused due to motion of eccentric weight with the help of prime mover. Generally these vibrations range from **6000 to 9000 vibrations per minute.**

Different vibrators may be used according to the type and size of the concrete element to be compacted. Surface vibrators are more suited for thin members and slabs, while **needle vibrators** (Fig. 5.4) are better suited for thick and deep members. Form vibrators are more suited where concrete can not be reached directly due to intricate shapes or heavy reinforcement. Selection of appropriate type of vibrators with suitable specifications is necessary for effective compaction to accomplish good quality concrete construction. Compaction of high strength rich-dry concrete by mechanical method becomes an essential condition for accomplishing good quality concrete. For effective compaction of some concrete elements combination of vibrators may be used. For laboratory tests vibrating tables may be used.

5.8.3 Precautions in Compaction

Concrete elements should be effectively compacted by use of suitable methods to achieve most optimum density. Special care should be taken to vibrate deep members cast in dry-rich concrete mixes using combination of vibrators if required. For highly plastic concrete compaction may be done manually by punning rods.

Vibrations should be continued till air **bubbles cease to come out from the concrete.** The inserted needle vibrator should be **withdrawn slowly so as to leave no hole marks** on the concrete surface. Needle vibrator insertions should be at **regular intervals** in either directions

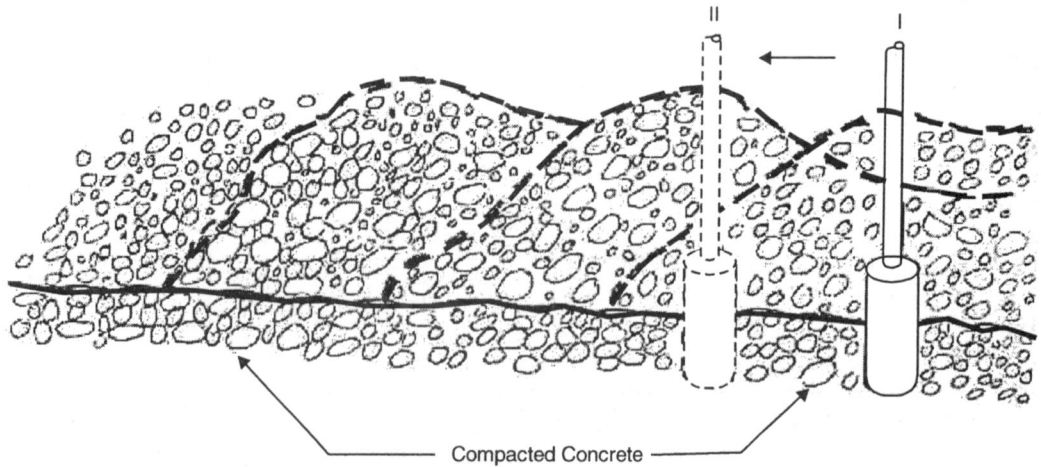

Fig. 5.4. Preventation of Segregation by Keeping Adequate Distance From Free Sloping Surface of the Concrete of the Needle Vibrator Moving in the Direction From I to II

such that each subsequent insertion lies with the vibration influence zone. Generally these intervals are at **300 to 600 mm spacing** depending on the size of the needle vibrator and the thickness of concrete element. Needle vibrators **should not touch forms.** Sloping edges of concrete slab should be carefully compacted. The compacted concrete should be finished to the required size of element.

Vibrating elements should be kept clean and in proper operating condition at the end of compaction job. The operation of correct compaction leads to good quality concrete construction without affecting economy. Compaction plays most critical role in accomplishment of high strength and other properties.

5.9 JOINT MAKING

5.9.1 Importance

It is essential to plan and design various types of joints in concrete constrictions for efficient and trouble free use of structures. The design should provide appropriate type, shape, location, and construction details of joints. Joint making becomes essential to compensate for expansion and contraction in large concrete structures without any undersirable effects. The main objective of joint making is to ensure adequate bond and continuity in different parts of concrete structure. Good joint making also eliminates weak spots, uneven surface and passage of dampness through concrete elements.

5.9.2 Types of Joints

There are two type of joints usually required in construction of concrete structures. These joints are:

- Construction joints formed to safeguard discontinuity in concreting operation;
- Movement joints necessary to allow movements due to expansion or contraction without damage or overstressing the concrete members.

The efficiency of these joints depends on their nature, workmanship and locations. The joint design and preparation depend on its alignment (vertical or horizontal), size (thickness, length, width and depth), locations, purpose and degree of restrain.

Construction joint is a surface between previously laid hardened concrete and freshly laid plastic concrete after certain time. The fresh concrete laid at these joints can not be integrally incorporated with previously hardened concrete by using normal method of construction without making joint. The structural design does not make any allowance of subsequent movement at these construction joints. Thermal and shrinkage movements cause stresses across these construction joints which need to be resisted appropriately. Design of construction joints is affected by its **size, alignment, presence of reinforcement, degree of restrain,** position or **location,** workmanship and surface finish of the joint. Bond between two portions of hardened concrete across the construction joint should be appropriately developed by proper workmanship. The quality of structural behaviour will be affected by workmanship in making the construction joint. Thermal stresses are usually caused at the construction joints due to **shrinkage** or **temperature variations** and depends on the **restraint** or lack of restraint between the two jointed parts. For example, tensile stresses are developed if the construction joint is located very near the fixed support.

Location of construction joints in walls and columns should be blended with the general architectural appearance. Suitable stopboards should be constructed in vertical joints for sealing purposes. Column-beam joints should be filled slightly below the junction. Since **shear and bond** requirement is **low in the middle portion of continuous slabs,** the construction joints may be located in **middle one third zone of continuous slab or beam** structures. **Construction joints** are preferred in **compression zones** of continuous structural members. In beam-slab construction joints should preferably be provided in the **centre or middle-one third** of the span where shear is minimum to avoid failure of the joint due to bond and shear. The construction joint between the slab and beam should provide a suitable **key and shear reinforcement** across the construction joint.

Movement joints in structures are provided to keep the length of each part less than 30 m so as to avoid damaging **cracks due to shrinkage and thermal variations.** The distance between the movement joints can be increased by providing temperature reinforcement to resist additional stresses caused by temperature variations. Expansion and contraction in roof slabs causes stress in parapet walls or ceiling plaster near their junctions and result in cracks. To create quality surface finish, there should be suitable grooves in finishings of these elements near the junctions. Further the continuity of surface finish should be broken at an appropriate interval based on shrinkage and thermal coefficients. Movement joints should be suitably formed to accommodate shrinkage and thermal movements in water retaining structures without breaking water seal. These movements should be capable of preventing leakage of water through the joint or the body of structural element. Movement joints must prevent collection of dust and debris by filling these joints with suitable flexible sealing materials.

There are variety of concrete surface finishings which may be selected based on the requirements of joint making, desired surface texture and other functional considerations. Joint making and concrete surface finishing should prevent unintentional and undesirable cracking due to shrinkage and thermal variations. The type and nature of joints in concrete structures affect the surface finishing and appearance. Concrete surface finish needs to be matching with the requirements of joint making for the desired quality of appearance and usage of the concrete structure. Thus formation and making of appropriate joints and surface finishing play a very important role in accomplishing quality concrete construction.

5.10 CURING OF CEMENT CONCRETE

5.10.1 Importance

The word curing means to treat or to bring back to health or to keep in good condition. In cement concrete, **curing refers to procedure which promotes the hydration of cement.** It may be defined as the act of **maintaining controlled conditions of moisture and temperature for freshly placed concrete for some definite period** after placing and finishing operations **to ensure the proper hydration of cement and hardening of cement concrete.** The process of hydration is dependent on the temperature and availability of adequate water for reaction in the mix. Loss of water from the body of concrete due to evaporation may cause the hydration process to stop, with a consequent reduced strength. Further, evaporation may cause rapid drying shrinkage resulting in tensile stresses and cracks in concrete body. Curing is therefore carried out to maintain the concrete in a **continuous moist condition for a certain period to prevent evaporation.** The period of moist curing depends on the type of structural member, placing conditions and environment. Intermittent drying and spraying water results in loss of strength in structural members. Initial curing for first few days plays critical role in quality of concrete specially **strength, impermeability** and **durability.**

5.10.2 Objectives of Curing

The main objective of curing cement concrete is to **promote proper hydration of cement** for achieving desired properties of concrete in its hardened stage such as strength, impermeability and durability. The factors of influencing the process of hydration and method of curing depends on:

- Availability and preservation of **adequate water** in the concrete mix;
- **Maintenance** of the most favourable **temperature** for optimum hydration;
- **Preservation of uniform temperature** throughout the body of concrete;
- **Protection of structural members** from mechanical disturbances, excessive loadings during curing period;
- Ensuring **adequate curing period** for safe carrying of service loads;
- Ensuring **adequate development of strength** and abrasion resistance in case of road concrete;
- Ensuring adequate **development of strength and impermeability** in case of water retaining structures.

The method, duration and conditions of curing are decided on the basis of specific requirements of the concrete element, environmental conditions and type of cement. Longer periods of curing required for slow-setting cements and important structural elements necessitating impermeability and strength.

5.10.3 Methods of Curing

Curing methods are broadly of 3 types:

- Those which interpose a direct source of water (ponding, wet material) to prevent evaporation such as ponding, covering with wet earth, wet straw, wet burlap and wet cotton mats.
- Those which minimize evaporation by **interposing impermeable medium** such as water proof paper, leaving shuttering and formwork, **spraying sodium silicate**, spreading calcium chloride and **spray of impervious membrane forming chemical.**
- Those which involve the application of **artificial heat** whilst maintaining concrete in moist condition such as low pressure **steam curing, high pressure steam curing, electrical** curing and **infrared radiation curing.**

The selection of method of curing will depend on the type, size, environment and requirements of structural element. The adequacy and appropriateness of curing influence the properties and quality of concrete elements. Although it is **difficult to measure curing effectiveness** at site but this can be assessed through various laboratory investigations. Interruptions, delay in curing and curing at high or low temperatures influence the properties and qualities of concrete structures to a great extent. While planning and analysing concrete construction, the curing aspect must be considered important in accomplishing the desired quality (Fig. 5.5).

To achieve optimum quality from curing, following factors should be considered:

- **Treatment and curing for first few hours** after placing the concrete are most critical for strength and other qualities of concrete;
- **Wet climate conditions** help in reducing variations in quality due to different approaches of curing;
- Initial **wet covering** with wet burlap for **first 24 hours** facilitate better results from subsequent curing by different methods;
- For better result **membrane forming compounds** should be **applied after setting and disappearance of surface water**;
- **Wetting of formwork** or subsoil base helps in avoiding loss of water from concrete body;
- **Avoid excessive service loads** and mechanical damage during curing period.

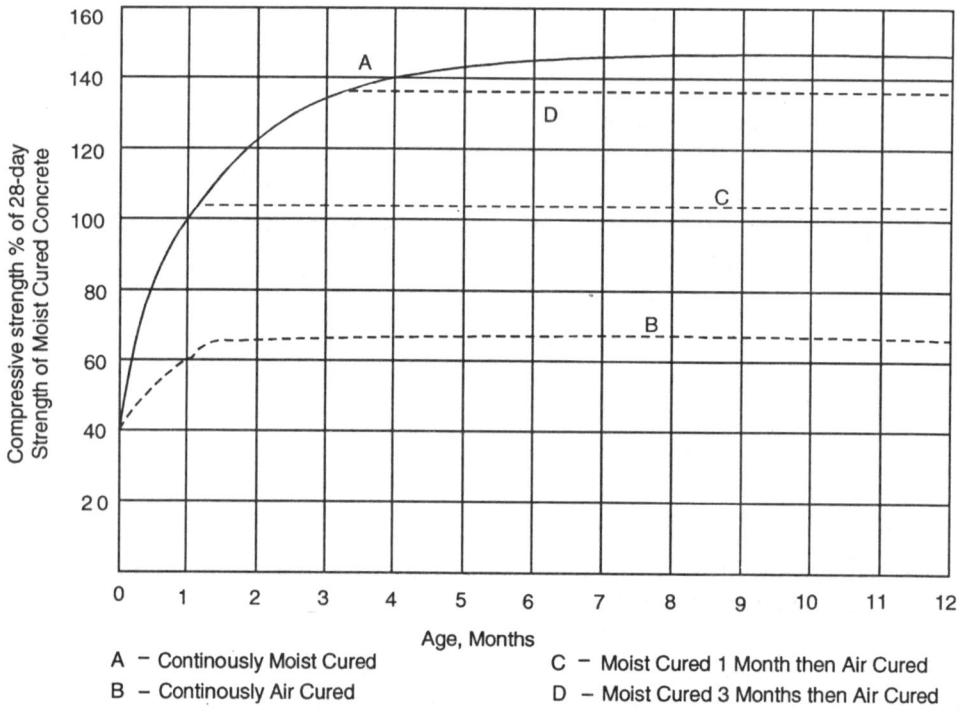

A – Continuously Moist Cured   C – Moist Cured 1 Month then Air Cured
B – Continuously Air Cured   D – Moist Cured 3 Months then Air Cured

Fig. 5.5. Curing and Strength Relationship for Portland Cement Concrete

5.11 FORMWORK REMOVAL AND PATCHWORK

For accomplishing good quality work, removal of formwork and shuttering is equally important. One of the aspects of cost and quality concrete construction is the duration after which the formwork needs to be removed. The **duration** of formwork depends on **type of cement used, nature and span** of structure, extent and pattern of **loading, environment** and expected occupancy. Another important consideration is **technique and sequence** of removal of formwork. Sometimes incorrect sequence of formwork removal may cause reversal of stresses in concrete elements as compared to those envisaged in the design and result in damage and cracking (e.g. removal of shuttering support in **cantilever elements must commence from the**

free end instead of middle of the span). Care should be taken to provide safety to workers and to avoid damage to adjoining elements during removal of shuttering and formwork. **Premature removal** of formwork may **damage the structure** badly due to **over stressing.**

Condition of cast concrete is observed clearly and defects, if any, are detected during removal of formwork. If the observed defects are of minor nature and do not warrant recasting of concrete, these minor defects need to be rectified immediately by special patchwork. This patchwork is necessary to avoid further damage to the structure and improve its appearance. Exposed reinforcement, if any, needs to be covered by water proof material. Suitable **coat of epoxy resin** is necessary for bonding and eliminating shrinkage cracks in the patchwork. Any **delay in patchwork** after removal of formwork will **enhance the damage or defect.** It may be understood that the **patchwork repair is no substitute for original good quality** concreting work during construction itself.

It is seen that for managing quality of concrete construction, it is important to **adopt correct technique of concreting operation at every stage right from selection of materials to removal of formwork** to put the structure to use. It is essential for every construction engineer and manager to understand various concreting techniques correctly before undertaking construction of cement concrete construction.

In this chapter we have studied the influence of different concreting operations on the quality of resulting cement concrete construction. Detailed descriptions of concreting operations and good practices can be studied from any standard book of Concrete Technology.

5.12 SUMMARY

Various concreting operations influence the properties and quality of fresh (plastic) and hardened concrete. These concreting operations at various stages are: **form work erection, storage, batching, mixing, transporting, depositing, compacting, joint making, curing, form work removal** and patchwork repair.

Formwork erection includes the **design, fabrication, alignment, size, shape, position** and **appearance** of finished surface. The objectives of formwork construction and erection are: to **support safely dead and live loads,** to achieve correct **size, shape** and **alignment,** and to **provide economical and efficient construction.** Materials used for form work are seasoned **timber, metal,** and **glass reinforced plastics.** Sometimes plaster waste moulds are used for special architectural/ornamental concrete elements.

Proper storage of materials **preserve and protect them from impurities and wastage** of materials. Storage of cement for long periods reduces strength of concrete. By using **12 months old stored cement, the strength reduces to 60 % of concrete** prepared with fresh cement. Storing of cement should ensure avoidance of contact of cement bags with moist walls, floors or air. **Avoid long storage** of cement.

Aggregate must be stored on dry, rigid and clean platform to avoid segregation and impurities and preserve uniformity of grading and moisture content. Generally aggregate stacks should be made for each size and to a height of 1.25 to 1.75 m. Water storage should avoid mixing of impurities and other organic matter. **Admixtures** must be stored properly **with labels** and **water proof containers.**

Essence of batching is **correct measurement of the ingredients.** Weigh batching is preferred for accurate measurement of materials specially cement and aggregate. Volume batching of sand must **compensate for bulking** effect due to moisture. Depending on the importance and the extent of project, suitable weighing device can be used. For quality concrete production, weighing should be accurate and avoid loss of batched materials. **Saturated surface dry aggregate should be used as a standard condition.**

Proper mixing of ingredients affects hydration process and development of desired characteristics and quality of concrete. Manual mixing can be done on small projects while machine mixing can be adopted for large and important construction projects. Machine **mixing ensures homogeneity, speed** and **quality** of concrete.

Transportation of concrete should be **timely and avoid loss of plasticity**, segregation and bleeding. Depending on the quantum and conditions of concreting job, a suitable method of transporting concrete can be selected. Common methods of transporting concrete are **manual, wheel barrows, dumpers, transit mixers, chutes, cableways, crane operated skips**, and **pumps.**

Concrete should be deposited after **preparing the sub base** or form work by proper cleaning and moistening. All **reinforcements** and other items to be embedded should be **checked prior to depositing concrete. Avoid segregation and bleeding** and formation of cold joints **while depositing concrete.**

Compaction of concrete after depositing affects various characteristics of concrete (**strength, durability** and **impermeability**). Presence of even **5 % air voids can result into 30 to 35 % loss of strength.** Compaction to achieve optimum density is most critical for quality of concrete construction. Depending on the situation and the extent of work, suitable method of compaction may be adopted. **Vibrators** should be used at **suitable intervals in either direction till air bubbles cease to come out** of concrete.

Joints are made to accommodate certain movements and to avoid weak cold joints. Joint sealing material must remain flexible to create effective barrier to moisture flow.

Curing promotes the hydration of cement under **controlled conditions of moisture and temperature. Curing** facilitates in improving **strength, durability** and **permeability** of concrete elements. Curing preserves the structure from mechanical or physical disturbances. Curing can be done by **ponding, wetting** or **membrane forming.**

Removal of form work should be done after the concrete element has developed adequate strength to sustain the expected loads. All minor defects observed must be set right immediately by appropriate method and use of epoxy resins. **Removal of shuttering** must be carried out in **suitable sequence.**

All equipments and tools used in concreting operations should be **cleaned, washed and oiled at the end of every day** work. This ensures quality performance of these equipments and tools for the production of quality concrete construction.

PRACTICE QUESTIONS

5.1 List concreting operations which affect the characteristics of concrete.

5.2 State influence of concreting operations in fresh concrete stage on hardened concrete in not more than 100 words.

5.3 Explain the importance and main design considerations of form work erection in achieving quality concrete

5.4 Explain how storage of ingredients affect the characteristics of cement concrete in not more than 200 words.

5.5 Define batching and explain importance of batching in producing quality concrete in not more than 200 words

5.6 State precautions in batching for controlled concrete production.

5.7 Describe importance of mixing in producing quality concrete in not more than 100 words.

5.8 List 5 methods of mixing and explain special features in not more than 50 words each.

5.9 List methods of transporting concrete.

5.10 State **precautions** in transporting concrete in not more than 100 words.

5.11 Explain **importance** of compacting in achieving quality of concrete in not more than 100 words.

5.12 List various methods of compacting and their suitability in different situations.

5.13 Explain importance of joint making in concrete structure.

5.14 State objectives of curing in not more than 100 words.

5.15 Explain 3 main methods of concrete curing and special factors to be considered for optimum results.

5.16 Explain the precuations in formwork removal in not more than 100 words.

5.17 Explain the need for patchwork repair in reinforced concrete elements in not more than 100 words.

UNIT III

CHAPTER 6

Basic Principles of Concrete Mix Design

LEARNING OBJECTIVES

The learner understands the basic principles involved in cement concrete mix design and will be able to:

- State **objectives** of cement concrete mix design;
- Explain **hydration** of cement;
- State **Duff Abram's W/C ratio law**;
- State **conditions of validity** of W/C ratio law;
- **Explain curves of W/C** ratio v/s strength of cement concrete;
- Explain the phenomenon of **workability**;
- Explain importance of **minimum fines** in concrete mix;
- Explain importance of **water content** in fresh cement concrete mix to achieve the ultimate characteristics.

6.1 INTRODUCTION

Proportioning of a concrete mix comprises of determining the **relative quantities of materials** to be used in production of concrete for a given purpose. The process of selecting proportions of these materials is called "**Concrete Mix Design**" and should not be misunderstood with structural design. Proportioning may be based on certain data obtained by practical experience and investigations of laboratory test results of various ingredients or any empirical data. The process of mix design involves the consideration of **properties** and **costs** of ingredients, conditions of placing and finishing the fresh concrete, and **expected properties of hardened** concrete such as **strength, durability, impermeability** and **volumetric** stability.

The main **objectives** of the concrete mix design can thus be stated as production of quality concrete which shall be:

- Satisfying the requirements **of fresh concrete (workability)**;
- Satisfying the properties of **hardened concrete (strength** and **durability)**;
- Most **economical** for the desired specifications and given materials at a given site; and
- **Performing most optimally** in the given structure under given conditions of environment.

The concrete mix design is thus based on the principles of:

- **Workability** of fresh concrete;
- Desired **strength and durability** of hardened concrete which in turn is governed by **water-cement ratio** law; and
- Material and **placing conditions** at the site which help in deciding workability, strength and durability requirements.

6.2 WATER-CEMENT RATIO LAW

The basic properties of hardened concrete mainly depends on "**hydration of cement**" (i.e. reaction of cement and water). Cement essentially consists of C_3S, C_2S C_3A and C_4AF which are compounds and react with water (C-CaO, S-SiO_2, A-Al_2O, F-Fe_2O_3).

$C_3A + H_2O =$ Hydrated Tricalcium Aluminate $+ Ca(OH)_2$

$C_3S + H_2O =$ Hydrated Tricalcium Silicate $+ Ca(OH)_2$

$C_2S + H_2O =$ Hydrated Dicalcium Silicate $+ Ca(OH)_2$

C_3A is the most active and reacts first with water. The reaction is mainly responsible for initial setting of cement. C_3S is the next active compound which reacts with water and its rate is slower than C_3A, but faster than C_2S. C_4AF is mostly **inert** compound.

The reaction of C_3S is responsible for **development of strength** for the first **28 days**. C_2S reacts slowly and continues for three years and contributes to the later strength development. **28 days strength** of concrete is about **80 % of its ultimate strength** after 2-3 years. For acceptance criterion 28 days strength is considered as standard strength.

The product of hydration of cement comprises of about **20 to 30 % calcium hydroxide** in crystalline form and are surrounded by hydrated calcium and aluminates which are colloidal in nature. The entire composition of calcium hydroxide, hydrated calcium silicates and aluminates is called "**Cement gel**". This cement gel on hardening gives strength to cement paste.

Cement paste develops full strength when cement particles hydrates fully. The volume of gel increases on hydration but the total volume of the **gel after full hydration is the same** as the volume of water and cement before hydration. After the cement paste becomes solid there is **no change in the volume of the cement paste** during hydration process although the volume of cement gel increases internally. The quantity of water added to cement plays an important role on the hydration and quality of gel formed. The quantity of **water required for full hydration is much smaller** than that is generally added. For understanding the effect of water on the strength consider three cases of different water- cement ratios of **0.20, 0.35,** and **0.60.** In the first case of 0.20 W/C ratio, water is insufficient to fully hydrate the cement particle to develop the maximum strength. In second case of 0.35 W/C ratio, water is just sufficient to fully hydrate the cement particle, and the cement gel occupies all the space without air voids or water spaces and thus develops optimum strength. In the third case of 0.60 W/C ratio, it is found that the **gel contains some water** even after full hydration of cement particles and the excess **water occupies some space (capillary pores)** in the cement paste which makes the paste **porous and weak,** and thus **reduction in the strength.** The W/C ratio and strength have been studied in detail and a relationship has been established.

The relationship of W/C ratio and strength of concrete were studied and stated by many scientists in different forms. Before statement of these relations, strength of concrete was incorrectly supposed to increase with the increase in the quantity of cement alone.

In 1919, Duff Abram established a law which states that **"with given materials and conditions of test, the ratio of the quantity of mixing water to the quantity of cement alone determines the strength of concrete so long as the mix is of a workable plasticity".** This law is known as water-cement ratio law or Duff Abram's water-cement ratio law. This law holds good when the specimens of concrete under test have the same **size, temperature, age,** and other conditions of **test.** According to this law, the strength of concrete was given in terms of W/C ratio as under:

$$S = \frac{984}{7^x}$$

Where,

S = 28 days crushing strength of concrete in Kgf/cm^2, and

X = water-cement ratio by volume

Above formula is valid for ordinary portland cement and the constants 984 and 7 may vary depending on the quality of cement, aggregate, conditions of curing, method of testing, age of concrete and size of specimen. This formula was found to hold only when concrete was fully compacted i.e. the mix being fairly **workable** to allow air voids to be practically eliminated.

Various factors such as quality of **cement** and **aggregate,** conditions of **curing,** method of **testing, age of concrete** and size of specimen affect the **W/C ratio** law. These factors are briefly explained below:

a) **Quality of Cement**: Concrete specimens having the same water-cement ratio but prepared with different cement such as OPC, RHPC, LHPC will give different strengths at 28 days and therefore, these strengths will not be comparable in the law.

b) **Quality of Aggregate**: Apart from the source, the shape, the maximum nominal size, and grading of aggregate should be the same. In high strength concrete grades, the specimens are likely to fail through bond between aggregate and cement paste rather than failure of cement paste. As such, strength of concrete specimens with the same W/C ratio but the different sizes and shapes of aggregates may not be comparable.

c) **Conditions of Curing**: At higher temperatures, the rate of hydration is faster producing more gel in a given time and thus developing strength faster especially during first 10 days (if the temperature is higher than 23°C). However, the **faster rate of hydration** at high temperatures produce **gel of poorer and porous quality** in comparison to the gel produced at **low temperatures.** It is, therefore, evident that for comparing the strengths, the concrete specimens should be cured under the same temperature conditions.

The cement paste is a **plastic network of cement gel.** Pores are filled with water and this water causes the hydration of remaining cement particles. The hydration continues so long as the pores contain water and therefore, it is essential to maintain the internal moisture conditions for the continuity of hydration. If the concrete is allowed to dry, the water in the pores evaporates and its place is taken by air. Subsequent watering cannot replace the entrapped air in pores with water and the hydration is interrupted, thus affecting the strength. Thus for comparing strengths of concrete specimens, the **internal moisture conditions must be kept uniform.**

d) **Method of Testing**: The strength varies with the size and shape of the specimens. Strength measured on cube specimens is higher than on cylinders. Also **smaller the size of the specimen, greater is the strength measured for the same concrete.** For example, strength of the concrete on 100 mm cube specimens may be about 105 % of the strengths measured on 150 mm cube specimens of the same concrete. Similarly 250 mm cube specimens may give about 90 % strengths for the same concrete. As such, for comparison of strengths, standard cube specimens of 150 mm size should be tested. Further the cubes should have the same standard moisture saturation state at the time of testing for comparison.

e) **Age of Concrete**: Strength of concrete increases with age. Water-cement ratio law, therefore, applies only when the specimen of concrete to be tested are of the same age. Indian standard specifies 28 days strength of concrete for various grades. Indian Standard

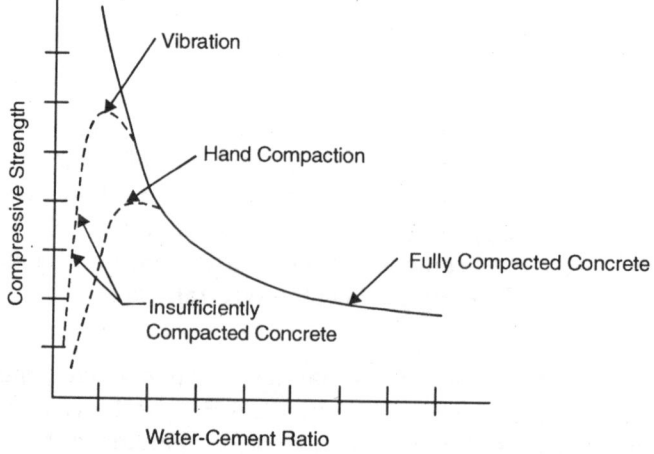

Fig. 6.1. Effect of Compacting on Compressive Strength of Concrete

(SP: 23-1982) specifies the compressive strength of concrete with respect to water-cement ratio for non-air-entrained and air-entrained concrete as given in Table 6.1. Further the strength developed at different ages is also specified in terms of 28 days in the Table 6.2 for ordinary portland cement concrete. For comparing strengths of concrete, the specimens should be tested after 28 days of standard curing.

Water-cement ratio law and understanding of its limitations, was a great break through in producing concrete of desired specifications and qualities. Water-cement ratio law for concrete strengths is specified by many investigators in different ways. One of the standard practice is to show the relationship of **strength of concrete and water-cement ratio by curves** for the convenience of concrete mix design.

Duff Abram had considered water-cement ratio by volume in his formula. Many research organisation in India and abroad have developed curves of relationship of strength of concrete versus water-cement ratio. Most of these curves are drawn considering water-cement ratios by weight (mass) i.e. the mass of water used for mixing divided by the mass of cement in the given concrete mix. Thus, in a given batch of concrete mix 25 kg. (25 litre) of water is added per 50 kg (one bag) of cement, the W/C ratio by mass shall be 25/50 = 0.50. Central Road Research Institute (CRRI) of India, New Delhi has prepared curves of 28 days concrete strength versus water-cement ratio by mass for six type of cements based on 7 days cement strength. Figure 6.2 shows CRRI based curves in which 28 days concrete strengths are shown as ordinate and W/C ratios by mass are shown as abscissa.

Table 6.1 Relationship between W/C Ratio And Compressive Strength*

28 Days Compressive Strength Kgf/cm² (N/mm²)	W/C Ratio By Weight	
	Non Air Entrained Concrete	Air Entrained Concrete
450 (45)	0.38	—
400 (40)	0.43	—
350 (35)	0.48	0.40
300 (30)	0.55	0.46
250 (25)	0.62	0.53
200 (20)	0.70	0.61
150 (15)	0.80	0.71

* (Ref: SP 23-1982)

Table 6.2 Age Factor for Strengths w.r.t. 28 Days Strength*

Age of Member (Months)	1	3	6	12
Age Factor Strength Ratio w.r.t. 28 Days	1.0	1.10	1.15	1.20

* (Ref: IS 456-1978)

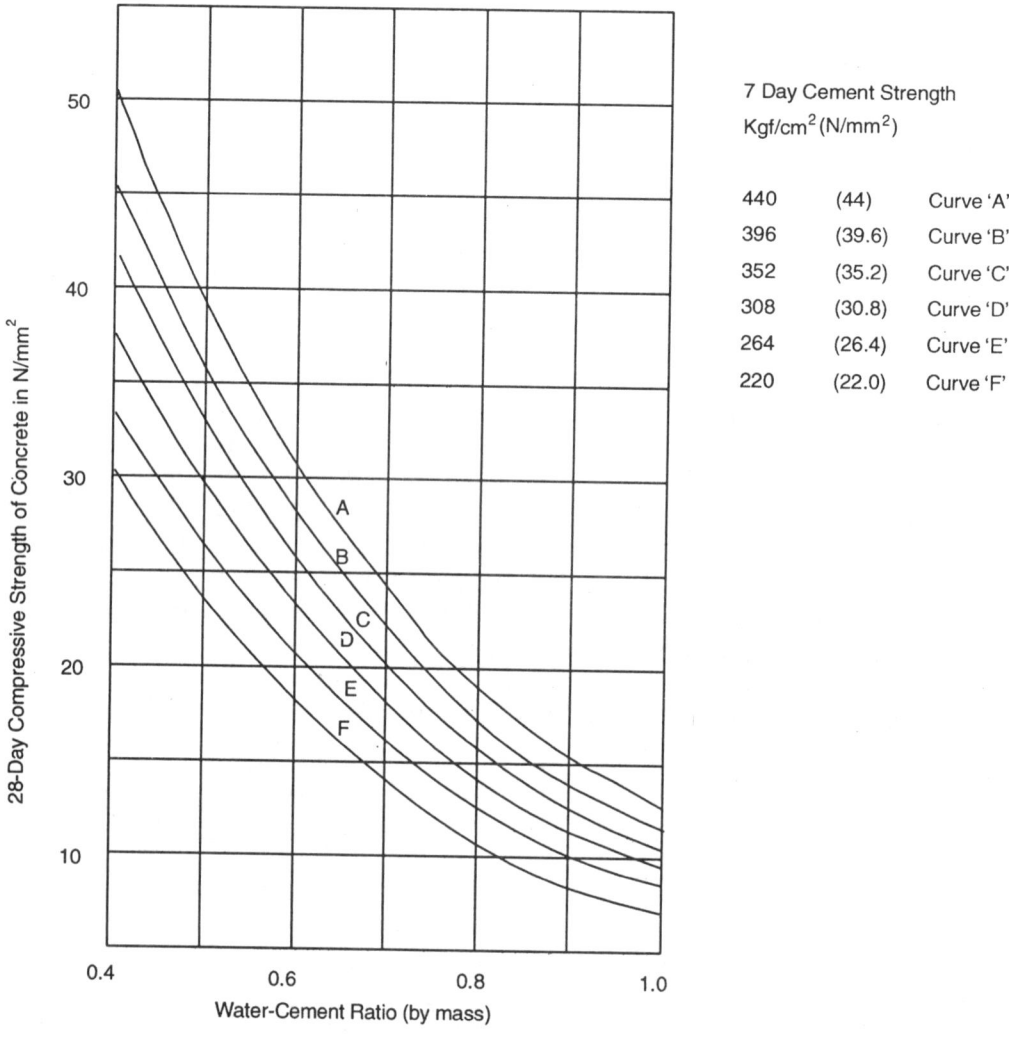

Fig. 6.2. Design Curve for **Cement Concrete Mixes** in Relation to 7-Day Compressive Strength of Cement (CRRI Based)

The type of cement manufactured in India may possess 7 days compressive strength in the range of 220, 264, 308, 352, 396 and 440 Kgf/cm². Depending on the type of cement (based on its 7 days compressive strength), the appropriate curve (A, B, C, D, E, and F) is chosen and W/C ratio obtained for the desired concrete strength. Indian standard IS:10262-1982 also provides a generalized relationship between W/C ratio and 28 days strength of concrete as shown in **Fig. 6.3.**

From the curve required W/C ratio can be obtained for the specified 28 days concrete strength. Water-cement ratio law applies equally well for the resistance of concrete to weathering, and water tightness. Lesser the water-cement ratio, greater is the resistance of concrete to weathering and greater is the impermeability and vice versa. In view of this, sometimes water-cement ratios are selected from the point of view of durability (resistance to weather/

chemical attack and impermeability) irrespective of the strength of concrete. For different conditions of exposure, standard SP:23-1982 specifies values of minimum W/C ratios required.

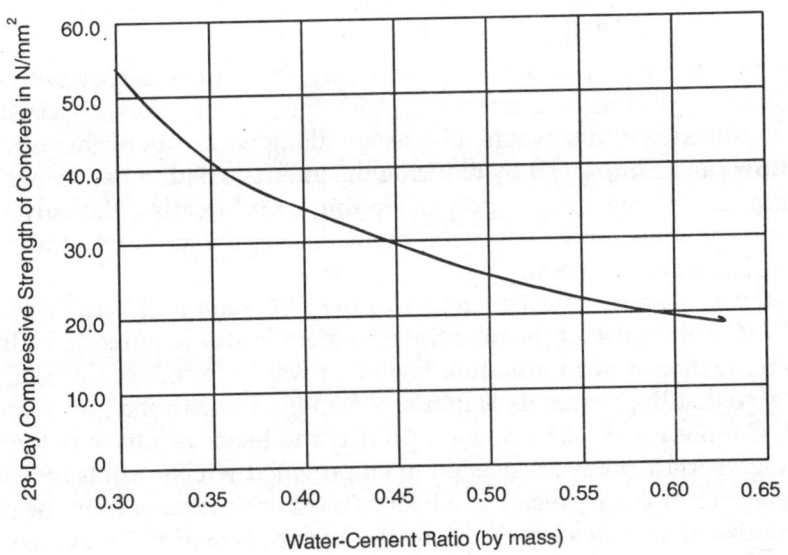

Fig. 6.3. Generalised relation between Free Water-Cement Ratio and Compressive Strength of Concrete

6.3 WORKABILITY

Workability of concrete plays vital role in achieving desired properties of concrete. Duff Abram's water-cement ratio law also gives importance to the workability of the mix. The principle of workability can be appreciated from the compaction point of view. The degree of compaction of concrete is directly related to the degree of workability for a given method of compaction. The **degree of compaction indicates the extent of elimination of air voids** and making concrete denser. From the graph of strength ratio versus density ratio of concrete, it is evident that even 5% reduction in density ratio leads to about 30 to 35 % reduction in strength. Thus, apart from employing appropriate method of compaction, the principle of workability is most important for the design of concrete mix proportions (Refer Fig. 5.3).

The process of compaction involves application of external force by tamping or vibration mainly to overcome internal friction between the concrete particles to move them closer. The portion of the work used in overcoming the internal friction of the concrete is called useful internal work. The workability can thus be defined as the property of concrete which determines the amount of useful internal work necessary to produce full compaction without causing segregation and bleeding. For reducing internal friction between the particles, appropriate quantity of water is required to lubricate the particle surface and to maintain homogeneity of the mix, a suitable grading shall be necessary.

6.3.1 Phenomenon of Workability

The phenomenon of workability is associated with the following concepts:

- Ease of flow
- Cohesiveness-movement without tendency of segregation
- Prevention of harshness
- Prevention of bleeding

A mix is said to be workable if it flows easily to become denser without tendency of segregation, bleeding and surface harshness. Workability is, therefore, important phenomenon and should be considered in concrete mix design along with water-cement ratio.

Ease of flow can be improved by reducing the internal **friction** between the particles. The internal friction can be reduced by proper **wetting or lubricating the surface of aggregate** particles. The lubrication can be improved by increasing the quantity of water for the given quantity of aggregate and cement.

The increase in quantity of water **increases the W/C ratio** and hence **decreases strength.** To keep the W/C ratio constant the quantity of **cement** is also required to be **increased** which results in **increase of cost of production.** The other way is to reduce the total surface area of the aggregate, so that the aggregate is lubricated optimally with the given quantity of water. For lower W/C ratios lesser surface area is needed and hence results in rich mixes. With high W/C ratios high surface areas are needed to be provided which results in lean mixes.

The surface area of the aggregate can be increased or decreased from the simple principle that **larger the size of particles lesser will be the surface area** of all the aggregate particles per **unit volume.** This can be understood by considering total surface area of a simple cuboid of 20 mm size and the total surface areas of 8 cuboids of 10 mm size obtained by breaking one cuboid of 20 mm size (Refer Fig. 3.5). From the surface area concept smaller sizes of coarse aggregate is required for concrete mixes with high W/C ratios but from economic considerations total surface area is increased by increasing the total quantity of aggregate of larger size C.A. Similarly for low W/C ratio concrete mixes from bond strength point of view, the coarse aggregate size is reduced to increase the surface area but the quantity of total aggregate is reduced to achieve the desired workability.

From the above concept, it is clear that for a given volume of all in aggregate, **coarser the grading, smaller is the specific surface area** and finer the grading, larger is the specific surface area.

Thus all in aggregate is obtained by combining coarse aggregate and fine aggregate in suitable proportions and considering appropriate shape and size of coarse aggregate. Fine aggregates i.e. sands are also grouped in 4 different grades as zones: I, II, III and IV by Indian Standards. Sand proportion in relation to a given coarse aggregate changes from maximum for zone I (coarsest) to minimum for zone IV (finest) for obtaining a suitable cohesive grading of all in aggregate (Refer Fig. 3.10 to Fig. 3.12).

The mix can be kept cohesive by providing sufficient **fines** in grading of aggregate, avoiding too much water and proper handling. Table 6.3 gives approximate **minimum quantities of fines per m³** of concrete for various sizes of coarse aggregate. If sufficient **fines** are not available from the normal grading of the aggregates and proportions of cement, then **extra cement** must be added or **additional quantity of fines** should be added.

When the concrete surface is finished with trowel, the surface may not become smooth due to lack of sufficient cement mortar to fill the voids in coarse aggregate-or due to the presence of excessive proportion of one particle size in aggregate grading. For avoiding harshness and bleeding, the aggregate grading should be so adjusted that the concrete contains

Table 6.3 Minimum Quantity of Fines

Nominal Size of Coarse Aggregate	10mm	20mm	40mm	63mm	80mm
Minimum Quantity of Fines in Kg per Cum of Compacted Concrete	525	400	350	325	280

Table 6.4 Approximate Water Contents (kg/m³) for Desired Workability for Different Sizes of Aggregate*

	Slump (mm) Vee Bee (S)	0-10 12	10-30 6-12	30-60 3-6	60-180 0-3
Aggregate Size (mm)	Type	(i)	(ii)	(iii)	(iv)
10	Uncrushed	150	180	205	225
	Crushed	180	205	230	250
20	Uncrushed	135	160	180	195
	Crushed	170	190	210	225
40	Uncrushed	115	140	160	175
	Crushed	155	175	190	205

* Ref: SP23-1982

Note: For different type of CA and FA, water content (W) is estimated by $W = 2/3\ W_{fa} + 1/3\ W_{ca}$, (where W_{fa} = water appropriate for F.A; W_{ca} = water appropriate for CA type).

Table 6.5 Approximate Mixing Water (kg/m³ of Concrete) for Different Slumps and Maximum Sizes of Aggregate

Max. Size of Aggregate (mm)	10	12.5	20	25	40	50	70	150
Slump (mm) ↓								
A. Non Air-Entrained Concrete								
30-50	205	200	185	180	160	155	145	125
80-100	225	215	200	195	175	170	160	140
150-180	240	230	210	205	185	180	170	—
Air in Concrete(%)	3.0	2.5	2.0	1.5	1.0	0.5	0.30	0.20
B. Air-Entrained Concrete								
30-50	180	175	165	145	140	140	135	120
80-100	200	190	180	175	160	155	150	135
150-180	215	205	190	185	171	165	160	—
Recommended Air Content %	8.0	7.0	6.0	5.0	4.5	4.0	3.5	3.0

sufficient fines and contains no particle size in excessive proportions. For determining the suitability of given mix proportions actual trowel test should be carried out to observe the tendency of harshness and bleeding. Table 6.4 and 6.5 specify the quantity of **water required** for certain **desired workability** for given **maximum size and type** of coarse aggregate.

Thus, for achieving the desired workability of **concrete mix without tendencies of segregation, harshness**, or **bleeding**, it is necessary to have **proper grading** of aggregate, certain **minimum quantity of fines**, and **enough quantity of water**. For the design of concrete mix proportion two principles involved are (i) **"Water-Cement ratio law"** and (ii) **"Workability"** apart from the **economy** considerations. These principles shall be used for the design of concrete mix proportions by selecting appropriate data and **values from the given tables and graphs** on the basis of the desired qualities according to a set of procedure given in different methods.

6.4 SUMMARY

Concrete **mix design refers to selecting proportions of various ingredients** to produce **concrete of desired characteristics** under given conditions of placement and environment. Design data is based on test results and past experience. The **objectives of mix design** is to produce appropriate **workability in fresh concrete, strength and durability of hardened concrete** with **optimum economy.** Mix design is based on **workability, W/C ratio,** and **placing conditions.**

Reaction of water with cement particles refers to **hydration of cement compounds** producing hydrates of **Tricalcium Aluminate, Tricalcium Silicate, Dicalcium Silicate** and **Calcium hydroxide. C_3A is mainly** responsible **for setting** while C_3S is for **early strength** and C_2S **is for later strength** development. 20 to 30 % of hydration products comprise of **crystalline calcium hydroxide surrounded** by C_3A, C_3S and C_2S and is **called gel.** Generally water required for hydration is less than 0.35 times mass of cement. **Lesser water reduces workability** and **higher water reduces strength** due to increased water capillary pores.

Duff Abram's law states that for given materials and conditions of test, the **ratio of** the quantity of **mixing water** to the quantity of **cement** alone determines the **strength** of concrete so long as the mix is of a **workable** plasticity. The law holds good under certain conditions of **type of cement, type of aggregate, curing, method of testing,** and **age of concrete.** Strength, v/s W/C ratio curves follow certain pattern within limits showing **increase in strength with reduction in W/C ratio.** Different curves can be drawn for different type of cement. Generally **28 days standard strength is about 80% of its ultimate strength.** Generally **7 days strength is about 60 to 70 % of its 28 days strength in cement concrete.**

The phenomenon of **workability** is associated with **ease of flow, cohesiveness, prevention of bleeding** and **harshness.** Workability is affected by **size, shape** and **grading** of aggregate. **Larger size particles have lesser surface area and results in better workability** when every thing else remains the same.

Total quantity of **fines play** important **role in workability and cohesiveness** of concrete mix. Certain **minimum fines** are used to produce concrete mixes of good workability and cohesiveness. Total quantity of mixing water depends on the size, grading and type of aggregate. Thus for producing concrete mix of desired characteristics and quality, basic principles of **workability and W/C ratio law** must be understood well and applied properly.

PRACTICE QUESTIONS

6.1   State main objectives of concrete mix design.

6.2   Explain hydration process of cement in relation to its basic compounds.

6.3   State Duff Abram's Water-Cement ratio law.

6.4   State conditions of validity of Duff Abram's W/ C ratio law.

6.5   Explain relationship of strength v/s water-cement ratio with the help of curves.

6.6   Explain CRRI curves of strength and W/C ratio.

6.7   Define workability in not more than 100 words.

6.8   Explain the phenomenon of workability in not more than 200 words.

6.9   Explain why the concrete needs certain minimum quantity of fines?

6.10   Explain how the concrete workability is affected by the size, shape and grading of aggregate?

6.11   Explain the effect of water content on workability and strength of concrete.

6.12   Explain how Economy, Quality, Workability and Strength can be achieved simultaneously.

CHAPTER 7

Management, Statistical Quality Control Procedures and Acceptance Criteria

LEARNING OBJECTIVES

The learner **understands** management and statistical quality control procedures and will be able to:

- Describe **inspection** procedures to check quality;
- List items covered in inspection to **check quality** of concrete;
- Explain **degree of quality control** for different site conditions;
- List factors causing **variability in quality** of concrete;
- Explain standard deviation and **coefficient of variation** for given data;
- Explain **target mean strength** with reference to site conditions and quality data;
- Explain **acceptance criterion** of concrete quality.

7.1 INSPECTION AND QUALITY CHECK

7.1.1 Need and Scope of Inspection

Inspection is necessary to ensure that the work is done in accordance with the plans, specifications, and good practices and, to prevent mistakes. Inspection is one of the several related actions which have a direct bearing on the results achieved in concrete construction. The following requirements are also considered necessary to the general problem of obtaining efficient and satisfactory construction.

i) Intelligent Design

Design involves not only the **proportioning** of the various parts of the structure but also the selection of an **appropriate grade** of concrete;

ii) Adequate specifications

This involves **appropriate** definitions of **quality** and **enforceable provisions**;

iii) Reliable construction

This involves the selection of an honest and dedicated contractor whose T&P (plant and equipment) are such that desired **control of quality** is possible;

iv) Competent Inspection

This involves the setting up of an organisation suitable to the type of construction involved, the selection of capable personnel and the delegation of sufficient authority.

The following items are commonly covered by the **inspection of concrete** construction at various stages:

- Sampling, identification, examination, and any **field testing** of materials;
- **Control** of concrete proportioning (within specification limits) and the **measurement** of materials;
- Examination of the foundation, forms and other work preparatory to concreting;
- Continuous inspection of the **batching, mixing, conveying, placing, compacting**, finishing and **curing** of concrete;
- Testing for **consistency of concrete** and preparation of suitable concrete specimens required for **laboratory testing**;
- General observation of contractor's **plant and equipment**, weather, working conditions and other items affecting the concrete; and
- Preparation of **records and reports**.

7.1.2 Choice of an Inspector

An ideal inspector is one who has both practical and technical know-how of the subject. The concrete inspector should be fully informed about all the phases of concrete technology. It is essential that the **inspector understands** the following:

- **Bulking influence** of moisture in fine aggregate (sand);
- Whether the mix is **well proportioned** or not;
- Concrete **consistency** is within permissible limits or not;
- Whether the particular concrete mix is producing the **desired slump** or not;

- Whether the **forms** are sufficiently **tight and substantially braced;**
- Whether the concrete being laid is **properly compacted** (by vibration) or not; and
- Early **removal of the forms** is avoided.

7.1.3 Authority of the Inspector

The quality inspector must be provided with the authority to:

- **Prohibit concreting** until preliminary conditions (such as completion of forms) have been fulfilled and the work inspected and until inspection report for the concreting has been provided;
- **Stop the use of materials, equipment,** which does not comply with the specification;
- **Stop any work** which is not being done in conformity with the plans and specifications; and
- Order the removal or repair of faulty construction or construction performed without inspection and not accessible to being inspected later.

Ordinarily the inspector is authorised to take direct action in the first three cases above and immediately reporting thereafter to his superior. However, he should stop work only as a last resort, when it is clear that continuing operations will result in unsatisfactory concrete.

7.1.4 Quality Check, Personnel and Equipment

The specification requirements regarding excavation, forms, reinforcement, embedded fixtures, and joints must be fulfilled and the work inspected before concrete is placed in a given section of the work. Forms should be of proper size, strength and located in their correct location.

Before concreting begins, the cement, aggregate, water and any other ingredient of the concrete mix should be inspected to see that they are satisfactory for making concrete of good quality. If the mix proportions are not given in the specifications, appropriate proportions should be determined by trial which will produce concrete of the specified quality. The batching equipment should be checked and adjusted, if necessary, to ensure proper quantities of materials in each batch. Concrete should not be mixed until the specification requirements have been met with regard to conditions and sufficiency of equipment for **mixing, transporting, placing, compacting, finishing** and **curing** of the entire unit of placement.

The inspector should pay particular attention to the **batching of materials,** the time of mixing, the consistency of the concrete and conditions or methods that might cause segregation in the concrete. He should also see that after concreting, proper curing of the concrete is being done or not as it influences the properties of the hardened concrete.

To check the strength of the concrete it is necessary to prepare test specimens for testing in the laboratory. Not less than three specimens should be taken for approximately each 200 cum of concrete, and in general, each sample should be taken from different points of the structure. It is advisable not to do the sampling from the conveying device as it may be segregated and may not represent the true sample of the entire lot. As far as possible the samples should be taken at irregular times and without prolonged preparations which may provide opportunity for the production of special batches for false sampling purposes.

The **sample** of concrete should be placed in a **watertight non-absorbent container**, re-mixed fast enough to make uniform mix and then moulded into **test specimens**. The test moulds should also be watertight and non-absorbent to avoid loss of water from the concrete. The entire operation of sampling remixing and moulding should be completed as promptly as possible, otherwise appreciable evaporation of water and stiffening may occur.

For compression test 150 mm cube moulds are commonly used. The moulds are filled in **three layers**, each layer being consolidated with **25 strokes of a 16 mm diameter and 600 mm** long round bullet pointed steel rod. After the top surface has been levelled, the specimen is covered with wet gunny bags to prevent evaporation.

Flexure specimens are usually 150 mm × 150 mm in section. The moulds are placed with their long axis horizontal and are filled in two layer, each layer being rodded 50 times. After rodding each layer, the concrete is spread along the sides and ends with a trowel. The top surface is finished with a wooden float and then covered with wet burlap, which is kept wet until the forms are removed. Test results so obtained should fall well within the specified limits, and if so needed, minor corrections may be applied to obtain the desired results.

Personnel and Equipment

To ensure the proper quality control, it is necessary to have a **well-trained team**, conversant with different quality procedure of **inspection, testing** and **data analysis**, and an **adequately equipped** field laboratory for carrying out routine control tests. For a construction project having a concreting production of 25-30 m^3 per day, the daily testing programme of the team and the composition of the team are given in Table 7.1.

Table 7.1 Daily Testing Programme and Team Composition

	Items of Tests	Daily frequency of Sampling/Testing	Minimum Team Required
1.	Seive Analysis for grading:		
	a. Coarse Aggregate	1	Supervisor – 1
	b. Fine Aggregate	1	Skilled Workmen – 3
			Unskilled Workers - 2
			Total personnel - 6
2.	Moisture content		
	a. Coarse Aggregate	1	
	b. Fine Aggregate	2	
3.	Workability of fresh concrete	8	
4.	Casting of cubes/beams	6	
5.	Strength test cubes or beams at		
	a. 7 days		
	c. 28 days	33 (Total)	
6.	Control at batching and mixing plants, Formwork placement, and compaction of concrete.	Consistently at regular interval	

Table 7.2 Degree of Quality Control Expected Under Different Site Conditions

Degree of Control	Conditions of Production at Site.
Very Good	Fresh cement from single source and regular tests, weigh batching of all materials, aggregates supplied in single sizes, control of aggregate grading and moisture content, control of water added, regular workability and strength tests, field laboratory facilities and frequent supervision.
Good	Carefully stored cement and periodic tests, weigh batching of all materials, controlled water, graded aggregate supplied, occasional grading and moisture tests, periodic check of workability and strength, intermittent supervision and experienced workers.
Fair	Proper storage of cement, volume batching of all aggregate allowing for bulking of sand, weigh-batching of cement, water content controlled by inspection of mix, occasional supervision and tests.

* Ref. IS 10262-1982.

7.2 MEASURES OF VARIABILITIES IN CONCRETE MIX DESIGN

7.2.1 Factors Contributing to Variability

It is found that the strength of concrete varies from batch to batch over a period of time. The strength of concrete may vary due to the following factors as already described in the previous units.

- Variation in the quality of constituent materials used;
- Variation in the mix proportions due to batching processes;
- Variations in the quality of batching and mixing equipments available;
- The quality of supervision and workmanship; and
- Variation due to sampling and testing of concrete specimens.

The above variations are inevitable during production to varying degrees. For example, cement from different batches may exhibit different strengths and the variability is more where cement from different sources is used. The grading and shape of aggregates may vary even from the same source and it is not economically feasible to eliminate such variations particularly when the aggregates are not factory made. Considerable variations occur in the mix proportions from batch to batch irrespective of whether the batching is by weight or volume. These can be attributed partly to the quality of plant available, moisture condition and partly due to efficiency of operation.

The propose of controlling the quality of concrete using the statistical means is to produce concrete of uniform quality, which can be achieved by good workmanship and maintaining the plant at peak efficiency. The compressive strength test results of cubes from random sampling of a mix exhibit variations, which are inherent in different operations involved in making and testing of concrete. If a large number of cube strength test results are plotted on a histogram, the results are found to follow a bell-shaped curve termed as "**Normal Distribution Curve**". The arithmetic mean or the average value of a number of test results gives no indication of the extent of variation of strength. However, this can be ascertained by relating the individual strength to the mean strength and determining the variation from the mean with the help of the characteristics of the normal distribution curve.

7.2.2 Standard Deviation

The root mean square deviation of the whole consignment is termed as the "**Standard Deviation**" and is defined numerically as:

$$S = \sqrt{\frac{\Sigma(x - \bar{X})^2}{n}}$$

Where, S = The standard deviation of the test results
 x = Any value in the test results
 $\bar{X}$ = The arithmetic mean of the test results
 n = The number of test results

The standard deviation increases with increasing variability and is expressed in the same units as the quantity. The characteristics of the normal distribution curve are fixed by the average value and the standard deviation. The spread of the curve along the horizontal scale is governed by the standard deviation, while the position of the curve along the vertical scale is fixed by the average value. It is possible to fix the limits below or above which the percentage of results can be expected to fall. The limits are set out as (X ± KS), where "K" is the probability factor having values shown in Fig. 7.1 and Table 7.3. For different values of "K" the

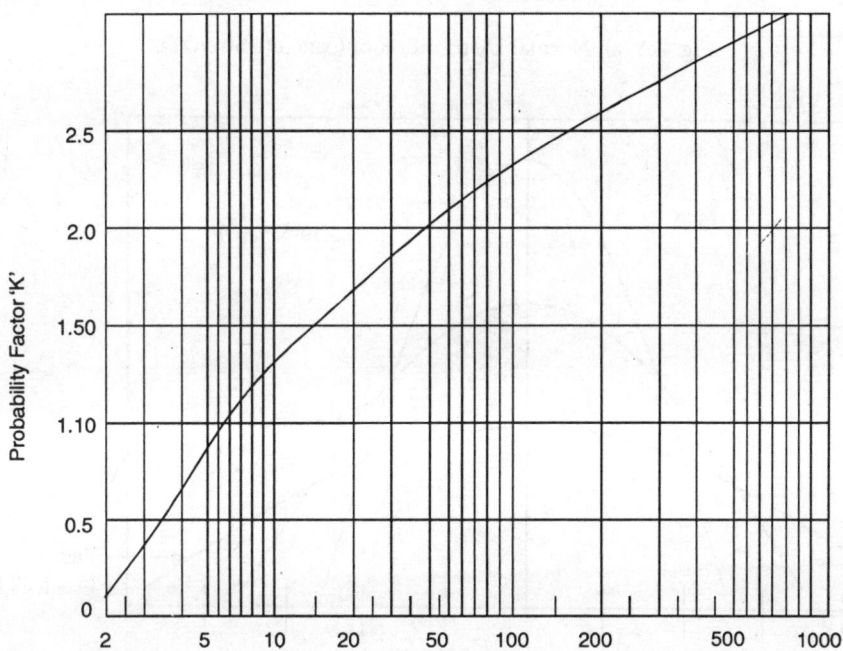

Fig. 7.1. Relation between the Factor 'K' and the proportion of Results
expected to be below the Minimum Strength

percentage of results falling above and below a particular value is illustrated in Fig. 7.2 (a, b) in relation to the area bounded by the normal probability curve.

Table 7.3 Values of Statistical Constant "K"

Percentage of Results Below the Characteristic Strengths	50	16	10	5	2.5	1.0	0.5	00
Constant "K"	0	1.0	1.28	1.65	1.96	2.33	2.58	00

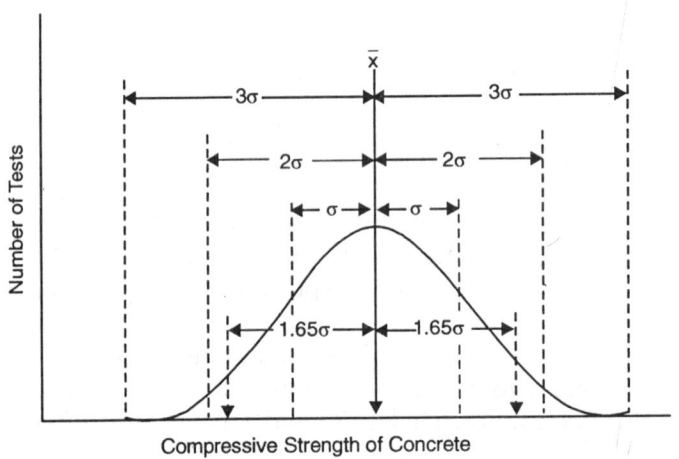

Fig. 7.2 (a). Normal Distribution of Concrete Strengths

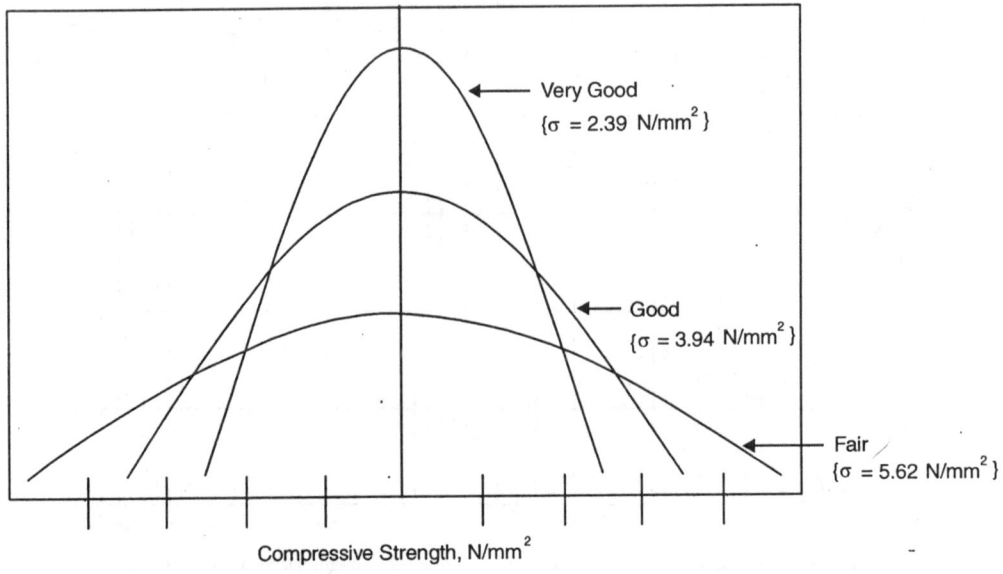

Fig. 7.2 (b). Typical Normal Frequency Curves for Different Control

Table 7.4 Assumed Standard Deviations

Grade of Concrete	M10	M15	M20	M25	M30	M35	M40
Standard Deviation Assumed (N/mm^2)	2.3	3.5	4.6	5.3	6.0	6.30	6.60

7.2.3 Coefficient of Variation

An alternative method of expressing the variation of results about the mean is by coefficient of variation, which is a non- dimensional measure of variation obtained by dividing the standard deviation, by the average value expressed as:

$$V = \frac{100\,S}{\overline{X}}$$

Where,

V = Coefficient of variation, S = Standard deviation in strength

$\overline{X}$ = Mean value of strength

When the coefficient of variation is constant, the standard deviation increases with the increase in strength of concrete. For higher grade of concrete the standard deviation is also higher.

Table 7.5 Suggested Values of Standard Deviation*

Grade of Concrete	Standard Deviation For Different Degree of Control in N/mm^2		
	Very Good	Good	Fair
M10	2.0	2.3	3.3
M15	2.5	3.5	4.5
M20	3.6	4.6	5.6
M25	4.3	5.3	6.3
M30	5.0	6.0	7.0
M35	5.3	6.3	7.3
M40	5.6	6.6	7.6
M45	6.0	7.0	8.0
M50	6.4	7.4	8.4
M55	6.7	7.7	8.7
M60	6.8	7.8	8.8

* Ref. IS 10262-1982

Note – Table 7.2 provides guidance regarding the different degrees of quality control to be expected, depending upon the infrastructure and practices adopted at the construction site.

7.3 STATISTICAL CONCEPTS OF MIX DESIGN

The advantages of applying the principles of statistics to the design of concrete mixes and to control the quality of concrete at site was first realised during the late 1940's by the early investigators. The design of the concrete mix shall be so done as to ensure that the **target mean strength** is achieved in certain percentage of cases during **preliminary trial mix testing** stage. Thus with the specified variation and accordingly taking the value of the constant K (from Table 7.3), the only parameter to be specified for obtaining the value of target strength (f_t) from the specified strength (f_{ck}) from the formula given below is the value of standard deviation (S).

$$f_t = f_{ck} + KS = f_{ck} + 1.65\ S$$

When in 95 % cases the desired target strength shall be realised.

The value of standard deviation for such a use shall be either an assumed value (Refer Table 7.4) or depending on the **degree of control** available at site (Refer Table 7.5) or a value based on past data using the similar plant, materials, and standard of supervision. The latter value, if available, it is to be preferred. Where the past data is not available during initial stages of preliminary trials the values given in Table 7.4 may be used. As soon as adequate number of test results (minimum 30) are available the actual calculated value of standard deviation shall be used for revising the mix design.

7.4 CHARACTERISTIC, TARGET MEAN STRENGTH AND ACCEPTANCE CRITERION

7.4.1 Characteristic Strength

As the cube test results follow the normal distribution there is always the probability that some results may fall below the specified strength. Recognising this factor, IS: 456-1978 (Latest IS: 456-2000) has brought in the concept of **characteristic strength.** The term characteristic strength means that value of the strength of the material **below which not more than 5 % of the test results are expected to fall.**

7.4.2 Target Mean Strength

Considering the inherent variability of concrete strength during production, it is necessary to design one mix to have a **target mean strength** which is **greater than the characteristic strength** by a suitable margin.

$$f_t = f_{ck} + K \cdot S$$

Where,

f_t = target mean strength
f_{ck} = characteristic strength
K = a statistical constant, depending on the definition of characteristic strength and is derived from the mathematics of Normal distribution and
S = standard deviation

The value of K is equal to 1.65 (see IS: 456-1978 or Latest IS: 456-2000) where not **more than 5 % of the test results are expected to fall below** the characteristic strength.

$$f_t = f_{ck} + 1.65\ S$$

7.4.3 Acceptance Criteria

1.   The concrete shall be **deemed to comply** with the strength requirements if:

a) every sample has a test **strength not less than the characteristic value;** or
b) the strength of one or more samples though less than the characteristic value, is in each case **not less than the greater of:**
   i) the **characteristic strength minus 1.35 times** the standard deviation, and;
   ii) 0.80 times the characteristic strength, and the average strength of all the samples is **not less than the characteristic strength plus** $[1.65- (1.65/\sqrt{n})]$ times the standard deviation.

2. The concrete shall be deemed **not to comply** with the strength requirements if: -
   a) The strength of any sample is **less than the greater of:**
      i) the **characteristic strength minus 1.35 times** the standard deviation, and
      ii) **0.80 times the characteristic strength,** or
   b) The average strength of all the samples is **less than** the characteristic strength plus $(1.65 - (3/\sqrt{n})$ times the standard deviation.

3. Concrete which does not meet the strength requirements as specified in "1" but has a strength greater than the required by "2" may, at the **discretion of the designer, be accepted** as being structurally adequate without further testing.
4. If the concrete is deemed not to comply pursuant to "2" the structural adequacy of the parts affected shall be **investigated and any consequential action as needed shall be taken.**
5. Concrete of **each grade** shall be assessed **separately.**
6. Concrete shall be assessed daily for compliance.
7. Concrete is liable to be rejected if it is **porous or honeycombed;** its placing has been interrupted without providing a proper construction joint, the reinforcement has been displaced beyond the tolerances specified; or construction tolerances have not been met. However, the hardened concrete **may be accepted after carrying out suitable remedial measures** to the satisfaction of the engineer-in-charge.

7.4.4 Example

In a construction work, concrete of grade M20 is to be used. The standard deviation of this grade of concrete has been established to be 4 N/mm^2. In testing of concrete cubes, the following results are obtained from a weeks production (average strength of 3 specimens tested at 28 days in each case) expressed in N/mm^2.

24.8, 27.0, 28.5, 23.6, 18.0, 21.6, 15.0 N/mm^2 .

Determine if the concrete is acceptable?

Solution

i) The first four results are straight way accepted since the sample strength is greater than the characteristic strength (20 N/mm^2) in each case.
ii) The 5th result of 18 N/mm^2 is less than the characteristic strength (20 N/mm^2) and is compared with

   (a) 0.8 × characteristic strength = 0.80 × 20 = 16.0 N/mm^2
   (b) (20.0 − 1.35 × 4) = 14.6 N/mm^2

Since $18\,\text{N/mm}^2$ is greater than **16 N/mm²**, recheck the average strength of first 5 samples, which is

$(24.8 + 27.0 + 28.5 + 23.6 + 18.0)/5 = \textbf{24.4 N/mm}^2$

For comparing average with:

$$20.0 + \left(1.65 - \frac{1.65}{\sqrt{5}}\right) \times 4.0 = \textbf{23.6 N/mm}^2$$

Since the average of 5 samples ($24.4\,\text{N/mm}^2$) is greater than **23.6 N/mm²**, the 5ᵗʰ sample is **acceptable**

(a) The 6ᵗʰ result is also acceptable.
(b) The 7ᵗʰ sample ($15\,\text{N/mm}^2$) is lower than $16.0\,\text{N/mm}^2$ (Refer to 'b' above).

The average strength of all the seven samples is

$(24.8 + 27.0 + 28.5 + 23.6 + 18.0 + 21.6 + 15.0)/7 = \textbf{22.6 N/mm}^2$

For comparing average with:

$$f_{ck} + \left(1.65 - \frac{3}{\sqrt{7}}\right) \times 4.0 = \textbf{22.1 N/mm}^2$$

Although the 7ᵗʰ sample ($15\,\text{N/mm}^2$) is **less than 0.8 times the characteristic strength** (i.e. $16\,\text{N/mm}^2$), but the **average strength of all the seven samples is greater than 22.1 N/mm²**, the sample may be accepted by the designer at his discretion.

7.5 SUMMARY

Inspection and testing are essential parts of quality control in concrete construction. Inspections ensure specifications and good practices in respect of desired mix design, enforceable specifications, reliability of contractors and competent inspectors. Inspections include sampling and field testing, measurement of mix proportions, evaluation of forms and bases, observations of batching, mixing, transporting, compacting (vibrations), curing, construction equipments, laboratory testing and preparation of records. Quality inspectors must be provided with an appropriate authority and team.

Degree of quality control for any construction project is decided on the basis of inspections, testing programmes, past experience and construction practices at project site. Quality of constituent materials, batching and mixing practices, performance of construction equipment, workmanship and sampling. Testing practices directly influence the variability in quality of concrete construction. Standard deviation increases with the increase in variability.

Target mean strength for the concrete mix is derived from the desired characteristic strength (f_{ck}), standard deviation for the desired grade of concrete and statistical constants.

$$f_t = f_{ck} + K \cdot S$$

Where,

f_t = target mean strength
f_{ck} = characteristic strength
S = standard deviation
K = statistical constant, depending on degree of quality control ($K = 1.65$, when not more than 5% test results fall below the characteristic strength)

Concrete is accepted if it satisfies various conditions laid down in Indian Standard.
(a) Every sample has strength not less than the characteristic strength
   - Samples less than 'f_{ck}' but not less than greater of

   $$(f_{ck} - 1.35\ S)\ \text{or}\ 0.80\ f_{ck}$$

   - Average of all the samples should not be less than

   $$f_{ck} + \left(1.65 - \frac{1.65}{\sqrt{n}}\right)S, \text{ where 'S' is SD and n is number of samples}$$

(b) Concrete shall be deemed not to comply if the strength of any
   - sample is less than greater of $(f_{ck} - 1.35\ S)$ or $0.80\ f_{ck}$, and

   - average strength is less than $f_{ck} + \left(1.65 - \dfrac{3}{\sqrt{n}}\right)S$

(c) Structural engineer can use his discretion if there is contradiction in criterion (a) and (b).

PRACTICE QUESTIONS

7.1   List main **requirements** for achieving good quality concrete construction.

7.2   List various stages of **concrete quality checks**.

7.3   Describe important **authority** of quality inspector.

7.4   Explain composition of **quality control team** for concrete construction.

7.5   Evaluate quality control levels from given site conditions and situations.

7.6   Define: **standard deviation, coefficient of variations** and **characteristic strength.**

7.7   List **factors** leading to **variability** of concrete quality.

7.8   Calculate **target mean strength** of concrete for a project from given site conditions and characteristic strength.

7.9   Evaluate if the given concrete is acceptable from given site conditions and test results.

7.10 Explain in not more than 200 words, how site conditions and testing helps in quality control of cement concrete construction.

7.11 Explain acceptance criteria of cement concrete quality.

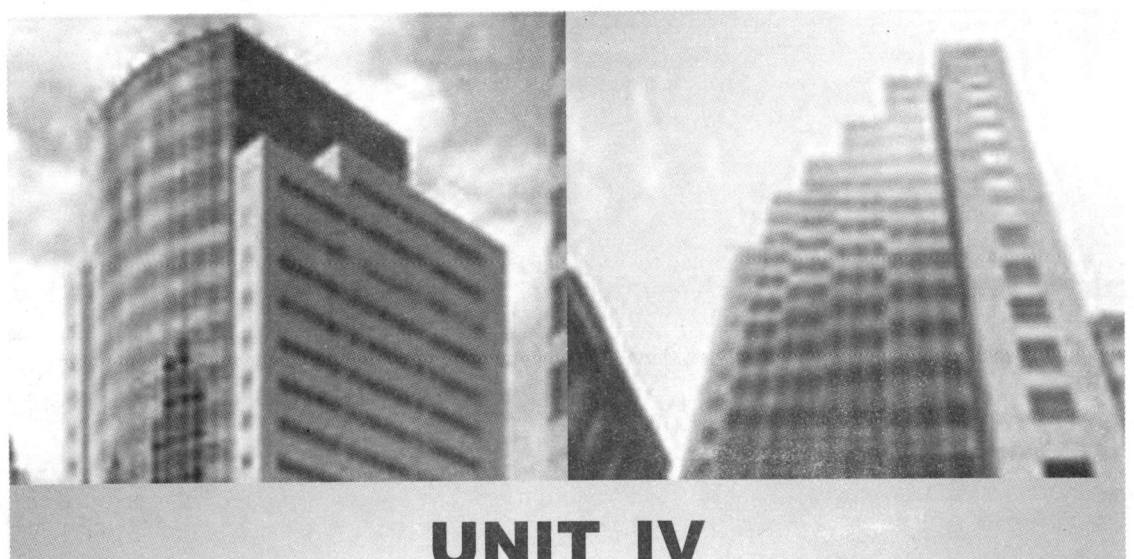

UNIT IV

CHAPTER 8

Cement Concrete Mix Design Procedures

LEARNING OBJECTIVES

The learner understands the cement concrete mix design procedures and will be able to:

- State **objectives** of cement concrete mix design;
- State basic **variables and characteristic data** considered in mix design;
- Explain the procedure to **assess moisture condition** of aggregate;
- Explain the effect of **bulking of sand** and its compensation;
- Explain steps of determining **proportions for nominal mix** concrete;
- List scientific **methods** of cement **concrete mix design**;
- Explain relationship of **W/C ratio with strength** of cement concrete;
- Explain **importance of accelerated strength** test for design of cement concrete;
- Explain **degree of workability** for various placing conditions;
- Explain **cement** requirements for different **exposure conditions**;
- Explain steps for the cement concrete mix design by **Indian standard guidelines**;
- Explain salient features of **ACI method** for cement concrete mix design;
- Explain briefly the **British method** of cement concrete mix design;
- Explain with example, cement concrete **mix design by any method**;
- **Compare specific aspects** of various methods of cement concrete mix design.

8.1 INTRODUCTION

Concrete is the most extensively used construction material and trends indicate that the supremacy of cement concrete as a construction material will continue for many decades to come. This is because of continuous improvement in the performance of concrete as a structural material due to scientific research and understanding of its properties. Now a days with the knowledge available it is quite convenient to produce cement concrete with compressive strength of 40 N/mm² and even it is quite feasible to produce special concrete of strength as high as 100 N/mm².

Concrete is a variable material, and it is realised by engineers and contractors that the design mix results into a better job at a lower cost. By suitable quality control, variations inherent with the cement concrete production are reduced and the concrete provides the desired properties of **workability, strength** and **durability** with **economy** in construction. The design mix satisfies stringent specification and requirements of concrete. The design of cement concrete mix in its simplest form is to **determine proportions of concrete ingredients** so that the concrete develops the **desired properties** both during fresh state and hardened state together with maximum **possible economy**. The main **objectives** of the cement concrete mix design can thus be stated as:

- To achieve the desired **workability** without segregation and bleeding for **full compaction** in its fresh state;
- To achieve the desired **strength** in its hardened state;
- To achieve the desired **durability** in the given placement and **environmental conditions;** and
- To achieve **optimum economy** in construction by using minimum possible quantity of cement in producing concrete satisfying above desired qualities.

In different methods of design of cement concrete mix proportions, there are following basic variables which must be considered:

- **Characteristics** of concrete such as **workability during fresh state** and **strength during hardened state;**
- Expected conditions of placing and exposure conditions of **weathering/chemical attack** and desired durability;
- **Characteristics of ingredients** such as quality and type of **cement,** size and grading of **aggregate** (coarse and fine);
- **Derived parameters** such as **water-cement ratio, water content** or **cement content proportions of fine** and coarse aggregates; and
- Site conditions regarding **degree of quality control, moisture,** and moisture state of aggregate.

From various principles studied in previous chapter, it is clear that **water-cement** ratio can be obtained for the **desired strength or durability** requirements. Suitable workability can be specified in terms of slump, compacting factor, or Vee-Bee degrees from the conditions of placement of structural concrete and the method of compaction. From the size, type, and grading of aggregates, and the desired workability, the quantity of **water per cubic** metre of concrete can be determined from the standard tables. Thus the important parameters needed to be specified or determined for the cement concrete mix design are:

- **Type of cement** and its compressive strength;
- **Strength** of concrete desired and degree of quality control at site; and
- **Degree of workability** required for placing conditions in terms of slump, compacting factor, Vee-Bee degrees;
- **Conditions of exposure** of concrete and concrete structure at site.
- **Size, shape, grading** and **physical data** (including specific gravity, bulk density, bulking of sand, moisture state of aggregate at site);

From these specified or observed basic data parameters, the design variables are computed using basic principles outlined earlier. These variables to be computed in the cement concrete mix design are:

- **Target strength** of the concrete mix;
- **Water-cement ratio** of the concrete mix;
- **Water** or cement **content** per unit quantity of the concrete mix for the **desired workability;**
- Proportions of **coarse and fine aggregates** and its grading;
- **Total aggregate content** per unit quantity of the concrete mix; and
- Quantity of **special admixture,** if any.

Thus the process of concrete mix design in any method shall involve specifying or observing of basic design data/parameters and computation of mix design variables using fundamental principles. Indian Standard Method of concrete mix design shall be dealt in detail while other methods shall be described briefly.

8.2 PROCEDURE FOR PROPORTIONING OF NOMINAL (ORDINARY) CONCRETE MIX

Concrete obtained by using nominal mix proportions, **without** stringent controls on the ingredients, processes, and laboratory trials of mixes, is called ordinary or **Nominal** concrete mix. For production of concrete for small and less important structures nominal mixes may be used. Good quality of even nominal mixes can be achieved if the proportioning is done on the basis of suitable guide lines laid in **IS: 456-2000** as given in Table 8.1. Nominal mix concretes are prepared only for concrete grades M5, M7, M10, M15 and M20. For higher grades, the Design Mix Concrete should be used.

In case of nominal mix concrete if it does not yield the specified strength, such concrete shall be classified as belonging to the appropriate lower grade. Nominal mix concrete proportioned for a given grade in accordance with **Table 8.1** shall not, however, be placed in higher grade on the ground that the test strengths are higher than the minimum specified.

The cement content of the nominal mix specified in **Table 8.1** for any grade shall proportionately be increased to overcome the difficulty of placement and compaction, so that the water-cement ratio as specified is not exceeded.

Water content should be adjusted depending on the condition of aggregate at site. If the aggregates contains any extra moisture on the surface of coarse aggregate or fine aggregate, it should be estimated and deducted from the mixing water (See Table 8.2).

Table 8.1 Proportions For Nominal Mix Concrete*

Grade of Concrete	Maximum Quantity of Water per 50 kg. of cement(litres)	Maximum Total Quantity of dry Aggregate by mass per 50 kg. of cement (kg.)	Proportions of FA to CA by mass
M5	60	800	Generally 1:2 subject
M7.5	45	625	to an upper limit of 1:1.5 and a lower
M10	34	480	limit of 1:2.5
M15	32	330	
M20	30	250	

* **Note-1:** The proportions of the fine to coarse aggregate should be adjusted from upper limit to lower limit progressively as the grading of the F.A. becomes finer (i.e. sand grading zone I to zone IV) and also as the maximum size of C.A. becomes larger (i.e. from 10 mm to 40 mm).

Thus for average sand zone II the **F.A.** to **C.A.** ratios for 10 mm, 20 mm, and 40 mm size C.A. shall be 1:1.5, 1:2, 1:2.5 respectively.

For 20 mm C.A. the ratio of F.A. to C.A. shall be 1:1.5, 1:2, 1:2.5 with sand of grading zone I, II and III respectively. These proportions are only tentative and can be modified by visual inspection.

Note 2: The quantity of water used in concrete mix for R.C.C. should be sufficient, but not more than sufficient to produce a dense concrete of adequate workability for the purpose. Workability of concrete should be controlled by maintaining water content that is found to give concrete which is just sufficiently wet to be placed and compacted without difficulty with the means available.

Table 8.2 Surface Water carried by Average Aggregate**

	Condition of Aggregate	Approximate Quantity of Surface Water	
		Litre/m^3	Percent by mass (litres/100kg)
1.	Very wet sand	120	7.50
2.	Moderately wet sand	80	5.0
3.	Moist sand	40	2.5
4.	Moist gravel or crushed rock	20-40	1.25 to 2.5

* Coarser the aggregate, less the water it will carry.
** Ref. IS: 456-1978 (456-2000).

Table 8.3 Water Absorption by the Aggregate

	Type of Aggregate	Approximate Quantity of moisture absorbed by mass of aggregate
1.	Average sand	1.0 %
2.	Pebbles and crushed lime stone	1.0 %
3.	Trap and Granite Rock	0.50 %
4.	Very light and Porous Aggregate	Upto 25.0 %

If the aggregate is too much dry at site, it shall **absorb** certain water, which should be **added extra** over and above estimated water depending on the absorption of water by the aggregate (See Table 8.3).

For approximately judging the surface moisture of sand in the field, take a handful of moist sand and squeeze in palm. When the lump is squeezed and then let free, the particles of moist sand fall apart. The moist sand appears slightly damp to the touch, but leaves little moisture on the palm. Moderately wet sand when squeezed in palm, it forms into a ball and leaves some quantity of moisture on hand. Very wet sand when squeezed in palm, leaves more moisture on hand and the sand does not form the lump but the particles fall apart and glisten.

After making necessary adjustments in quantity of water, the workability of concrete should be checked to ensure fairly correct adjustments of moisture. After obtaining nominal proportions, if slump of the concrete mix is less than the desired then reduce the proportion of total aggregate to cement. If in addition to less slump the mix has the tendency of harshness and non-cohesiveness then in addition to **reduction of aggregate-cement ratio**, increase the proportion of **fine to coarse** aggregate by judgement. If in addition to less slump, the concrete mix shows over sanding, then in addition to reduction of Aggregate-Cement ratio, also **reduce the proportion of fine aggregate** to coarse aggregate.

If the mix shows more slump than desired, **then reduce water-cement ratio** by judgement. If in addition to more slump, there is bleeding, then reduce water-cement ratio and increase the **proportion of sand to coarse aggregate** to make the mix more cohesive. Various quantities given in Table 8.1 are for general guidance and adjustments can be made within the limits shown on the basis of actual properties of concrete.

8.3 PROCEDURE FOR PROPORTIONING OF DESIGN MIX CONCRETE

8.3.1 Introduction

When the concrete proportions are obtained by stringent control on ingredients, and processes, and preliminary laboratory trial tests are carried out to ensure the desired minimum strength and other properties on the basis of well established principles, then such a concrete is called **Design Mix (Controlled) Concrete**. In design mix concrete various parameters are considered as accurately as possible from various design data and variables as discussed earlier and shall be computed for the desired strength and other properties.

The mix design methods used in different countries are mostly based on empirical relationships, charts, and graphs developed from extensive experimental investigations. Most of these methods are based on the principles explained earlier. There are only minor variations in the process of selecting the mix proportions in different methods. Some of the following methods shall be discussed here:

i) Mix Design Method according to Indian Standard Recommended Guidelines;
ii) The ACI Mix Design Method;
iii) The USBR Mix Design Practice; and
iv) The British Mix Design Method.

In all the four methods the **water-cement ratio** is chosen for the **target mean strength** from the empirical strength versus water-cement ratio relationships, and water content is chosen for required workability for aggregates in saturated surface dry condition. So far as the aggregate volume is concerned, the methods differ to some extent. In the ACI method, the volume of coarse aggregate in the concrete mix is first determined on the basis of the maximum size of aggregate and the grading of fine aggregate. Whereas in the British method, the proportion of fine aggregate is determined first depending on the maximum size of aggregate, the degree of workability, the grading of fine aggregate and the water-cement ratio of the concrete mix.

In USBR method, the proportion of dry-rodded coarse aggregate is determined corresponding to the maximum size of aggregate, a fixed fineness modulus of sand and a fixed workability in terms of slump. The ACI method also determines the proportions of dry- rodded coarse aggregate in the concrete mix. It is based on the concept that in dry-rodded void content, the differences in the amount of mortar required for workability with different aggregates due to differences in particle shape and grading are automatically compensated for.

The latest British mix design method does not consider the combined aggregate grading curves. This implies admission to the use of aggregates of any grading as long as they are within the grading limits specified by the appropriate codes/specifications.

In the ACI and USBR methods, the air content of concrete is considered to achieve unit absolute volume of the mix ingredients. The batch weight of the materials per unit volume of concrete is calculated from the absolute volumes. In the British method, the quantities of the ingredients are calculated directly from the wet density of concrete which is dependent on specific gravity of the combined aggregate (on saturated surface dry condition).

The Indian Standard recommended guidelines for the mix design includes design of normal concrete mixes (non air entrained), both for medium and high strength concrete. In this method of mix design, the **water content** and **proportion of fine aggregate** corresponding to a **maximum size of aggregate** are first determined with reference to **standard values of workability, water-cement ratio** and grading of the fine aggregate (**sand grading zones**). The water content and the proportion of fine aggregate are then adjusted for any difference in **given workability, water-cement ratio** and **grading of fine aggregate** (sand grading zone) in a particular case from the reference standard values. The batch mass of materials per unit volume of concrete is calculated on the **absolute volume basis.** Various tables and charts have been developed by exhaustive tests at the Cement Research Institute of India and on the basis of data on concrete being designed and produced in this country. Although various tables and charts for design are based on the use of ordinary portland cement (OPC) but they can also be applied for concrete prepared with portland pozzolanic cements (PPC). The final mix proportion, selected after trial mixes may entail some minor changes in each case. Such variations may be necessitated due to change of type of cement or aggregate or their source.

For selection of **water-cement ratio for the target compressive strength** at 28 days generalized relationship shown in Fig. 6.3 may be used both for OPC and PPC with sufficient accuracy. For accurate selection of water-cement ratio **Fig. 8.1** may be used depending on the 28 days strength of cement and hence does not make any difference with OPC or PPC. **Figure 8.2** may be used instead of **Fig. 8.1** when cements are classified on the basis of **accelerated curing** strengths.

Fly-ash cement concretes for comparable strengths using 3 to 5 % fly-ash and the proportion of fine aggregates are also reduced by 2 to 4 %. These generalization may not be applicable directly for concrete made with PPC and the precise variations in these proportions should be established by trials with actual materials available at site. Indian Standard method uses similar approach to that of USBR and specified in IRC 44-1976 for mix design for concrete pavements.

In IRC 44-1976 approach water-cement ratio is based on relationship of 28 days concrete strengths for different cements classified on 7 day's strengths (**Refer Fig. 6.2**).

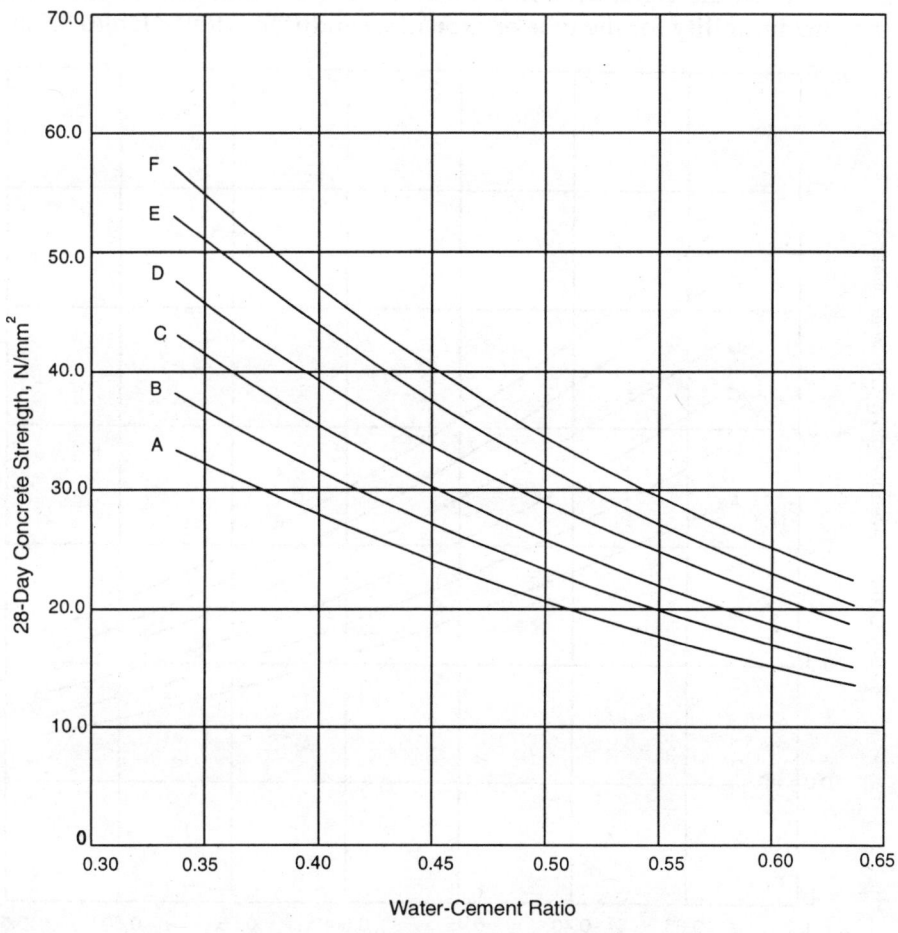

28 - Day Strength of Cement, Tested According to IS: 4031 - 1968

A – 31.9-36.8 N/mm^2 (325-375 Kg/cm^2)   D – 46.6-51.5 N/mm^2 (475-525 Kg/cm^2)
B – 36.8-41.7 N/mm^2 (375-425 Kg/cm^2)   E – 51.5-56.4 N/mm^2 (525-575 Kg/cm^2)
C – 41.7-46.6 N/mm^2 (425-475 Kg/cm^2)   F – 56.4-61.3 N/mm^2 (575-625 Kg/cm^2)

Fig. 8.1. Relationship between Free Water Cement Ratio and Concrete
Strength for different Cement Strengths (28 days)

8.3.2 Indian Standard Guidelines for Mix Design (IS: 10262-1982)

For mix design the following **data** is required to be **specified** or determined from various **site conditions** and materials:

i) **Characteristic Strength** at 28 days (f_{ck}) or desired grade of concrete (not more than 5 % results shall be allowed to fall below this limit;

ii) **Workability desired** for given placing conditions at the site (for guidance Refer Table 8.4 (a) and 8.4 (b));

iii) Limitations on the **water-cement ratio** and the **minimum cement content** to ensure adequate **durability** for the type of **exposure conditions** (Refer Tables 8.5 and 8.6);

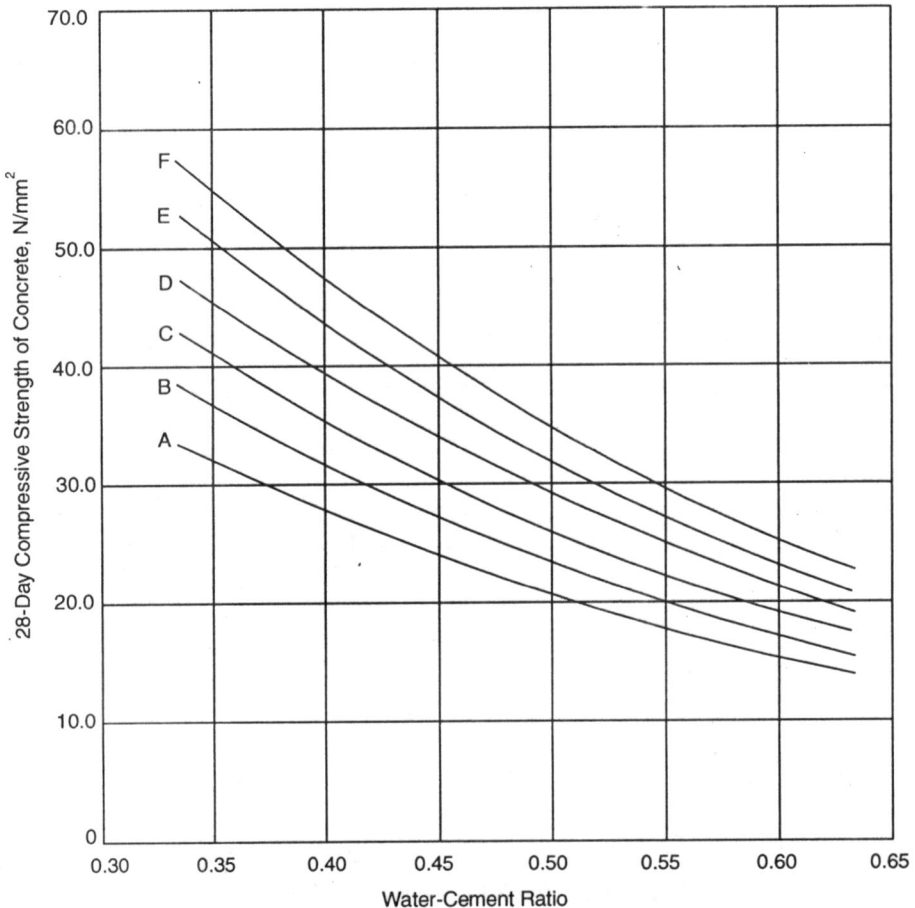

Acclerated Strength (Tested According to IS: 9013-1978) of Reference Mix

A – 12.3 - 15.2 N/mm^2 (125 - 155 Kg/cm^2) D – 21.1 - 24.0 N/mm^2 (215 - 245 Kg/cm^2)
B – 15.2 - 18.1 N/mm^2 (155 - 185 Kg/cm^2) E – 24.0 - 27.0 N/mm^2 (245 - 275 Kg/cm^2)
C – 18.1 - 21.1 N/mm^2 (185 - 21 5 Kg/cm^2) F – 27.0 - 29.9 N/mm^2 (275 - 305 Kg/cm^2)

Fig. 8.2. Relationship between Free Water Cement Ratio and Compressive Strength of Concrete for different Cement Strengths Determined on Reference Concrete Mixes (Acclerated Test-Boiling Water Method)

iv) The **type and maximum size of coarse aggregate** to be used/available at site;

v) The **type and compressive strength of cement** to be used for concrete production;

vi) Standard deviation (S) for compressive strength on the basis of previous quality control conditions at site or expected degree of **quality control** based on the infrastructure and practices adopted at the construction site (Refer Tables 7.2, 7.4, and 7.5).

Table 8.4 (a) Suggested Ranges of Workability of Concrete for Different Placing Conditions**

Placing Conditions	Degree of Workability	Values of Workability
1. Concreting shallow sections with vibration	Very Low	20-10 seconds. Vee-Beetime or 0.75-0.80 compacting factor
2. Concreting of lightly reinforced sections with vibration	Low	10-5 seconds. Vee-Beetime or 0.80-0.85 compacting factor
3. Concreting of lightly reinforced sections without vibration or heavily reinforced section with vibration	Medium	5-2 seconds. Vee-Bee time or 0.85-0.92 compacting factor or 25-75 mm slump for 20 mm CA*
4. Concreting heavily reinforced sections without vibrations	High	Above 0.92 compacting factor or 75-125 mm slump for 20mm CA

* For smaller aggregate the values will be lower
** Ref. SP: 23-1982

Table 8.4 (b) Comparison of Consistency Measurements by Various Methods*

Workability Description	Slump (mm)	Vee-Bee Time (Seconds)	Compacting Factor
1. Extremely dry	—	32-18	—
2. Very stiff	—	18-10	0.70
3. Stiff	0-25	10-5	0.75-0.80
4. Stiff Plastic	25-50	5-3	0.85
5. Plastic	50-100	3-0	0.90
6. Flowing	150-175	—	0.95

* Ref. SP: 23-1982

The **step-by-step** procedure of **mix proportioning as per IS-guidelines**.

Step I: Calculate the target mean strength with respect to the expected quality control at site and the desired characteristic strength at 28 days by the equation given below:

$$f_t = f_{ck} + K \cdot S \qquad \qquad ...(8.1)$$

Where

f_t = target mean compressive strength at 28 days (N/mm^2).

f_{ck} = characteristic compressive strength at 28 days (N/mm^2).

S = standard deviation (Refer Tables 7.4 and 7.5) depending on degree of control and the grade of concrete (N/mm^2).

K = a statistical constant depending on the accepted proportion of low results and the number of tests (Refer Table 7.3).

As per Indian Standard IS: 456-1978 (Revised 2000), the characteristic strength is defined as that value below which not more than **5 % of the test results are expected to fall.** As such the value of:

$$K = 1.65$$
$$f_t = f_{ck} + 1.65\,S \qquad\qquad ...(8.1a)$$

Table 8.5 Minimum Cement Required in Concrete to Ensure Durability under Specified Conditions of Exposure*

Exposure condition	Plain Concrete		Reinforced concrete		Prestressed concrete	
	Min. cement Kg/m^3	Max W/C Ratio	Min. cement Kg/m^3	Max W/C Ratio	Min. cement Kg/m^3	Max W/C Ratio
1. Mild – Completely, protected against weather, or aggressive conditions, except for a brief period of exposure to normal weather conditions during construction.	220	0.70	250	0.65	300	0.65
2. Moderate – Sheltered from heavy and wind driven rain and against freezing, whilst saturated with water, burried concrete in soil, and concrete continuously under water.	250	0.60	290	0.55	300	0.55
3. Severe – Exposed to sea water, alternate wetting and drying, and to freezing whilst wet, subject to heavy condensation of water of corrosive fumes.	310	0.50	360	0.45	360	0.45

* Ref. : SP23-1982.

NOTE-1 : When the maximum W/C ratio can be strictly controlled, the cement may be reduced by 10 %.

NOTE-2 : The minimum cement is based on 20 mm C.A. It should be increased by about 10% for 12.5 mm C.A. and reduced in cement by 10% for 40 mm C.A. In case of pre-stressed concrete 10 % reduction in cement for 40 mm C.A. applies only in severe exposure while 10 % increase **in cement** for 12.5 mm applies to severe and moderate conditions of exposure.

Step II: Determine the **water-cement ratio** required for the computed value of **target strength** using curves given in Fig. 6.3 for Indian cements if compressive strength and type of cement is not known. In case the 28 days or accelerated compressive strength of cement is known then **water-cement ratio** is determined from the Fig. 8.1 or Fig. 8.2 selecting **characteristic cement curve** respectively. In case 7 days compressive strength of cement is known then water-cement ratio is determined from the target strength of concrete by selecting appropriate characteristic cement curve shown in Fig. 6.2 (developed by CRRI).

This water-cement ratio is verified for the desired target strength by actual trial test in laboratory. It may be noted that relationship of 28 days concrete strength and W/C ratio is more precise in case of 28 days characteristic cement compressive strength curves shown in Fig. 8.1.

Such trials will need 28 days for determining the compressive strength of cement and another 28 days for the trial mixes of concrete. In order to cut down the time, curves drawn for reference accelerated concrete mixes as shown in Fig. 8.2 should be used.

Table 8.6 Requirements for Plain, Reinforced and Prestressed Concrete
Exposed to Sulphate Attack*

Concentration of Sulphates Expressed as SO_3			Type of Cement	Requirements for Fully Compacted Concrete With Aggregate Complying To IS: 383-1970	
In Soil Total SO_3 %	SO_3 in 2:1 Water (gm/l)	Ground Water PPM		Minimum Cement (Kg/m^3)	Maximum Free W/C Ratio
1. Less than 0.20	—	Less than 30	OPC, Portland Slag Cement or PPC	280	0.55
2. 0.20 to 0.50	—	30 to 120	OPC, Portland Slag Cement or PPC	330	0.50
			Supersulphated Cement	310	0.50
3. 0.50 to 1.0	1.9 to 3.1	120 to 250	Supersulphated Cement	330	0.50

* Ref: SP: 23-1982

NOTE-1 : The table applies to concrete made with 20 mm C.A. placed in neutral ground waters (PH 6 to 9), containing naturally occurring sulphates but not contaminants, such as ammonium salts. For 40 mm C.A., the value may be reduced by about 15% and for 12.5 mm C.A., the value may be increased by about 15%. OPC is not recommended for acidic conditions (PH 6 or less). Super sulphated cement gives an acceptable life in mineral acids, down to PH 3.5 provided that the concrete is dense and prepared with W/C ratio of 0.40 or less.

NOTE-2 : For severe conditions in thin sections under hydrostatic pressure on one side only and partly immersed, W/C ratio may be further reduced, and cement increased to ensure the degree of workability needed for full compaction and minimum permeability.

Step III: Water-cement ratio determined above is checked against the limiting water-cement ratio for the requirements of durability in the given condition of exposure and type of concrete such as plain, reinforced, or prestressed concrete (Refer to Table 8.5 and 8.6). Lower of the two values of water-cement ratios required for strength and durability is adopted.

Step IV: Depending on the conditions and method of placement a suitable workability is assumed for the given job from Table 8.4 (a) and 8.4 (b).

Step V: Determine the water content and percentage of sand in total aggregate by absolute volume from Table 8.8 or 8.9 for medium and high strength concrete corresponding to the maximum aggregate size and the following standard conditions:

(i) Crushed/angular coarse aggregate;

(ii) Fine aggregate consisting of natural sand conforming to grading zone II (as per IS: 383-1970), in saturated surface dry conditions;

(iii) Water-cement ratio of 0.60 and 0.35 for medium and high strength concrete respectively; and

(iv) Workability corresponding to compacting factor of 0.80.

Table 8.7 Approximate Entrapped Air Content*

Nominal Maximum sizeof Aggregate (mm)	10	20	40
Entrapped air as percentof volume of concrete	3.0	2.0	1.0

* Ref. IS: 10262-1982.

Table 8.8 Approximate Sand and Water Content per cum of Concrete for Grades upto M35*
(W/C = 0.60, Workability = 0.80 C.F.)

Maximum Size of Aggregate (mm)	Water Content/m^{3**} of concrete (kg)	Sand as Percent of Total Aggregate by Absolute Volume
10	208	40
20	186	35
40	165	30

* Ref. IS 10262-1982
** Water content corresponding to saturated surface dry aggregate

Step VI: Determine the variations from standard values in conditions of workability, water-cement ratio, grading zone of fine aggregate, and type of aggregate likely to be used for the preparation of concrete at the site. Depending on the selected water-cement ratio and compacting factor of concrete, the variations in these values from the standard values in Table 8.8 and 8.9 for medium or high strength concrete is determined. From the variations in standard conditions of W/C ratio, compacting factor, grading zone of sand and type of aggregate, the water content and sand content adjustments are determined from Table 8.10.

Table 8.9 Approximate Sand and Water Content per m3 of Concrete for Grades Above M35
(W/C = 0.35, Workability = 0.80 C.F.)

Maximum Size of Aggregate (mm)	Water Content/$m^{3**}$of concrete (kg)	Sand as Percent of Total Aggregate by Absolute Volume
10	200	28
20	180	25

* Ref. IS 10262-1982
** Water content corresponding to saturated surface dry aggregate.

Step VII: The cement content is now calculated from the selected water-cement ratio and the final water content arrived after adjustments for various factors as in step VI. Cement content is obtained by dividing net water content by water-cement ratio [C=W/ (W/C) Kg/m³].

The cement content so calculated is checked against the minimum cement content for durability and sulphate attack prevention requirements (Refer Tables 8.5 and 8.6) and the greater of the two values adopted.

Step VIII: Determine estimated air entrapped in concrete mix on the basis of the maximum size of aggregate from the Table 8.7.

Step IX: With the quantities of water and cement per unit volume of concrete and the percentage of sand in the total aggregate already determined, the coarse and the fine aggregate content per unit volume of concrete are calculated from the absolute volume equations given below:

$$V = [W + C/S_c + 1/p \times f_a /S_{fa}] \times 1/1000, \text{ and} \qquad ...(8.2)$$

$$V = [W + C/S_c + 1/(1\text{-}p) \times ca/S_{ca}] \times 1/1000 \qquad ...(8.3)$$

Where,

V = Absolute volume of fresh concrete.

= gross volume (1 m³) minus the volume of entrapped air (Refer Table 8.7).

S_c = Specific gravity of cement (generally 3.15).

W = Mass of water (Kg) per m³ of concrete.

C = Mass of cement (Kg) per m³ of concrete.

p = Ratio of fine aggregate to total aggregate by absolute volume.

f_a, ca = Total masses of fine aggregate and coarse aggregate (Kg) per m³ of concrete respectively, and

S_{fa}, S_{ca} = Specific gravity of saturated surface dry fine aggregate and coarse aggregate respectively.

Step X: From various quantities of ingredients per m³ of concrete, the batch proportions are calculated in terms of unit mass of cement on the basis of mixer capacity at site. Various quantities of ingredients per m³ are divided by mass of cement per m³ to obtain the **proportions** in terms of **batch of unit mass of cement.** For obtaining batch proportions, **unit cement mass proportions** are **multiplied** by mass of cement per batch according to the mixer capacity.

Step XI: The designed mix proportions are verified by actual trials in the laboratory for the desired qualities and properties using actual materials to be used in the construction.

The procedure of mix design shall be illustrated by an example.

8.3.3 Example 8.1: Concrete Mix Design (Grade M25)

a) Design stipulations

i. Characteristic compressive strength at 28 days : 25 N/mm²
ii. Degree of quality control expected at site : Good
iii. Maximum size of aggregate and type. : 20 mm (angular)

iv.	Degree of workability desired	: 0.85 (C.F)
v.	Type of exposure	: Mild and no sulphate attack
vi.	Concrete use	: RCC Structure

b) *Test data of materials expected to be used*

i.	Cement OPC satisfying the requirements of IS 269-1976	
ii.	Sp. gravity of cement	: 3.15
iii.	Sp. gravity of C.A. (Saturated surface dry condition)	: 2.65
iv.	Sp. gravity of F.A. (Saturated surface dry condition)	: 2.60
v.	Water absorption coarse aggregate	: 0.50 %
vi.	Water absorption fine aggregate	: 0.50 %
vii.	Surface moisture-coarse aggregate	: Nil (and absorbed nil)
viii.	Surface moisture-fine aggregate	: 2.0 %
viii.	Sand grading zone	: III

Table 8.10 Adjustment of Values in Water Content and Sand Percentage for other Conditions*

Sr. No.	Change in condition Stipulated for Tables 8.8 to 8.9(i)	Adjustment required in Water Content (ii)	Percent sand in total aggregate (iii)
1.	For sand conforming to grading		
	Zone I	0	+ 1.5 %
	Zone III	0	− 1.5 %
	Zone IV	0	− 3.0 %
	Of IS 383-1970		
2.	Increase or decrease in the value of Compacting factor by 0.10	± 3 %	0
3.	Each 0.05 increase or decrease in Free W/C ratio	0	± 1.0 %
4.	Rounded aggregate	− 15 kg/m²	− 7.0 %

* Ref: IS: 10262-1982

MIX DESIGN CALCULATIONS

i) Target mean strength (Table 7.5 for S)
: $f_t = f_{ck} + K \cdot S$
= 25 + 1.65 (5.3)
= 33.75 N/mm²

ii) Water-cement ratio (Fig. 6.3) : W/C = 0.42

iii) For mild exposure reinforced concrete (table 8.5) : Max. W/C = 0.65
Adopt W/C = 0.42 (lower of the two)

iv) Workability desired is specified in terms of **0.85 C.F.**

v) Water content and sand percentage for concrete upto M35 grade for standard W/C = 0.60 and C.F. = 0.80, for 20 mm angular aggregate are (Table 8.8): W = 186 Kg/m³, sand percent p = 35% of total aggregate.

vi) Difference in W/C ratio = 0.60 – 0.42 = 0.18 (decrease), Difference in workability in terms of C.F. = 0.85 – 0.80 = 0.05 (increase).

Grading Zone of fine aggregate = III

Aggregate Type – Crushed (angular)

Sr.	Change of Condition (Table-8.10)	Adjustments Required in Water Content	Percentage Sand
1.	Sand Zone-III	0	– 1.5
2.	Increase in C.F. by 0.85 – 0.80 = 0.05	+ 3 × 0.05 / 0.10 = + 1.5%	0
3.	For decrease in W/C ratio by 0.60 – 0.42 = 0.18	0	– 1.0 × 0.18 /0.05 = – 3.6
4.	Total	+ 1.5 %	– 5.1 %

Required Water Content W = 186 + 1.5/100 × 186 = 188.8 = 189 l

$$W = \textbf{189 Kg/m}^3$$

Required sand content percent = 35 – 5.1 = 29.9 = **30.0 %** of total aggregate

p = 0.30

vii) Cement content C = 189/0.42 = **450 Kg/m³**

Minimum C (for mild exposure from Table 8.5) = **250 Kg/m³**

Thus **adopt** (greater of the two values) C = **450 Kg/m³**

viii) For 20 mm aggregate expected entrapped air = 2%

(Ref. Table 8.10) = 0.02

ix) From absolute volume equations 8.2 and 8.3, we have

Mass of F.A. : $(1 - 0.02) = (W + C/S_c + 1/P \times f_a/S_{fa}) \times 1/1000$

$980 = (189 + 450/3.15 + 1/0.3 \times f_a/2.6)$

$f_a = \textbf{506 Kg/m}^3$

Mass of C.A. : $(1 - 0.02) \times 1000 = (189 + 143 + 1/0.7 \times C_a / 2.65)$

$C_a = \textbf{1202 Kg/m}^3$

x) The Mix Proportions by mass are:

Unit of Batch	Water W	Cement C	F.A. f_a	C.A. C_a	Total Aggregate $(f_a + C_a)$
1 m³ Concrete	189 kg	450 kg	506 kg	1202 kg	1708 kg
Unit Cement	0.42	1.0	1.124	2.671	3.795
1 Bag Cement	21 litres	50 kg	56.2 kg	133.6 kg	189.8 kg

xi) Actual quantity of materials to be added at site in one bag batch are:

Water

a. For W/C ratio = 0.42, water = 21 litres
b. Extra water to be added for absorption in:
   F.A. = Nil as already saturated
   C.A. = 0.50/100 × 133.6 = 0.67 kg = **0.67 litres**
c. Excess surface water to be deducted in F.A. = 2/100 × 56.2 = **1.124 kg**
   C.A. = Nil
d. Net quantity of water to be added at mixer:
   = 21 + 0.67 – 1.124 = 20.546 = **20.55 litres**

Sand Allowing for mass of free moisture:
   56.2 + 1.124 = 57.324 kg = **57.33 kg**

Coarse Aggregate
   133.6 – 0.67 = 132.93 kg = **133 kg**
   Actual quantities to be added for one bag batch:

Water	**20.55 litres**	
Cement	**50 kg**	
Sand	**57.33 kg**	
Coarse Aggregate	**133 kg**	(This may be obtained by combining different fractions of various sizes available to make good grading as provided in IS: 383-1970)

Further adjustments may be carried out as per actual observations at site for the desired specifications of fresh concrete. Various adjustments shall be further discussed in subsequent chapter.

8.4 THE ACI MIX DESIGN METHOD
(Absolute Volume Method)

8.4.1 Approach

The ACI 211.1-1977 provides the recommended practice for selecting the proportions for normal and heavy weight concrete by **American Concrete Institute.** In this method of mix design, the concrete mix is designed for different maximum size of aggregate. The bulk volume of coarse aggregate per unit volume of concrete is determined for different fineness modulii of sand. The water-cement ratio is determined in a usual procedure to satisfy both **strength and durability requirements.** The volume of fine aggregate is determined for unit volume of concrete from the difference in solid (absolute) volume between the concrete and other ingredients. Allowance for entrapped air content in concrete is made prior to calculating the volume of fine aggregate. The step by step procedure is as follows:

i) Compute target mean 28 days compressive strength on the basis of **degree of quality control** at work site (Refer Table 8.11).
ii) Select water-cement ratio from the target mean 28 days compressive strength (Refer Table 8.12). Check W/C ratio for durability (refer Table 8.5a and 8.6). Adopt smaller of the two water cement ratios.
iii) Select suitable workability in terms of slump or compacting factor on the basis of conditions and method of placing (Refer Table 8.4a and 8.4b).

iv) Select water content for the desired workability and maximum size of aggregate (Refer Table 8.13).

v) Compute cement content from the water content and the selected water-cement ratio. Check minimum cement content for the durability requirement (Refer Table 8.5 and 8.6).

vi) Compute fineness modulus of sand from the given sieve analysis data. Also determine the unit mass and specific gravity of coarse and fine aggregates at site if not specified.

vii) The bulk volume of coarse aggregate content is estimated for the maximum size of aggregate and the fineness - modulus of sand (Refer Table 8.14). From the bulk volume, mass of coarse aggregate is calculated from the unit mass of C.A. and then solid volume calculated from specific gravity of C.A.

viii) The solid volume of the fine aggregate (sand) is determined by subtracting the sum of the solid (absolute) volumes of coarse aggregate, cement, water and entrapped air content from the unit volume of concrete.

ix) Mass of fine aggregate is calculated from the absolute volume and specific gravity of fine aggregate.

x) Various quantities are given on dry condition for 1.0 m³ of concrete. Surface moisture or absorption at site is observed and various quantities are modified for the site conditions.

Table 8.11 28 Days Compressive Strength required for a Degree of Quality Control*

Degree of Quality Control	Required Minimum site Strength at 28 days in N/mm² (Concrete Grade)			
	M10	*M20*	*M30*	*M40*
Good	15.0	26.7	37.5	50
Fair	16.7	30.0	43.3	53.3
Poor	20.0	33.3	Not to be attempted	

* Ref: ACI Manual of Concrete Practice, Part-I-1979

For stiffer concrete mixes ACI-211-65 recommended practice for selecting proportions for the "no slump concrete" should be used. In this method, the workability measurement is done by compacting factor, Vee-Bee consistency, or flow Table (drop table) test instead of slump test. Another difference is that the C.A. content is higher for more workable mixes. Thus the tables for water requirement for different degrees of workability and coarse aggregate volume per unit volume of concrete are modified. Rest of the approach remains the same.

Table 8.12 Relationship between Water-Cement Ratio and Compressive Strength of Concrete*

Compressive Strength at 28 days (N/mm²)	15	20	25	30	35	40	45
W/C Ratio Non-air entrained	0.80	0.70	0.62	0.55	0.48	0.43	0.38
Air entrained	0.71	0.61	0.53	0.46	0.40	—	—

* Ref: ACI Manual of Concrete Practice, Part-I-1979.

8.4.2 Example 8.2: Design concrete mix for the interior R.C.C. components for the following requirements:

Grade of concrete – M20, workability – medium
Degree of quality control at site – Fair
Aggregates available at site are:
C.A. : Max, size 20 mm, 52% passing through 10 mm sieve,
Sp. gravity = 2.80, unit mass = 1800 Kg/m^3, surface moisture = 1% by mass.

F.A.: Sieve size	percent passing
4.75mm	100
2.36mm	90
1.18mm	76
600 micron	58
300 micron	22
150 micron	4

Sp. gravity = 2.65, unit mass = 1700 Kg/m^3, surface moisture = 2%

Table 8.13 Approximate Mixing Water for Different Slumps and Maximum Size of Aggregate*

Workability Slump (mm)	Water (litres) per m^3 of concrete for maximum size of aggregate (mm)						
	10	12.5	20	25	40	50	80
Low 25-50	250	235	220	210	195	180	170
Medium 75-100	270	260	240	230	220	200	190
High 150-175	290	270	250	240	225	210	200
ApproximateEntrapped Air percent	3.0	2.5	2.0	1.5	1.0	0.50	0.30

* Ref.: ACI Manual of Concrete Practice, Part-I, 1979.

Table 8.14 Bulk Volume of Dry-Rodded Coarse Aggregates Per Unit Volume of Concrete*

Maximum size of aggregate (mm)	Fineness Modulus of Sand			
	2.4	2.60	2.80	3.00
10	0.50	0.48	0.46	0.44
12.5	0.59	0.57	0.55	0.53
20	0.66	0.64	0.62	0.60
25	0.71	0.69	0.67	0.65
40	0.76	0.74	0.72	0.70
50	0.78	0.76	0.74	0.72
70	0.81	0.79	0.77	0.75
150	0.87	0.85	0.83	0.81

* Ref. ACI Manual of concrete practices Part-I, 1979.

Mix Design Ex.: 8.2

(i) From Table 8.11, the target mean strength required for M20 grade of concrete under fair degree of quality control = 30 N/mm^2.

(ii) From Table 8.12, the required **W/C ratio** for the (28 days) target strength of 30 N/mm^2 = 0.55 (non-air-entrained concrete). For interior R.C.C. elements (very mild exposure) max W/C = 0.65 (Table 8.5). **Select W/C = 0.55**

(iii) For medium workability-slump may be considered as **75 mm** or compacting factor = **0.90,** say (Refer Table 8.4a and 8.4b).

(iv) Fineness modulus of sand and coarse aggregate are found as under:

Sieve Analysis F.A.

Sieve Size	Percentage Retained	Passing
4.75 mm	0	100
2.36 mm	10	90
1.18 mm	24	76
600 micron	42	58
300 micron	78	22
150 micron	96	4
Total	250	F.M. = 250/100 = 2.50

Coarse Aggregate

Sieve Size	Percentage Retained	Passing
20mm	0	100
10mm	48	52
4.75mm	100	0
2.36mm	100	0
1.18mm	100	0
600 micron	100	0
300 micron	100	0
150 micron	100	0
Total	648	F.M. = 648/100 = 6.48, Say 6.50

(v) Select water content for 20mm maximum size of aggregate and the desired medium workability (slump say 75mm) from Table 8.13, W = **240 litres/m^3.** Entrapped air content = 2%

(vi) Mass of cement content C = W/(W/C) = 240/0.55 = **436 Kg/m^3** Minimum cement content for durability (mild exposure) (Table 8.6) = **250 Kg/m^3** Adopt cement content C = **436 Kg/m^3**

(vii) Select bulk volume of C.A. for 20mm maximum aggregate size, fineness modulus of 2.5 for sand from Table 8.14.

Volume of C.A. = **0.65 m³/m³** of concrete, Unit mass of C.A. = **1800 Kg/m³**
Mass of C.A. = 0.65 × 1800 = **1170 Kg/m³** of concrete

(viii) Absolute (solid) volumes of ingredients :

Cement = 436/3.15 × 1/1000 = **0.1384 m³**
Water   = 240/1 × 1/1000 = **0.24 m³**
C.A.   = 1170/2.80 × 1/1000 = **0.418 m³**
Entrapped air = 2% = 0.020 m³
Total Volume = **0.8164 m³**
Solid volume of **sand** required = 1 − 0.8164 = **0.1836 m³**
Mass of dry sand required = **0.1836 × 2.65 × 1000 = 487 Kg**

(ix) Masses of various ingredients in 1 m³ of concrete are:

Water (W)   =   **240 Kg**
Cement (C) =   **436 Kg**
Sand (dry)   =   **487 Kg**
C.A. (dry)   =   **1170 Kg**
Surface water in aggregate for site conditions:
Coarse aggregate : 1170 × 1/100 = **11.70 Kg**
Fine aggregate : 487 × 2/100 = **9.74 Kg**
Total surface water = **21.44 Kg**

(x) Quantities adjusted for site conditions:

Mixing water (W) = 240 − 21.44 = **218.56 = 218.6 Kg**
Fine aggregate (F.A.) = 487 + 9.74 = 496.74 = **497 Kg**
Coarse aggregate (C.A.) = 1170 + 11.70 = 1181.70 = **1182 Kg**
Field Proportions

Water(Litre)	Cement (Kg)	F.A.(Kg)	C.A(Kg)	Unit Batch
218.6	436	497	1182	1m³ concrete
0.50	1.0	1.136	2.71	Per Kg. Cement
25	50	56.8	135.5	Per bag of cement

8.5 THE USBR MIX DESIGN METHOD

In this method, the water content of air-entrained concrete and the proportions of fine and coarse aggregates are determined for a fixed workability and grading of fine aggregate. The water-content and percentages of sand or coarse aggregate are adjusted for changes in the materials and mix proportions. The water-cement ratio for compressive strength is determined in usual way. The step-by-step procedure of mix proportions is as follows:

(i) Select suitable required compressive strength of concrete for the job and type of quality control by usual approach.

(ii) Select water-cement ratio for the target mean 28 days compressive strength of concrete from Table 8.15 for either air-entrained concrete or for air-entrained concrete with water-reducing set-controlling admixtures.

(iii) Approximate air and water content and the percentages of sand and coarse aggregate per cubic metre of concrete are determined from **Table 8.16** for air-entrained concrete (AEC or for air-entrained concrete with water-reducing, set controlling admixtures (AEC,

WRSCA) for concrete containing natural sand with a fineness modulus of 2.75 and workability of 75 to 100 mm slump.

Table 8.15 Probable Minimum Average Compressive Strength of Concrete from Various Water Cement Ratios*

Water-cement Ratio	Compressive Strength at 28 days N/mm^2	
By Mass	Air Entrained Concrete	Air Entrained with water-reducing set-controlling Admixtures
0.40	39.9	45.5
0.45	34.3	39.2
0.50	29.4	33.6
0.55	25.2	29.4
0.60	21.7	25.2
0.65	18.2	21.7
0.70	15.4	18.9

* Concrete Manual Eighth Revision - 1981 Bureau of Reclamation, U.S.A (SP 23-1982).

Table 8.16 Approximate Air and Water Content Per m^3 of Concrete and the Proportions of F.A. and C.A.*. (Sand FM = 2.75, Concrete Slump 75-100mm)

Max. size of C.A.(mm)	Air content percent	F.A. percent of total aggregate by solid volume	Percent dry rodded unit mass of C.A. per unit vol. of concrete	Average water content in A-E-CKg/m^3	Average water content A-E-C, W-R, S-C-A,Kg/m^3
10	8	60	41	189	177
13	7	50	52	180	168
20	6	42	62	165	156
25	5	37	67	156	147
40	4.5	34	73	145	136
50	4.0	30	76	136	122
75	3.5	28	81	121	112
150	3.0	24	87	97	91

*Ref. Concrete Manual Eighth Edition – 1981, Bureau of Reclamation, USA (SP: 23-1982).

(iv) Adjustments of values in water content and percentages of sand or coarse aggregate are made as provided in **Table 8.17** for changes in the fineness modulus of sand, slump of concrete, air content, water-cement ratio and sand content other than the reference values in **Table 8.16.**

(v) The cement content is calculated using the selected water-cement ratio and the final water content of the mix after the adjustments.

Table 8.17 Adjustment of Values of Water Content, Percent Sand and Percent of Dry-Rodded Coarse Aggregate*

Sr. No.	Changes in Material or Mix Proportions	Adjustment Required in		
		Water content percent	Sand percent	Percent Dry-rodded C.A.
1.	Each 0.10 increase or decrease in F.M. of sand.	—	± 0.50	± 1.0
2	Each 25mm increase or decrease in slump.	± 3.0	—	—
3.	Each 1 percent increase or decrease in air content.	± 3.0	± 0.5 to 1.0	—
4.	Each 0.05 increase or decrease in water-cement ratio.	—	± 1.0	—
5.	Each 1 % increase or decrease in sand content.	± 1.0	—	± 2.0

*Ref. Concrete Manual Eighth Edition – 1981, Bureau of Reclamation, USA (SP: 23-1982).

(vi) Proportions of aggregate are determined by estimating the quality of coarse aggregate from Table 8.16 (dry-rodded unit mass coarse aggregate method) or by computing the total solid volume of sand and coarse aggregate in the concrete mix and multiplying the final percentage of sand after adjustment. Either of approach shall produce approximately the similar proportions.

8.6 THE BRITISH MIX DESIGN METHOD (Department of Environment Method)

8.6.1 Procedure

This method is replacement of traditional British mix design method of Road Note No. 4. It discards the use of specific grading curves of the combined aggregate, uses the relationship between water-cement ratio and compressive strength of concrete depending on the type of cement and type of aggregate to be used. It replaces the mix design tables correlating water-cement ratio, aggregate-cement ratio, maximum size of aggregate, type of aggregate differing in shapes (rounded and irregular), degree of workability and overall grading curves of the combined aggregates in earlier Road Note No. 4. Instead, water content required to give various levels of workability is determined for two types of aggregates, namely, crushed and uncrushed.

The degree of workability "**very low**", "**low**", "**medium**" and "**high**" have now been referred in terms of specific values of slump and Vee-Bee time. The method of mix design results in expressing the mix proportions in terms of quantities of materials per unit volume of concrete in line with European and American practice. The step-by-step procedure of mix design is given as under:

i) The target mean compressive strength is found in usual way for the given quality control at site for the given grade of concrete.

ii) The water-cement ratio for the target mean compressive strength is determined using Table 8.18 and Fig. 8.3 and compared with the maximum water-cement ratio specified for durability and the lower of these two values adopted. From the values of strength corresponding to 0.50 W/C ratio as given in Table 8.18, the relevant curve in Fig. 8.3 is selected. Then for the target mean strength W/C ratio is selected from this curve. For intermediate values the curves may be interpolated.

iii) The water content depending upon the type and maximum size of aggregate to give a concrete of the specified slump or Vee-Bee time is selected from Table 8.19.

iv) The cement content is calculated from the water-cement ratio and water content of the mix.

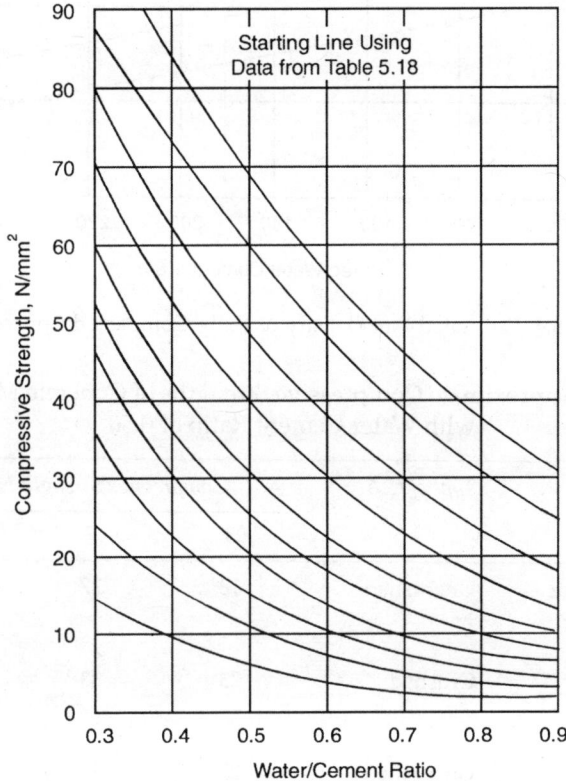

Fig. 8.3. Relationship between Compressive Strength and Water-Cement Ratio

v) The total aggregate content (saturated and surface dry) is determined by subtracting the cement and water content from the wet density of concrete, the wet density being obtained from Fig. 8.4 depending upon the water content and the relative density of the combine aggregate.

vi) Finally, the proportions of fine and coarse aggregates are determined from Fig. 8.5, depending on the water-cement ratio, the maximum size of aggregate, the workability level and the grading zone of the fine aggregates.

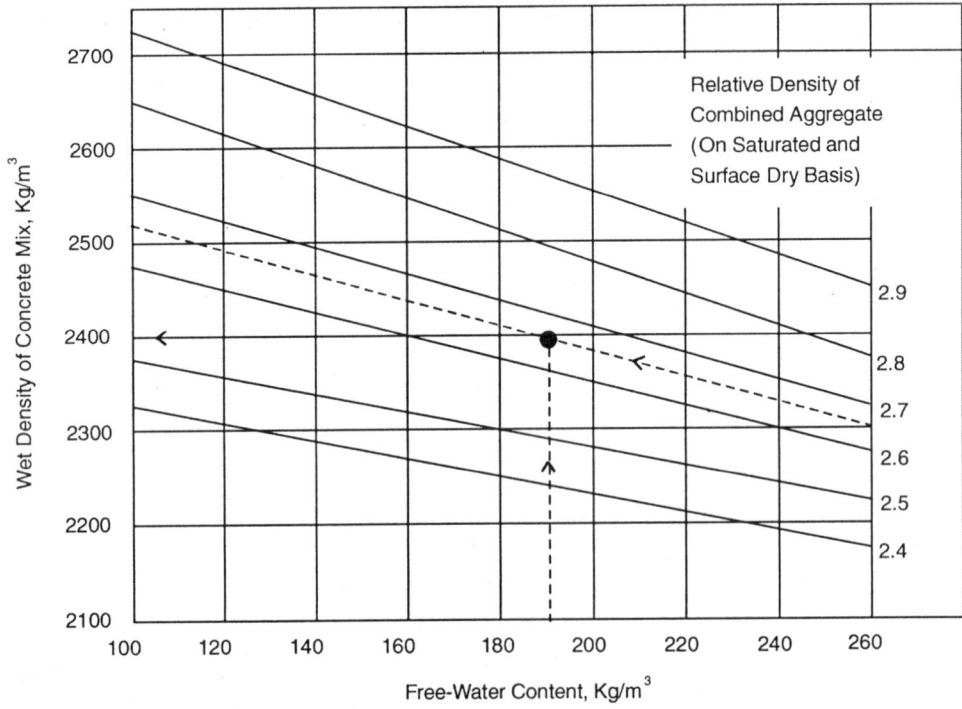

Fig. 8.4. Estimated Wet Density of Fully Compacted Concrete

Table 8.18 Approximate Compressive Strengths of Concrete Mixes made
with Water Cement Ratio of 0.50

Type of cement	Type of C.A.	Compressive Strength N/mm², Age (days)			
		3	7	28	91
Ordinary Portland cement or	Uncrushed	18	27	—	—
Type 40	—	—	—	—	48
Sulphate Resisting Portland cement	Crushed	23	33	—	—
Type 47	—	—	—	—	55
Rapid Hardening Portland Cement	Uncrushed	25	34	—	—
Type 46	—	—	—	—	53
	Crushed	30	40	53	60

*Ref. HMSO "Design of Normal Concrete Mixes –1975 (SP: 23-1982).

8.6.2 Example: 8.3

Design concrete mix **M20** grade of concrete by British method. The quality control at the site is expected to be fair and the placing conditions demand low degree of workability (slump about 25 mm). Use 20 mm maximum size crushed aggregate in saturated surface dry condition and sand of grading zone III and ordinary portland cement. Mild exposure is expected. Relative density of combined aggregate = 2.65.

Table 8.19 Approximate Water Contents (Kg/m³) Required to Give Various Levels of Workability*

Slump (mm)		0-10	10-30	30-60	60-180
Vee-Bee (S)		12	6-12	3-6	0-3
Max-size of Aggregate (mm)	Type of Aggregate				
10.	Uncrushed	150	180	205	225
	Crushed	180	205	230	250
20.	Uncrushed	135	160	180	195
	Crushed	170	190	210	225
40.	Uncrushed	115	140	160	175
	Crushed	155	175	190	205

* Note 1: For different type of C.A. and F.A., water content is estimated by

$W = 2/3\ W_{fa} + 1/3\ W_c$, Where, W_{fa}, W_c are water contents appropriate to type of F.A. and C.A. respectively & W is total water content.

Note 2: Ref. "Design of normal concrete mixes", 1975 HMSO (SP 23-1982)

Design of Mix

i. Target mean 28 days strength f_t = 20 + 1.65 × 5.6 = 29.24 = 30 N/mm² say.

ii. Referring Table 8.18, 28 days strength for O.P.C. is 47 N/mm² for W/C = 0.50, the curve falls between 3rd and 4th and is nearer to curve 3 from top in Fig. 8.3.

From this tentative curve, we get the W/C ratio 0.67. For mild exposure max W/C ratio = 0.65 (refer Table 8.5). Thus adopt W/C ratio = 0.65.

iii. Water cement for 20 mm crushed aggregate and low workability, 30 mm slump, say (Table 8.22) = 190 Kg/m³.

iv. Cement content = W/(W/C) = 190/0.65 = 292 Kg/m³.

v. From Fig. 8.4 and relative density of combine aggregate = 2.65 and free water content of 190 Kg/m³, the wet density = 2400 Kg/m³.

Total aggregate = 2400-190-292=1918 Kg/m³.

vi. For 20 mm max. size, 30 mm slump, sand grading zone III, and W/C = 0.65, percent of F.A. = 30% (Fig. 8.5).

Mass of F.A. = 0.30 × 1918 = 575 Kg.

C.A. = (1-0.30) × 1918 = 1343 Kg.

Thus the proportions of ingredients per m³ of concrete are:

Water	= 190 Kg	0.65
Cement	= 292 Kg	1.00
Sand (F.A)	= 575 Kg.	1.97
C.A.	= 1343 Kg	4.60

8.7 OTHER METHODS

8.7.1 Fineness Modulus Method

W/C ratio is determined for the required strength and the workability from various tables or curves.

The water needed is determined as in ACI method for the required workability and maximum size of aggregate.

Fineness modulus F_c for CA and F_f for fine aggregate is obtained by sieving. If the fineness modulus required for combined aggregate is F_m, the ratio of fine aggregate to coarse aggregate X can be determined as

$$X = \frac{(F_c - F_m)}{(F_m - F_f)}$$

Fine and coarse aggregates are combined in the ratio X:1 and the required F_m may be checked by sieving. Rest of the method is similar to ACI method.

The main **emphasis** in the method is on **proportioning of fine and coarse** aggregate for the minimum voids and desired workability.

8.7.2 Murdock's Surface Area Coefficient Method

This method is based on workability related to grading and shape of aggregate measured in the form of angularity index. The mix design is based on **surface area coefficients**. It is based on empirical index and lays too much **emphasis on small size aggregate particles.** No effect of shape is separately considered in workability but various parameters of mix are assumed to depend on angularity index. Other details remain the same as normal approaches.

8.7.3 Trial And Error Method

The values of average target strength, water-cement ratio and the desired workability are determined in the same way as in ACI or British or **Indian standard method.** The proportions between the constituents are determined by trial and error.

For finding the proportions of given **sand and coarse aggregate** to obtain the **maximum density** of combined aggregates, take a container of constant volume and fill it full by dry mix of sand and C.A. in certain proportions. **Find the unit mass** (weight) of mixed aggregates for various percentages of sand (say 25, 30,35, 40, 45 and 50) to total aggregate by mass. Shaking or vibration of the container shall be avoided to eliminate segregation of coarse aggregate. The dry mix should be as uniform as possible. The filled top of the container should be struck off level by means of a straight edge. Repeat the experiment till the sand percentages give maximum density of combined aggregates.

Plot a curve between the unit weight and the percentage of sand. Adopt the percentage of sand giving maximum unit mass (weight) for preparing concrete.

Adopt the proportions giving dense mixed aggregate and mix it with cement and water slurry (having required water-cement ratio corresponding to the specified target strength) till the desired workability (interms of slump, compacting factor, or Vee-Bee time) is obtained. Weigh the residue slurry and fix the proportions of materials. Prepare 10 cubes of 150 mm size and test them after 7 days standard wet curing. In this method three set of trial mixes may be

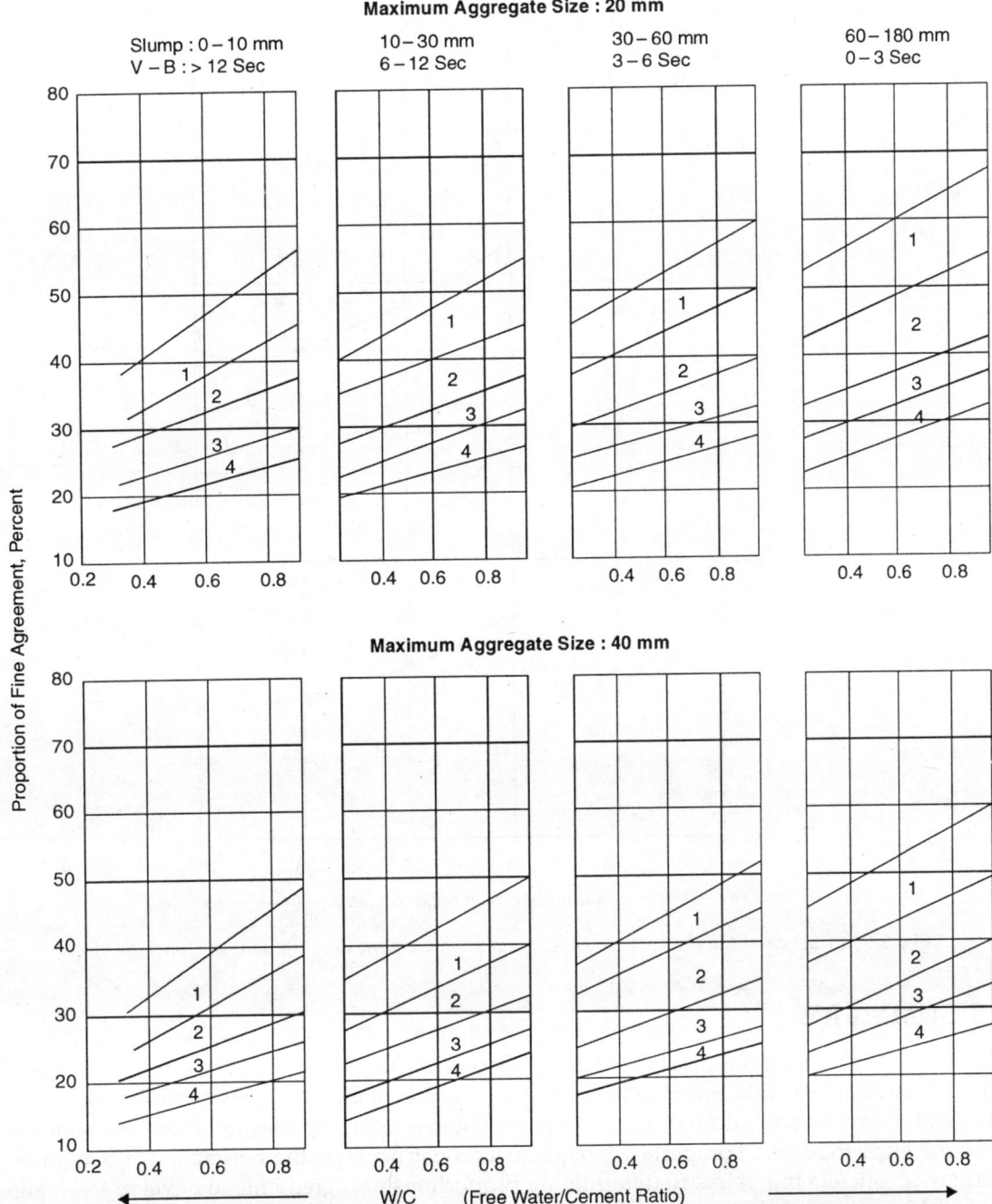

Fig. 8.5. Recommended Proportions of Fine Aggregate for Grading Zones 1, 2, 3 and 4 and 20 mm CA

prepared with higher and lower water-cement ratios to establish the desired strength as accurately as possible and corresponding W/C ratio. For economy in mix with different size of C.A. refer to Fig. 8.6 for guidance with reference to different W/C ratio curves.

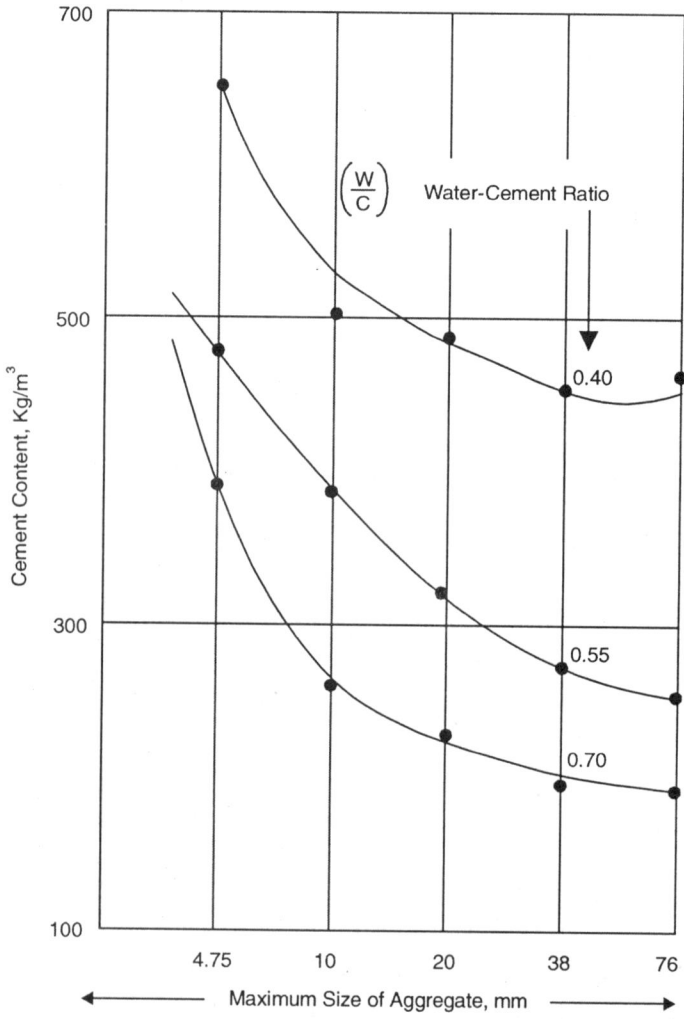

Fig. 8.6. Influence of Maximum Size of Aggregate on Cement Requirement of Concrete Mix

8.8 SUMMARY

Cement concrete has been most extensively used construction material due to research and technology developments. The developments have further made it a **highly controlled** scientific **material.** Cements and admixtures are now available in many grades to design the concrete mix to achieve the desired properties and quality of construction to the satisfaction of customers. Design of concrete mix refers to determining proportions of ingredients to develop fresh and hardened concrete of desired properties and quality.

Objectives of mix design are to achieve desired **workability, strength, durability** and **economy.** Mix design requires data on **workability, strength, placement** and **exposure** conditions, ingredients, quality levels at sites and derived data on W/C ratio, water or cement contents and aggregate grading and FA/CA ratios.

Based on various parameters, determine target mean strength, W/C ratio, water or cement content for the desired workability at site, FA/CA ratio and grading, total aggregate, and special admixtures, if any. Nominal mix proportions are assumed from the table of approximate proportions using basic principles and experience without elaborate calculations.

Various methods of cement concrete mix design are based on Indian standard guidelines, ACI, USBR, British and other approximate approaches. Most of the approaches require calculations of **target mean strength** based on expected quality **levels** at site, **W/C ratio** for the desired **target strength and durability, water content** for the desired **workability** at site for the available **size, shape, grading** and **type of aggregate**, ratio of FA/CA based on **fineness** of F.A. and **size and type of CA, cement content** from **water and W/C ratio**, total aggregate based on **absolute** volume of concrete mix including air solids and adjustments at site to compensate for actual site conditions.

Water content and **minimum sand contents** are decided from the tables based on **fineness of sand and size** and type of **coarse aggregate.** Minimum cement content and maximum W/C ratios are also compiled in tables on the basis of **durability** and past experience. Best method of **cement concrete mix design** is to adopt any of the scientific approaches and confirm the proportions by trials using available materials and actual site conditions using various principles of quality and mix design.

PRACTICE QUESTIONS

8.1   State objectives of **cement concrete mix design.**

8.2   List basic **variables** and **characteristic** data required for **mix design.**

8.3   Considering fine aggregate to coarse aggregate ratio as 1:2.5 and determine nominal cement concrete proportions for M I5 grade of concrete according to IS 456-1978 if minimum quantity of water required is **32 litres per bag** of cement and the **total aggregate should not be more than 400 kg/per bag** of cement.

8.4   Explain the effect of **bulking of sand** on its batching and moisture content.

8.5   List 4 common methods of concrete mix design.

8.6   Explain with graph **relationship of concrete strength, cement type and W/C ratio.**

8.7   Calculate **target mean strength** of concrete from required **characteristic strength** of 30 N/mm^2 for a site where **not more than 5% results should fall below** characteristic strength. Degree of quality expected at site is **very good.**

8.8   Describe briefly **11 steps** involved in **cement concrete mix design** considering **Indian Standard guidelines.**

8.9   Using various tables and curves from Indian standard guidelines, design a M30 cement concrete mix for a project expected to have **good quality control** and cement having 28 days **strength of** 53 N/mm^2. **Workability** desired is of **0.80 CF** using graded aggregate with **20 mm max. Size** and FA falls in **sand grading zone II.** Assume suitable conditions of aggregate and other data necessary.

8.10   Compare basic differences in **ACI approach and Indian standard guidelines** for cement concrete mix design.

8.11   Compare basic **differences in British approach and Indian standard guidelines** for cement concrete mix design.

8.12   Explain method of modifying water content required for a given **size of coarse aggregate** when coarse and fine aggregates are of different types.

8.13   State trend and effect of following factors on **concrete properties whether increasing or decreasing** when other factors remain the same.

Variable Factor	Workability of fresh concrete	Strength of hardened concrete
(a) Increase of water content	—	—
(b) Increase of fineness of cement	—	—
(c) Increasing of fineness modulus of combined aggregate.	—	—
(d) Increasing of W/C ratio	—	—
(e) Increasing the size of coarse aggregate	—	—
(f) Sand of grading Zone I instead of Zone II	—	—
(g) Sand of grading Zone IV instead of Zone II	—	—
(h) Use of plasticizers v/s no plasticizers	—	—
(i) Use of river shingle instead of crushed type CA	—	—
(j) Temperature rises beyond 40°C	—	—
(k) Temperature falls beyond –4°C	—	—

CHAPTER 9

Concrete Mix Quality Assurance Trials

LEARNING OBJECTIVES

The learner understands the importance of cement concrete trial tests for quality assurance and will be able to:

- Describe importance of trial tests in assuring desired properties;
- Interpret trial test results and variations from desired results;
- Determine field adjustments in mix proportions based on variations in trial test results using CRRI adjustment chart;
- Calculate adjustments and decide final proportions of cement concrete mix for the desired properties.

9.1 TRIAL MIXES

Material properties and the factors, which influence the properties of concrete, have already been studied in detail in previous chapters. There are large numbers of material parameters involved in achieving the concrete of specified quality and hence it is difficult to control these parameters theoretically on precise lines. It is therefore, very common practice to prepare trial mixes to ensure the desired quality of concrete (specially **workability** and **strength**). These trial mixes are planned and prepared in laboratory prior to production of design mix concrete at the job site. The materials for the trial-mixes in laboratory shall be the same as for the work site. The process of preparing trial mixes shall be also similar to that which is likely to be followed for the production of actual concrete at site.

Two most important properties of concrete to be ensured are (a) **workability** in fresh state and (b) compressive **strength** in hardened state. Both these properties shall be verified by preparing trial mixes from the designed proportions. For a given set of materials and conditions of test, both these properties are influenced directly by water content and water-cement ratio.

For ensuring the specified workability with the given ingredients, three trial mixes are prepared with three different water contents. First trial mix is prepared with the correct quantity of water content as determined in the basic design mix proportions, second trial mix is prepared with 5 % more water content than the design mix proportions and the third trial mix is prepared with 5 % less water content than the basic design mix proportions. Mix proportions are worked out for all the three trial mixes and concrete samples are prepared. From all the three samples, workability is measured in terms of slump, compacting factor, or Vee-Bee time. These workabilities are compared with the desired workability for the design mix. It may be noted that the aggregate used in the trial samples shall be in **saturated surface dry condition** prior to preparation. From the workabilities of these 3 trial samples, the exact quantity of water content shall be estimated by interpolation corresponding to the specified value of workability in terms of slump or compacting factor or Vee-Bee time. It is important to note that the workability increases with increase in water content. The quantity of water in basic design mix depends on the shape and maximum size of aggregate and is obtained from the empirical tables.

Having assessed the correct value of water content for the desired workability in the fresh state, **three trial mixes are prepared with the selected water content** and **three different water-cement ratios** to achieve the specified **target strength** at the site. The first trial mix shall be prepared on the basis of the correct water-cement ratio corresponding to the desired target strength using relevant **design strength curves** in any method. The second trial mix is prepared with 0.05 higher water-cement ratio and corresponding proportions with respect to the **same selected water content**. The third trial mix is prepared with **0.05 lower water-cement ratio** and corresponding proportions with respect to the same selected water content.

Second and third trial mixes are prepared to design proportions more precisely for the required strength at the site. At least 6 test cube specimens shall be prepared in each of 3 trial samples. It may be again noted that all aggregates shall be the same as those likely to be used at the actual work site and all these aggregates shall be brought to saturated surface dry condition prior to use.

9.2 INTERPRETATION OF TEST RESULTS OF TRIAL MIXES

When trial mixes are prepared by using the specified quantity of water and other basic ingredients and design mix proportions give workability as specified, then there is no need for other two trial mixes. In case the basic trial mix proportions give lesser workability, next fresh trial mix shall be prepared with 5% higher water content and workability shall be observed. This may provide the desired or higher workability. Water content may be interpolated for the desired workability.

In case **workability is more** than the specified value then prepare the next fresh trial mix with **5% less water** and measure the workability. **Interpolate** the correct water content from these observed values of workabilities and water contents. Thus estimate the correct value of **water content for the desired workability** for the available shape, size and grading of aggregates. The mix should also be observed for the tendency of segregation or bleeding, if any, and sand proportion also modified to make the mix cohesive.

With selection of correct water content for the specified workability of 3 trial mixes are prepared with **0.05 higher and 0.05 lower water-cement ratios** than the designed W/C ratio corresponding to target strength and 6 cubes are filled with each trial mix. Cubes are tested after suitable period of curing of 7 days or 28 days. Strength test results of all the 3 sets of trial concrete mixes are plotted with respect to corresponding water-cement ratio. The **water-cement ratio** required for providing the **target strength is interpolated** from this graph. Water content selected **for workability** and **W/C ratio for desired target strength** are adopted at site for the design mix.

9.3 DIRECT FIELD ADJUSTMENT (CRRI CHART)

9.3.1 Introduction to Chart

Owing to large variation in the quality of cement from different factories and supply of cement from different sources at the job site, the concrete technologist and site engineers find it extremely difficult to maintain the design strength of concrete mixes. Facilities to redesign the mixes with every fresh lot may not exist at all the sites. In such cases the direct field adjustment chart for concrete becomes very handy in making the adjustments in mix proportions on the basis of ratio of actual strength, with fresh cement and design field strength. Central Road Research Institute of India (CRRI) has developed adjustment chart shown in Fig. 9.1. This method envisages field adjustment of concrete mixes through alteration in water-cement ratio and aggregate-cement ratio with the use of a directly usable unique chart.

The adjustment **chart will be useful to counteract variations** in concrete compressive strength due to change in cement quality, and other factors such as human control and climatic factors.

The **aggregate-cement ratio adjustment curves** are based on **sp. gravity of 2.65 and 3.15** for aggregate and cement respectively. If there are marked variations in the specific gravity of either aggregate or cement, these curves may be suitably modified using the following equation of absolute volume –

$$V = (W + C/S_c + A/S_a) \cdot 1/1000$$

Where,

V = Absolute Volume of the wet mix (say 1 m^3)

W = Mass (weight) of water in Kg/m^3

C   = Mass (weight) of cement in Kg/m^3
A   = Mass (weight) of total aggregate in Kg/m^3
S_c, S_a = Sp. gravity of cement and aggregate respectively.

The chart provides adjustments in terms of increments and decrements necessary with reference to the original mix proportions. In the use of the chart, it is assumed that the original mix is known and is correctly designed and there are no subsequent changes in aggregate type, size and grading. The chart is applicable for all types of aggregates such as crushed (angular) and rounded (gravel). The chart is of direct utility to site engineer for controlling and maintaining better quality of concrete construction.

9.3.2 Method of Using the Chart

It is assumed that the site engineer obtains data of laboratory mix design proportions on the basis of average cement sample to be used. On getting new cement sample, the site engineer should prepare the concrete cube **specimens using the same proportions** including water-cement ratio according to the earlier designed mix. The strength of the specimens with new cement should be tested. If there is **substantial difference in strength**, the mix proportions shall be **redesigned** on the basis of an adjustment chart shown in Fig. 9.1. The chart is **applicable both for 7 days and 28 days** concrete **strength**. It may also be used for **adjustments with strength ratios** obtained on the basis of **accelerated curing by boiling.**

Determine the ratio of **actual field strength** and the **design field strength** (target strength) for concrete at **7 days**, 28 days or by **accelerated boiling test.** Mark this ratio on bottom horizontal axis and draw vertical projection line to intercept with **water-cement ratio adjustment curve.** From this point of intersection draw horizontal projection line to the left. The interception of this projection line with the vertical axis gives the **reduction or increment required in water-cement ratio.** If the **ratio of strength is less than 1.0** the **water-cement ratio shall be decreased.** Similarly for **strength ratio more than 1.0** the **water-cement ratio can be increased.**

From this point of intersection on water-cement ratio curve **move horizontally to intercept appropriate aggregate-cement (A/C) ratio adjustment curve** selected on the basis of total water content/m^3. From this point of interception on the selected curve, **move vertically upto** adjustment reading (increase or decrease) in **aggregate-cement (A/C) ratio on top horizontal axis.** For strength ratios of more than 1.0, the adjustment in A/C ratios increase, while for strength ratios less than 1.0, the adjustment in A/C ratios decrease. The A/C ratio adjustment curves are given for 5 different water contents ranging from 154 Kg to 202 Kg per m^3 of concrete. For intermediate values of water content, the **adjustment of A/C may be estimated by interpolation.**

The proportion of sand and coarse aggregate, the **grading and proportions of various fractions shall remain unchanged** and shall be kept the same as that of the original design mix.

9.4 FIELD (SITE) PROPORTIONS

From laboratory testing the mix proportions are finalized on the basis of the test results of preliminary trial mixes. The condition of the aggregates is assessed at site and necessary adjustments in the quantities of water and aggregate are made. If the aggregate is **dry and**

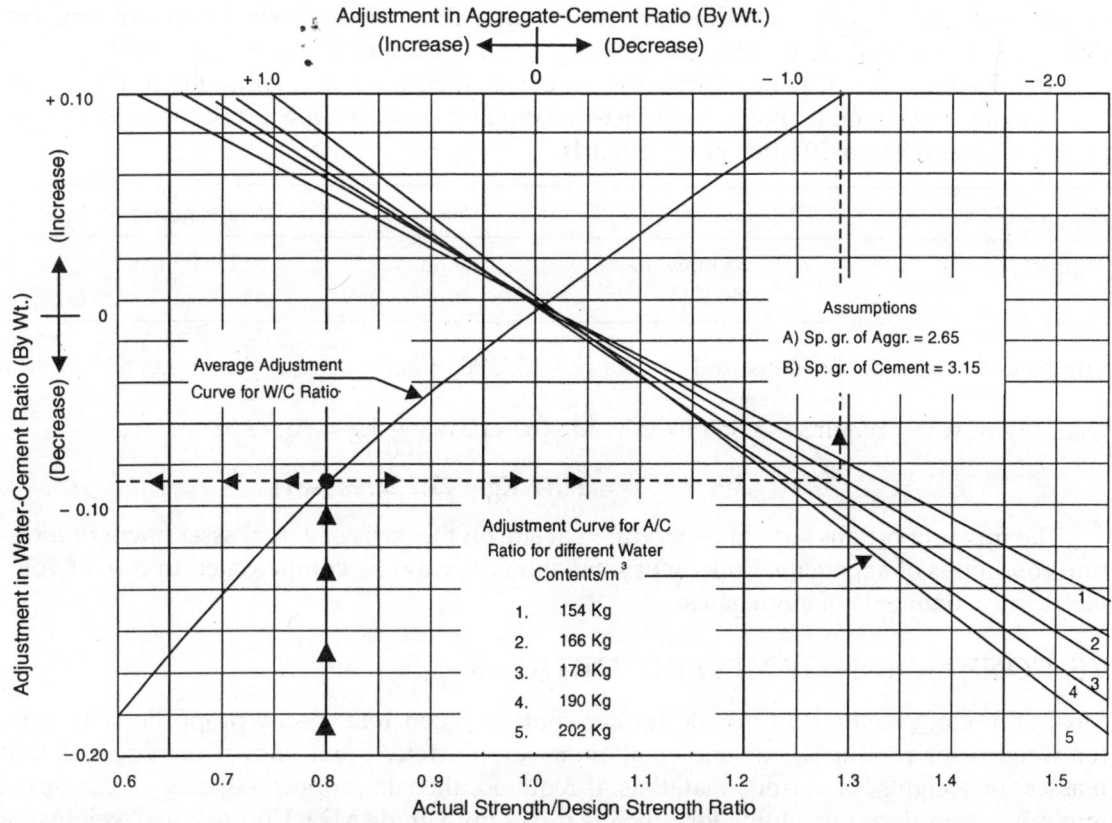

Fig. 9.1. Direct Adjustment Curves for Cement Concrete Mixes Using Different Indian Cement
to Counteract Variations in Compressive Strength in the Field

absorbs water, the quantity of **water to be mixed is increased** by the equivalent mass of water likely to be absorbed. If the aggregate contains the extra surface moisture over and above saturated surface dry condition, the quantity of mixing **water should be reduced** by this extra amount of surface water. Condition of coarse aggregate (CA) and sand should be assessed separately and necessary allowance in water be made accordingly. To ensure the correctness of moisture assessment, the **workability** of fresh concrete is **measured** at site and a necessary adjustment in water content is made so as to achieve the desired workability.

Another method of arriving at field proportions is to first bring the aggregate to standard conditions of moisture. The aggregate stacks are **sprinkled with enough water 24 hours before** use to make aggregate to absorb water and **bring it to saturated surface dry condition.** In such a case the water quantity to be added shall be same as computed in the mix design. Control on the quantity of water can always be kept by regular assessment of workability. Any deviation in **workability** from the desired value, should be **corrected by an appropriate change in the quantity of mixing water.**

Depending on the job, there may or may not exist suitable facility for weigh batching. In such a case, the batch proportions by mass are converted from mass to bulk volume and measurement of aggregate is carried out by volumetric containers (boxes) of appropriate size. The bulk density of aggregate is also regularly checked at site from time to time. Care should be taken to make appropriate **allowance for bulking effect of sand** when measurements are done by volume. Bulking effect in sand, if any, can be observed by using graduated glass jar at site daily 2 to 3 times. For example, at certain site, the sand has 1.50 Kg/litre bulk density, 15% bulking and contains about 3% surface moisture by mass. The mix proportions by volume are 1:1.8:3.8 and water 30 litres per bag batch.

Required Water	Cement	Sand	Coarse aggregate
30 litres	35 litres (50 Kg.)	63.0 litres	133.0 litres

The bulked sand shall be measured $= 63 \times (1 + 15/100) = 72.50$ litres at site (to get 63 l actual).

$$\text{Quantity of water required to be mixed} = 30 - (63 \times 1.5) \times \frac{3}{100} \text{ kg} = 30 - 2.83$$

$$= 27.17 \text{ Kg} = 27.2 \text{ litres (say)}$$

The mix proportions should be modified at site on the basis of actual assessment of moisture conditions of aggregate. Bulking of sand should always be compensated in case of volumetric measurements of aggregates.

9.5 CONVERSION OF MIX PROPORTIONS

Most of modern methods of mix design for controlled concrete specify proportions by mass (or weight) for production of one cubic metre of concrete. From known or observed unit masses (or weights) of various materials, if required, the site proportions may be converted into mix proportions by volume for concrete mixes upto grade M25. Unit masses (weights) of fine and coarse aggregates can be obtained at the site. Let, for example, designed mix proportions by mass are as under:

Water (Kg)	Cement (Kg)	Sand (Kg)	Coarse Agg. (Kg)	Unit of Concrete Batch
240	436	487	1170	1.0 m³

The bulk densities (unit masses) of cement, sand and coarse aggregates are respectively observed to be 1.44, 1.50 and 1.60.

The mix proportions by bulk volume shall be derived as under:

	Per 1 m³ concrete	Per bag cement batch (Vol)
Water	240/1 = 240 litres	27.73 litres
Cement	436/1.44 = 303 litres	35 litres
Sand	487/1.50 = 325 litres	37.5 litres
C.A.	1170/1.60 = 731 litres	84.5 litres

It may be noted that for quality work proportions by mass (weight) are preferred and when volume batching is used at site, proper precautions are taken to **compensate for bulking** effects.

Having discussed various adjustments in ingredients, their conversions and implications in case of normal concrete, the concreting under some of special conditions shall be briefly discussed to ensure quality control in special conditions of placement.

9.6 SUMMARY

For any cement concrete mix main concerns are **workability, strength, durability** and **economy.** Workability is measured during fresh state and compared with expected values. Adjustments are made in water content, aggregate grading, and cement content keeping W/C ratio constant. Three sets of samples are prepared for the desired workability and with W/C ratios equal, higher and lower than the calculated values. Accelerated strengths of all these samples are determined and W/C ratio corresponding to the desired strength is selected and adopted in the field.

In the field different materials (moisture conditions, grading and cement, etc.) vary in their conditions and require suitable field adjustments. For this CRRI had developed a field adjustment chart for adjusting cement concrete proportions at site. These charts facilitate continuous monitoring of quality of cement concrete at site by field engineers.

Measurement of various properties and trials of cement concrete at site facilitate assurance of desired properties and quality.

PRACTICE QUESTIONS

9.1 Describe importance of wokability trial tests in cement concrete construction.

9.2 Explain variations in expected strength results in field.

9.3 Explain main basis of field adjustments with the help of sketch.

9.4 Explain why there is need of measurement of workability at site.

9.5 Explain with simple example, how cement concrete mix proportions are converted from mass to bulk volume.

9.6 If change of cement at site has resulted in actual site strength to target strength ratio as 0.9, suggest likely change in W/C ratio and A/C ratio.

9.7 If the change in ingredients has resulted in 20% increase in strength ratio at site, suggest the likely changes in W/C and A/C ratios.

9.8 At certain site design mix proportions by volume are 1:1.6:3.0. Sand supplied at site has 5% moisture by mass and 25% bulking effect. If the mix uses 30 litres water per bag batch, find the correct proportions of the mix at site. Bulk densities of cement, sand and CA are respectively 1.44, 1.52, and 1.6 respectively.

9.9 Convert following mix proportions by mass into mix proportions by volume if the materials available are on surface saturated conditions. The bulk densities for cement, sand and CA are as 1.45, 1.5 and 1.60 respectively. Mass proportions are:

W	C	S	CA
270kg	435 kg	525 kg	1128 kg

CHAPTER 10

Quality of Cement Concrete Under Special Conditions of Placement

LEARNING OBJECTIVES

After studying this chapter, the learner understands quality considerations under special conditions and will be able to:

- Explain effect of high temperature on quality of cement concrete;
- Explain effect of low temperature on quality of cement concrete;
- Explain special **precautions and practices** to achieve good quality concrete construction in hot weather;
- **Calculate** temperature of **ice cooled concrete** mix from the given proportions and characteristics of ingredients;
- Explain special **precautions and practices** to achieve good quality concrete construction in **extreme cold weather** conditions;
- Explain basic **principles** to be adopted for achieving good quality concrete construction placed **under water**;
- Explain method of placing concrete under water;
- Explain special features of quality of **polymer impregnated** and **fibre reinforced concretes**;
- Explain **high performance concrete (HPC)**;
- Explain **self compacting concrete (SCC)**;
- Describe testing of SCC.

10.1 EFFECT OF TEMPERATURE ON CONCRETE

10.1.1 High Temperatures

Concreting during hot weather requires good planning. **High temperatures accelerate** the setting and hardening of concrete. At higher temperatures, **more mixing water** is generally required for achieving certain consistency. The concreting done at a temperature **beyond 40°C** is placed under this category and is usually not recommended without necessary precautions. The climatic factors affecting concreting in hot weather are **high ambient temperature** and **reduced relative humidity**. The effect of these conditions get enhanced with **increase in wind velocity**. The effects of hot weather are more critical during periods of **rising temperature**, falling relative humidity, or both. The various effects of hot weather on concreting are stated in subsequent paras.

(i) Accelerated Setting

A higher temperature of the fresh concrete results in the **rapid loss of water** from it, which leads to **accelerated setting**, reduces the **handling time** of concrete and **results into lowering of quality** of cement gel which ultimately **lowers the strength** of hardened concrete. Loss of water from the fresh concrete would necessitate the addition of more quantity of water, which if not added proportionately would **adversely influence the workability** and hence placement, consolidation and finishing. With the increasing in concrete temperature, the slump (workability) decreases and hence the water demand increases for constant consistency. Another disadvantage from the accelerated setting of concrete is the **formation of cold joints**.

(ii) Reduction in Strength

Concrete mixed, placed and cured at higher temperatures normally develops higher early strength than concretes produced and cured at normal temperatures. Tests have shown that for concretes placed and cured upto 28 days at various temperatures ranging from 49°C to − 4°C at 100% relative humidity, the initial strength was generally greater at higher temperatures. However, the difference in strengths at various temperatures tends to be narrower as the age of concrete increases, and indeed at an age of nearly one year the **low temperature concretes develop higher strengths** than those of higher temperatures (Fig. 10.1).

High temperature results in **greater evaporation loss** and hence necessitates increase of mixing water and consequently results in reduction of strength. It was also found that longer the delay between production of concrete and placing, greater is the strength reduction mainly due to **stiffening and inadequate compaction**.

(iii) Increased Tendency to Cracking

Rapid evaporation of water from the surface of the concrete may lead to **plastic shrinkage and cracking** and subsequently cooling of the hardened concrete would introduce tensile stress. It is generally believed that **plastic shrinkage occurs when the rate of evaporation is more than the rate at which bleeding water rises** to the surface. However, it has been recently found out that cracks also form under a layer of water and merely become apparent on drying. It has also been found from the results of road construction that intensity of **cracking is proportional** to the **maximum day temperature** during construction. **Cracking** is mostly confined to the **morning work**, which is attributed to the exposure of concrete to **direct sun and higher air temperature** for a longer time, than the concrete laid during the afternoon.

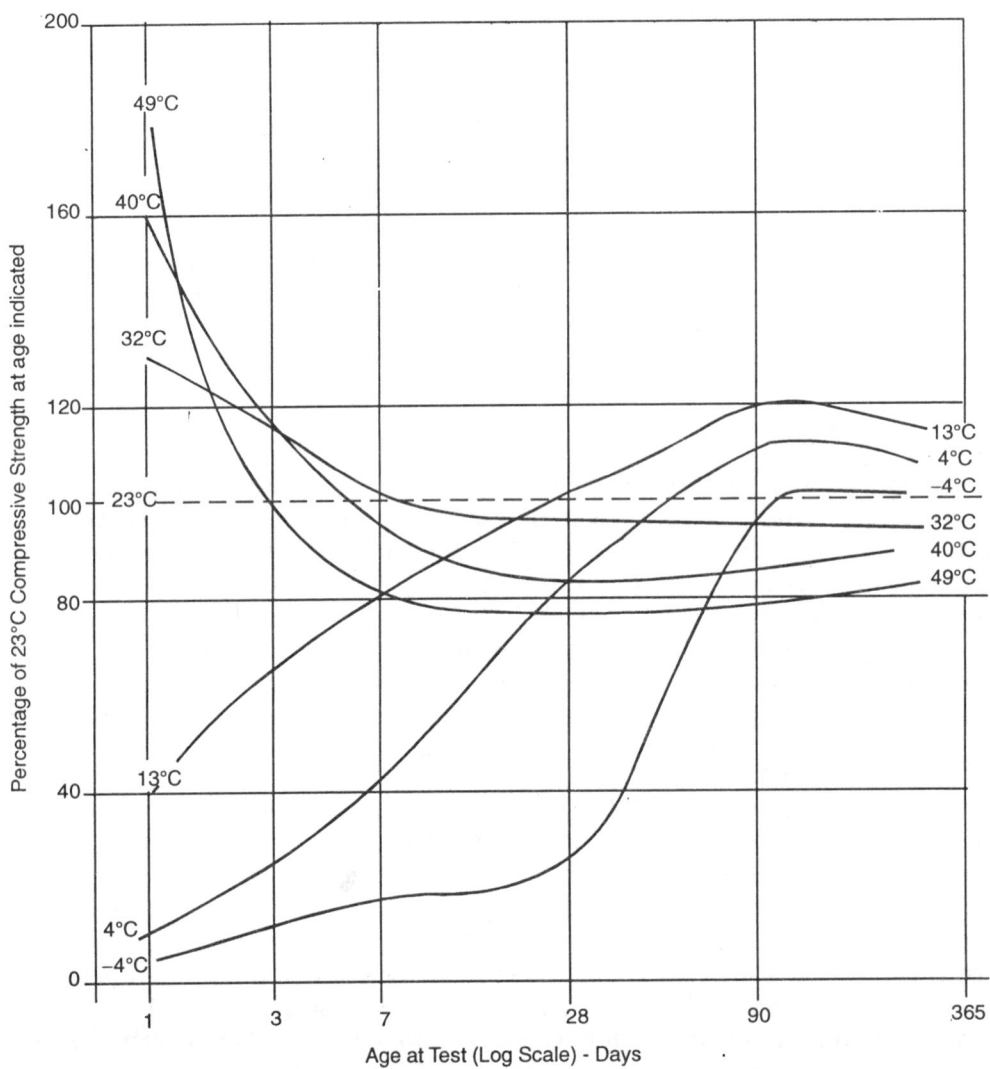

Fig. 10.1. Effect of Temperature During the First 28 Days on the Strength of Ordinary Portland Cement Concrete (W/C Ratio 0.41, Air Content 4.5%)

(iv) Rapid Evaporation of Water during Curing Period

In order to obtain concrete of higher strength and good physical structure (non-porous), it is necessary that the concrete be properly cured. The necessity of curing arises from the fact that hydration of cement can take place only in **water-filled capillaries**, so the water loss by **evaporation** from the capillaries must **be prevented**.

(v) Difficulty in Control of Air Content in Air-Entrained Concrete

It is more difficult to control air content in air-entrained concrete at higher temperatures. This also adds to the difficulty of controlling workability. For a given amount of air-entraining agent, the concrete at higher temperature will entrain less air than concrete at normal temperature (**Table 6.5**).

In the I.S. 7861 (Part I)-1975 on hot weather concreting, there is a graph to determine the loss of water by evaporation (to be compensated for keeping the **workability** same) as a **function of temperature, relative humidity and wind velocity**. This may be used for adjusting water content as per prevailing site conditions.

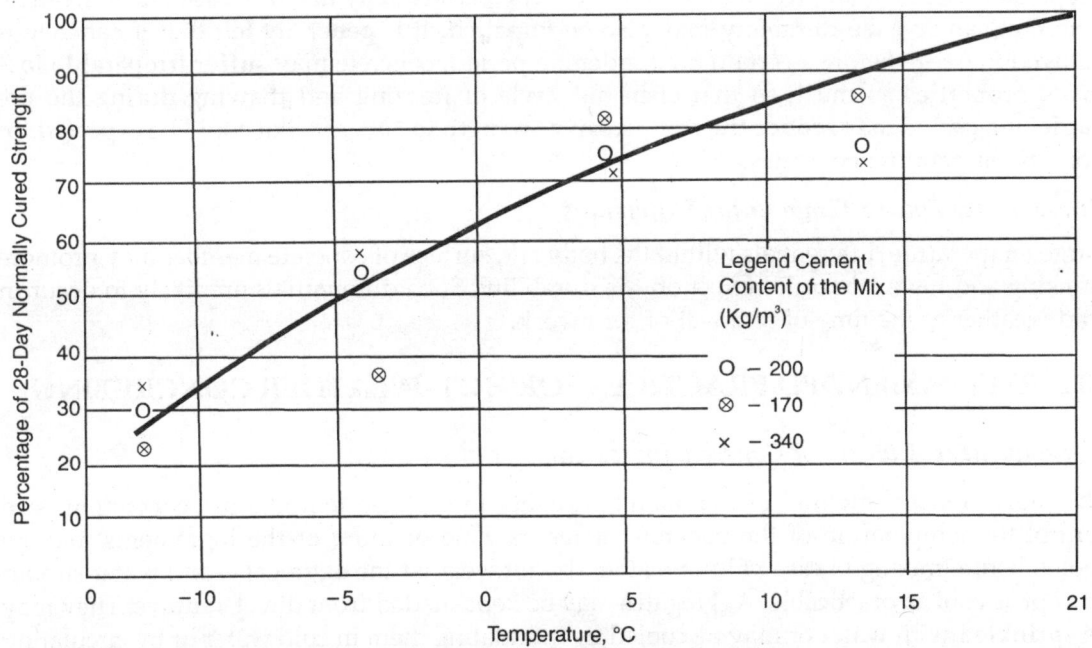

Fig. 10.2. Effect of Low Temperature on Compressive Strength of Concrete

10.1.2 Low Temperatures

Production of concrete in cold weather introduces special problems, which do not arise while concreting at normal temperatures. The problems are mainly due to slower development of concrete strength. If the concrete in the plastic stage is exposed to low temperature, **ice crystals** are formed which cause **expansion in the pore** structure. This subsequently causes damage and disintegration of concrete due to **alternate freezing and thawing** of the hardened concrete. Thus concreting operations carried out **below 5°C are termed as cold-weather concreting**. The effects of cold-weather concreting are described in subsequent paras.

(i) Delayed Setting and Hardening

When the temperature is falling to about **5°C or below**, the hydration of cement and development of strength is retarded compared with the strength development at normal temperatures (15-20°C). The hardening period necessary before removal of formwork is thus increased and the practices of concreting at normal temperature cannot be transformed to low temperature directly. Effects of temperature of concrete on the strength development can be expressed as in Fig. 10.2. Although the initial strengths of concrete placed at low temperature are lower, it has now been established that the long-term strengths of concrete are **not severely affected** provided that the concrete has been prevented from freezing during the fresh state and early life.

(ii) Freezing of Concrete at Early Stages

When concrete is exposed to freezing temperatures there is the risk of concrete suffering irreparable loss of strength and other qualities. The permeability may increase due to freezing temperatures and the durability may also be impaired. It is generally felt that if concrete is allowed to freeze before a certain pre hardening period, concrete may **suffer irreparable loss in its properties** so much so that even one cycle of freezing and thawing during the pre hardening period may **reduce the compressive strength to 50%** of what could be expected for normal temperature concretes.

(iii) Stresses Due to Temperature Differences

Large temperature differentials within the body and surface of concrete member may promote cracking and have a harmful effect on the durability. Such differentials are likely to occur in cold weather at the time of removal of formwork.

10.2 RECOMMENDED PRACTICES FOR HOT–WEATHER CONCRETING

(i) Temperature Control of Concrete Ingredients

The most practical method of maintaining quality in extreme temperature concreting is to control the temperature of the concrete materials. One or more of the ingredients may be cooled before mixing in case of hot weather. In hot weather the aggregates and water should be kept as cool as practicable. Aggregates may be kept shaded from direct sunrays. They may be **sprinkled with water** or may be cooled by inundating them in cold water or by circulating refrigerated air through pipes or other suitable methods. **Mixing water has the greatest effect on lowering the temperature** of concrete, as its specific heat is nearly 5 times more than that of common aggregate. Reduction in water temperature is generally caused by mechanical refrigeration or by mixing with crushed ice. Ice can also be directly used as part of the mixing water provided it is completely melted by the time mixing is completed. Cement temperature has only a minor effect on the freshly mixed concrete because of cement's low specific heat and relatively small quantity in the mix. Wherever mass concreting has to be done, circulation of refrigerated water through network of pipes should be done to **dissipate heat from the body** of the freshly mixed mass concrete.

Temperature of the aggregates, water and cement shall be maintained at the lowest practical levels so that the **temperature of resulting concrete is below 40°C** at the time of placement. The temperature of the concrete is calculated by the following formula:

(a) Cold water as mixing water (without ice)

$$T = [S(T_a W_a + T_c W_c) + T_w W_w + T_{wa} W_{wa})]/\{S(W_a + W_c) + W_w + W_{wa})\}$$

b) With ice added to mixing water

$$T = [S(T_a W_a + T_c W_c) + (W_w - W_i)T_w + W_{wa}T_{wa} - 79.6 W_i]/\{S(W_a + W_c) + W_w + W_i + W_{wa}\}$$

Where, T = Temperature of freshly mixed concrete (°C)

T_a, T_c, T_w, T_{wa} = Temperature of aggregate, cement, added mixing water, and free water on aggregate, respectively (°C)

W_a, W_c, W_w, W_{wa}, W_i = mass of aggregate, cement, added mixing water, free water on aggregate and ice respectively (kg), and

S = Specific heat of cement and aggregate (K Cal/Kg°C)

'S' = for aggregate and cement are generally taken as 0.22 K cal/kg°C, while for water it is 1.0 K cal/kg°C

(ii) Transporting, Placing, Compacting, Finishing, Curing and Protection

Transporting, placing and compaction should be done as quickly as practicable during hot weather. The containers should be as close as possible to minimize evaporation loss. Delays contribute to loss of slump and an increase in concrete temperature. Enough workmen should be available to handle and place concrete immediately upon delivery. All steps in **finishing should be taken promptly** but not before the water sheath disappears or before the concrete can support the weight of workmen. Prolonged mixing, even at agitating speed, should be avoided.

Since concrete hardens more rapidly in hot weather, extra care in placing techniques is required to **avoid cold joints and plastic shrinkage cracking.** Wooden form should be sprayed with water, so that it may not absorb part of the mixing water. Continuous moist curing is preferred during hot weather. Curing should commence as soon as the surface is finished and covered with gunny bags to continue for at least 24 hours. If moist curing is not continued beyond 24 hours, the concrete should be covered with curing paper or with heat reflecting plastic sheets while the surface is still damp.

In case of hot, dry and windy days, temporary sunshades and windbreakers must be erected. **Cover the fresh concrete** surface with wet hessian cloth.

(iii) Admixtures

Retarding admixtures are sometimes used during hot weather to **delay the setting time** of concrete and lessen the need for an increase in mixing water. **Water-reducing retarders** may be beneficial provided they do not interfere with strength development and other properties of the concrete and provided their use is carefully controlled. They should only be used to supplement other hot-weather concreting procedures. Sugar upto a maximum of **0.05% by mass of cement** may be used to effectively retard setting. It also improves strength. Higher percentage of sugar may prove to be **harmful** and **fatal.**

(iv) Time of Concreting

Concreting should be done during **late evenings to minimize the exposure** of fresh concrete to direct sun.

(v) Prior Preparation

All machines such as mixers, chutes and hoppers should be painted white or covered with wet hessian cloth as far as possible.

All materials and equipment must be well prepared before use. Forms, reinforcing steel, and subgrade must be sprinkled with water. The whole environment where concrete is to be laid must be wetted to improve relative humidity around the concreting site.

(vi) Type of Cements

It is preferable to use low heat portland cement, portland pozzolanic cement and portland blast furnace slag cement.

10.3 RECOMMENDED PRACTICES FOR COLD-WEATHER CONCRETING

(i) Temperature Control of Concrete Ingredients

The best method to keep concrete temperature above the permissible minimum is by controlling the temperature of the ingredients. All available means shall be used for maintaining these materials at as high temperature as practicable. Heating of the aggregates shall be such that frozen lumps, ice and snow are eliminated and at the same time, overheating is avoided. The average temperature of an aggregate for an individual batch **shall not exceed 65°C.** The mixing water shall be heated under such a control and in sufficient quantity as to avoid appreciable fluctuation in temperature from batch to batch. The required temperature of mixing water to produce specified concrete can be obtained from Fig. 10.3 for the given cement content and W/C ratio. The heated water shall come in direct **contact with aggregate first** and not in contact with cement. Water having temperature upto the boiling point may be used provided the aggregate is cold enough to reduce the temperature.

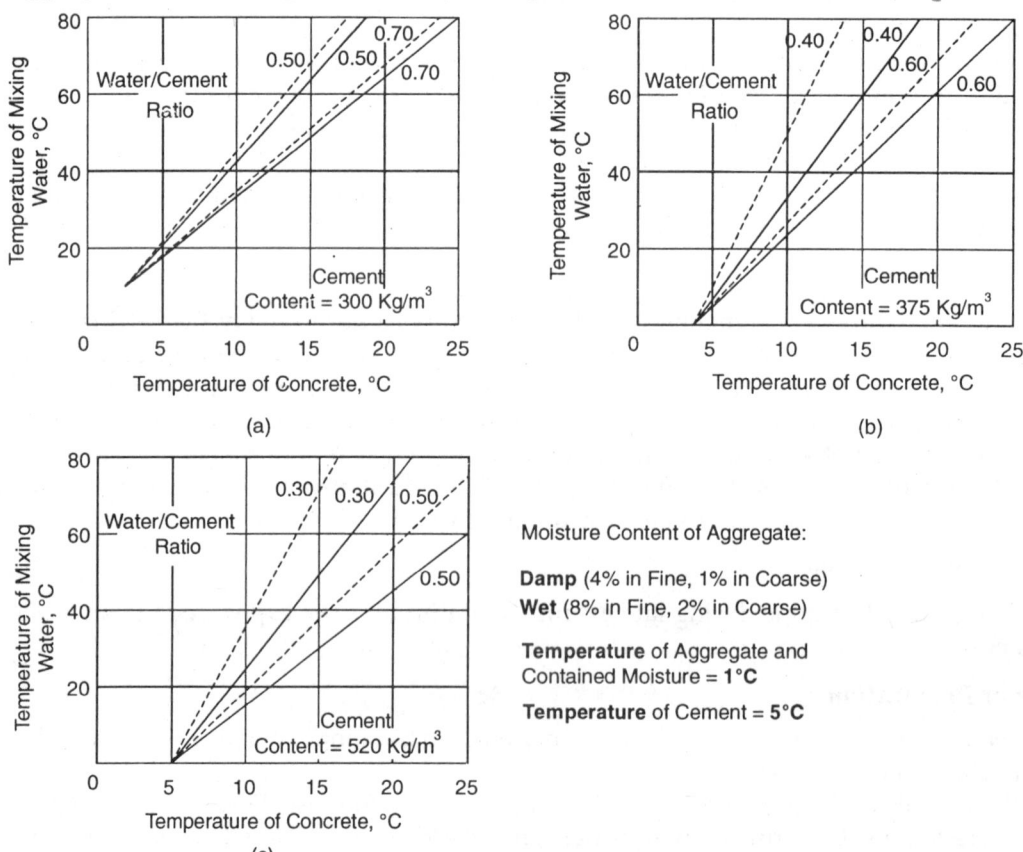

Moisture Content of Aggregate:

Damp (4% in Fine, 1% in Coarse)
Wet (8% in Fine, 2% in Coarse)

Temperature of Aggregate and
Contained Moisture = **1°C**

Temperature of Cement = **5°C**

Fig. 10.3. Required Temperature of Mixing Water to Produce Heated Concrete

(ii) Use of Insulating Form Work

Sufficient amount of heat is generated during hydration of cement. Such heat can usefully be conserved by having insulated formwork with covers which may maintain the concrete temperature above the desirable limits for the first 3 days and even when the ambient temperatures are lower. The formwork covers can be of timber, clean straw, blankets, sacking, tarpaulins and plastic sheeting, in conjunction with air gap as insulation. The efficiency of insulating material depends upon the thermal conductivity of the medium as well as on the ambient temperature conditions. For moderately cold weather, timber formwork alone is sufficient and is preferable to steel formwork. The following comparison of different insulating materials (refer ACI 306-1966) indicates how the efficiency of different combinations varies due to their coefficient of thermal conductivity. The insulating values are indicated with reference to 25 mm commercial blankets as reference.

	Insulation Materials IS 7861 (Pt. II) -1981		Equivalent Thickness 'mm' (with reference to 25 mm thick commercial blankets)
a)	25 mm commercial blanket	—	25
b)	25 mm loose-fill insulation of fibrous type	—	25
c)	25 mm insulating board	—	20
d)	25 mm saw dust	—	15
e)	25 mm timber	—	8
f)	25 mm damp sand	—	0.6

(iii) Proportioning of Concrete Ingredients

Since the quantity of cement in the mix affects the rate of increase in temperature, additional quantity of ordinary portland cement, rapid hardening portland cement or accelerating admixtures used with proper precaution can help in getting the required strength in a shorter period. Air-entraining agents are generally recommended for use in cold weather. Air-entrainment increases the resistance of hardened concrete to freezing of fresh concrete. In India, calcium chloride has not been successfully used due to non severe winter conditions and its **corrosive action** on the reinforcement.

(iv) Placement, Curing and Protection

All ice or snow and frost should be removed from the base or formwork before the concrete is placed. It should be kept in mind that no amount of insulation can supply heat at below freezing point and that the surface on which the concrete is to be placed along with the steel to be used should be kept sufficiently warm. Whenever it is proposed to place concrete at or **below 2°C**, it is essential to know that the time taken for the concrete temperature to fall to freezing point is atleast equal to the minimum pre-hardening period. During periods of freezing or **near-freezing conditions, water curing is not necessary. Freezing and thawing in the concrete should be avoided.** It should be kept in mind that the temperature of the concrete should **not fall below – 4°C**, otherwise reaction and subsequent hardening shall not take place at all.

(v) Delayed Removal of Formwork

Because of **slower rate** of gain of **strength** during the cold weather, the formwork and props have to be kept in place for longer time than in usual concreting practice. The appropriate time for removal of formwork may be ascertained from the **strength of test cubes left at site under** the **same conditions** of temperature and humidity as the structural element is concerned. The time of removal of formwork, thus arrived should not be less than the duration of necessary protection required by the formwork. As a general guidance, the minimum time limits for stripping of formwork of members carrying only its own weight, and at air temperature of about 3°C are given below:

Elements	Using OPC (Days)	Using RHC (Days)
Beam sides, walls, columns	5	3
Slabs (props left under)*	7	4
Beam soffits (props left under)	14	8
Removal of props to slabs	14	8
Removal of props to beams	28	16

(vi) Admixtures

Admixtures used in cold-weather concreting should be such that they accelerate the setting and strength development of concrete. **Calcium chloride, sodium silicate** and sodium hydroxide are some of the well-known admixtures. However, calcium chloride is the most commonly used admixture, but the **excess use** of this may cause a **flash set**. The use of **calcium chloride is avoided** in lightweight insulating concrete and where **sulphates are present** in the aggregate.

(vii) Type of Cement

It is preferable to use Rapid Hardening cement, High Alumina cement which produces more heat during hydration and also accelerates the process of hardening.

10.4 UNDER-WATER CONCRETING

10.4.1 Basic principles

Following principles should be adopted for under-water concreting:

(i) It is necessary that the concrete shall contain **10% extra cement** than that required for the same mix placed in the dry conditions to compensate the part of the cement which may be lost in water. This may also change water-cement ratio and hence change in its strength.

(ii) As **compaction** of the concrete under water **cannot be carried** out by vibrators, so it becomes essential that concrete to be laid should have the **self-compacting properties** under its own weight. For this it is required that concrete should be sufficiently workable. IS: 456-2000 limits the minimum and maximum slump for the concrete to be laid under water at **100mm** and **180mm** respectively.

 (iii) Concrete prepared for under water operation should be **placed in the final position** without any disturbance by providing cofferdams around the element casted.
 (iv) **Admixtures** may also be used at times, so as to **accelerate** the setting and hardening of the concrete.
  (v) Rate of hardening of the concrete can also be increased by using **cements of different quality** (for example **High Alumina** cement) depending upon the placing conditions.
 (vi) The area to be concreted should be **enclosed by cofferdams,** to ensure that water does not seep into the enclosure and also the **velocity of water is reduced** to less than 3 m/min. **Dewatering by pumping shall not be done while concrete is being placed or until 24** hours thereafter.
(vii) Concrete shall be **deposited continuously** until it is brought to the required height and size.
(viii) Concrete in seawater shall be atleast **M15 grade** in the case of plain concrete and **M20 or M25 in case of reinforced concrete.** The use of slag or pozzolana cement is advantageous under such conditions.
 (ix) The concrete to be used in sea-water should be designed so as to get the **densest possible concrete** in which the use of **slag, broken-brick, soft limestone, soft sand stone** or other porous or weak aggregates are avoided.
  (x) As far as possible precast members unreinforced, well cured and hardened, without sharp corners be used in case of concreting in seawater.
 (xi) No construction joints shall be allowed within 600 mm below low water level or within 600 mm of the upper and lower planes of wave action in sea.
(xii) For reinforced concrete structures to be used in seawater, care shall be taken to protect reinforcement from exposure to saline atmosphere during storage and fabrication.

10.4.2 Placing Methods

The methods to be used for depositing concrete in or under water shall be one of the following:

(i) Tremie

When concrete is to be deposited under water by means of a tremie the top section of the tremie shall be a hopper large enough to hold one entire batch of the mix or the entire contents of the transporting bucket, if any. The tremie pipe shall be not less than 200 mm in diameter and shall be large enough to allow a free flow of concrete and strong enough to withstand the external pressure of the water to which it is subjected, even if a partial vacuum develops inside the pipe. A separate lifting device shall be provided for each tremie pipe with its hopper at the upper end. Unless the lower end of the pipe is equipped with an approved automatic check valve, the upper end of the pipe shall be plugged with a wadding of the gunnysack or other approved material before delivering the concrete to the tremie pipe through the hopper. When the concrete is forced down from the hopper to the pipe, it will force the plug down the pipe and out of the bottom end, thus establishing a continuous stream of concrete. It will be necessary to **raise the tremie slowly** in order to cause a **uniform flow** of the concrete, but the tremie shall not be emptied so that water enters the pipe. At all times after the placing of concrete is started and until all the concrete is placed, the lower end of the tremie pipe shall be below the top surface of the fresh concrete. This will cause the concrete to build up from below instead of flowing out over the surface, and thus avoid formation of laitance layers. If the charge in the tremie is lost while depositing, the tremie shall be raised above the concrete surface, and unless sealed by a check valve, it shall be replugged at the top end, as at the beginning before refilling for depositing concrete.

(ii) Drop bottom bucket

The top of the bucket shall be covered with a canvas flap. The bottom doors shall open freely downward and outward when tripped. The bucket shall be filled completely and lowered slowly to avoid backwash. The bottom doors shall not be opened until the bucket rests on the surface upon which the concrete is to be deposited and discharged. The bucket shall be withdrawn slowly until well above the concrete.

(iii) Bags

Bags of atleast 0.028 m^3 capacity of jute or other coarse cloth shall be filled about two-thirds full of concrete. The spare ends are turned under so that bag is square ended and securely tied. They shall be placed carefully in header and stretcher courses so that the whole mass is interlocked. Bags used for this purpose shall be free from deleterious materials.

(iv) Grouting

A series of round cages shall be prepared and made from 6mm steel bars and 50 mm mesh extending over the full height laid vertically over the area to be concreted so that the distance between centres of the cages and also to the faces of the concrete shall not exceed one metre. Stone aggregate of not less than 50 mm nor more than 200 mm size shall be deposited **outside the steel cages over the full area and height** to be concreted with due care to prevent displacement of the cages.

A suitable **1:2 cement: sand grout** with a water-cement ratio of **not less than 0.6 and not more than 0.8** shall be prepared in a mechanical mixer. This grout is set down under pressure through 38 to 50 mm diameter pipes terminating into **steel cages**, about 50 mm above the bottom of the concrete. As the grouting proceeds, the pipe shall be **raised gradually** up to a height of not more than 600 mm above its starting level after which it may be withdrawn and placed into the next cage for further grouting by the same procedure. The process is continued till the completion of concrete.

10.5 SPECIAL CONCRETE

10.5.1 Polymer Concrete (A New Emerging Building Material)

Polymer concrete consist of 5 to 10% polymer binder, of total weight of polymer concrete and mineral filler such as aggregate, gravel and crushed stone. Polymer concrete also known as resin concrete is composite material in which the binder consists of **synthetic organic polymer**. Polymer binder can be thermo plastic or more frequently thermo setting.

I Salient Features of Polymer Concrete

Polymer concrete as compared to ordinary Portland cement concrete has:

  (i) **Higher Strength**
 (ii) Greater **resistance to chemical** and corrosive agents
(iii) Has **lower water absorption**

(iv) Higher **freeze thaw stability**
(v) **High density**
(vi) **High dielectric** characteristic
(vii) **Damping** characteristic
(viii) **Less shrinkage**

Polymer Concrete is used in selected situations and has very restricted use due to high cost.

II Properties of Polymer Concrete v/s Cement Concrete

The compressive strength of polymer concrete is near about **four times** that of conventional cement concrete. Its **tensile strength** is also about four times and the **modulus of elasticity** is about two times. The **modulus of rupture** has been found to be almost **four times**. The flexural modulus of elasticity is about 1.5 times and the **creep deformation is about 10% only**. The hardness and impact values have been observed to be more than two times. The water **permeability becomes negligible** and water absorption is greatly reduced.

III Application of Polymer Concrete

Polymer concrete is basically used for the following purposes:

(i) Construction of **underground** engineering **structures.**
(ii) **Covering facades.**
(iii) **Covering floors** and **making roads** (Particularly for **air strips** where construction time is of prime importance).
(iv) Sound **insulating** materials.
(v) Agriculture and Horticulture in pipe construction.
(vi) **Load bearing supports** like props and ties in industrial structures.
(vii) Polymer plaster for **interior decoration.**
(viii) Residential and other civil engineering constructions.

IV Limitation of Polymer Concrete

It is well known that one of the vital drawbacks of Polymer Concrete is their comparatively **low thermal stability** (80°C – 120°C). Some binders like organisilicon and other binders, can be used to produce Polymer Concrete with a thermal stability up to **600°C** or more.

V Classification of Polymer Concrete

Special Concrete based on Polymers are divided into four main categories depending on the composition and the method of production

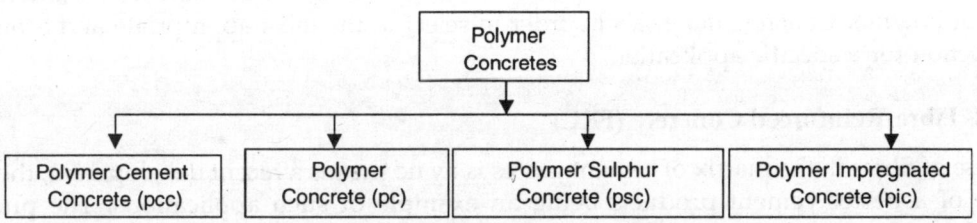

Fig. 10.4. Classification of Polymer Concretes

(i) Cement Concrete modified by polymers – **polymer cement concrete (PCC)**
(ii) Concrete based on polymer binders – **Polymer concretes (PC)**
(iii) Concrete modified by sulphur polymers – **polymer sulphur concrete (PSC)**
(iv) **Cement Concretes impregnated with monomers** or oligomers – **Polymer Impregnated Concrete (PIC)**

(i) **Polymer Cement Concrete (PCC) :** These cement concretes are prepared by adding **water soluble polymers** during the production. **Poly vinyl acetate** or water soluble epoxide resins are added to the mix to the extent of **2 – 20% by mass.**

(ii) **Polymer Concrete (PC) :** Polymer Concrete denote compositions based on **synthetic resins** or **monomers** with chemically stable fillers and aggregates. The Polymer Concretes consists of large quantity of fillers and aggregates. The main properties of Polymer Concretes are exhibited by the chemical nature of synthetic resins and the finely dispersed filler fraction.

(iii) **Polymer Sulphur Concrete (PSC) :** These concretes are prepared with a sulphur binder during the production of which modifying additives such as dicyclopentadiene or **chloroparaffin** are added to the **molten sulphur in proportions of 1 – 15%.**

(iv) **Polymer Impregnated Concrete (PIC) :** These are cement concretes which after placing, drying and curing are impregnated with various monomers which fill the pores of concrete to produce a non porous concrete.

The **impregnation** of cement concrete with monomers or oligomers ensures the possibility of producing **high density and strength** characteristics.

Polymer-impregnation has made it possible to develop high-grade concrete. The conventional concrete of **35 N/mm^2 strength**, the polymerization raises the compressive strength to **140 N/mm^2**. With high silica cement, stone aggregates and high temperature steam curing, the compressive strength can be developed upto **270 N/mm^2** and tensile strength of **25 N/mm^2**. Water absorption is reduced by 80 to 90 percent, and the freeze-thaw resistance is substantially improved. Further, polymerized concrete exhibits essentially **zero creep properties.** It transforms conventional concrete from plastic material to almost an elastic one with almost twice the modulus of elasticity.

VI *Scope of Polymer Concrete in INDIA*

Furan Polymers are based on for furfuryl alcohol, which is derived from agricultural residues such as corn hobs, rice hulls, oat hulls or sugar cane bagasse. In India, being an agricultural country, there is sufficient amount of raw material available for furan resin.

Polymer should be used only in application in which the higher cost can be justified by superior properties or low energy requirements during processing and handling. It is therefore important that architects and engineers have some knowledge of the capabilities and limitations of Polymer Concrete materials in order to select to the most appropriate and economic production for a specific application.

10.5.2 Fibre-Reinforced Concrete (FRC)

The use of fibres in the matrix of cement pastes is by no means a recent development, the long usage of asbestos cement products being an example of such applications. The purpose underlying the introduction of fibres into the matrix of a cement paste has been to **increase tensile and flexural strengths**, to **improve impact** resistance and to **control cracks.** Recently several materials other than asbestos such as steel, carbon, plastics and glass have been tried as fibres in concrete, and dramatic improvement in properties have been observed with some of them. For example, steel fibres, upto about 4 percent by volume, were found to increase both the first crack flexural strength and splitting tensile strength of concrete by upto 2.5 times. The dynamic strength of concrete reinforced with different types of fibres and subjected to explosive charges and dropped weights increases atleast fivefold. Improvements are also recorded in **fatigue strength, resistance to corrosion** by salt water, thermal conductivity and abrasion resistance.

Fibre-reinforced concrete has potential for applications in **pavement overlays, blast resistant structures, marine structures, shell structures** and **refractory linings** due to superior qualities/properties.

10.6 MODERN TRENDS IN CEMENT CONCRETE*

10.6.1 Introduction

With the use of **chemical admixtures** and development of cement concrete, the construction technology is getting revolutionized. Investigations and **research** are going on to develop **high performance structures** to resist **extreme conditions**. Work on **high performance concrete (HPC)** and **self compacting concrete (SCC)** was first started systematically in **Japan in 1988** to construct **high durability** concrete structures under extreme conditions. The new materials (**HPC & SCC**) are being used in many countries including India. The R & D in **HPC** and **SCC** along with cement manufacturing technology are going to **revolutionize the concrete technology**. In time to come **HPC** and **SCC** are likely to become **standard normal concrete**.

With the development of **chemical admixtures** and HPC, it is now possible to construct structures in **extreme conditions**. HPC is required to display high quality properties both in **fresh (plastic) state** and **hardened state**. Important properties during fresh state pertain to **workability** and **compatibility**. During **hardened state** main properties of **performance** are related to **strength** (resistance to **forces**), **Durability** (resistance to **deterioration**) and **Impermeability** (resistance to **moisture movement** and dampness).

10.6.2 High Performance Concrete (HPC)

As explained earlier **HPC** (high performance concrete) can be produced by using **suitable admixtures**. **Super plasticizers** in suitable proportions are used to achieve high degree of **workability** leading to **self compacting concrete (SCC)** with high performance in other properties also.

Another important aspects of **HPC** are its **mechanical properties** (viz **strength**, bond, dimensional changes, etc), **durability** and **impermeability. Strengths** as high as **200 Mpa** in case of **ultra high performance concrete** can now be achieved by use of suitable admixtures and modern technology. Present conventional technology can produce concrete grades of M50 to M60. Ultra high performance concrete (**UHPC**) can even **eliminate** the use of **reinforcement** in some structures. **Durability** of concrete structures suffer due to **corrosion of reinforcement** caused by **penetration and movement of air and moisture** through **permeable concrete cover**.

---

* EFNARC, Association House, 99 West Street, Farnham, Surrey GU97EN, UK. www.efnarc.org
  Fax: +441252739140 (O) Tel: +441252739147 (O)

Therefore, high performance concrete requires – **high strength, high impermeability** and **high durability** at the same time. These properties are developed by using admixtures having **high microsilica** slurry in suspension form (5 – 20 %) and suitable percent of **stabilizers** to control **viscosity** of concrete.

There are numerous type of admixtures for specific purposes viz – high **early strength**, water reducing **retardars, Air Entraining Agents (AEA), corrosion inhibiting** admixtures, injection grout admixtures, **water proofing** agents, etc. We shall now discuss emerging technology for high performance.

10.6.3 High Workability In Fresh Concrete

High **workability** is achieved by use of modern **super plasticizers** based on **metamineformaldehyde** (0.80 to 3.0 % by mass of cement) such as those developed by **MC** Bauchemie Muller GmbH & Co. KG, Germany and number of Indian Partners (www.mc-bauchemie.com).

New generation **super plasticizers** are based on **multicarboxylatether (MCE), polyacrylate** (0.2 to 3.5% by mass of cement). **Microsilica** slurry and stabilizer can be used (5 to 20 % by mass of cement) for **high performance concrete (HPC)** and **self compacting concrete (SCC)**. These admixtures are available by their trade name **centrilit fumes** in slurry suspension form.

Numerous admixtures are available in the market with different trade names and variations in specifications to suit **specific requirement** viz high early strength, high ultimate strength, retardation in setting to **retain workability** for the longer period as desired in case of **ready mix concrete (RMC)**. Air entraining agents (AEA) such as MC plast AEA (liquid) are available to improve **workability** (cohesiveness) and **resistance** to **freezing and thawing** at low dosage (0.15 to 0.30 % by mass of cement).

The **corrosion inhibiting** admixtures (**2 to 6 %** by mass of cement) such as **MC-Corrodur** may be used for **protection and control of corrosion** to improve **durability** without affecting **strength**.

Integral **water proofing** compound (powder 1 % by mass of cement) or high performance **water proofing agent** (liquid 0.3 to 0.50 % by mass of cement) which allows **higher dispersion** and better workability making concrete more **watertight** by building **water barrier**.

10.6.4 Self Compacting Concrete (SCC)

Self compacting concrete (SCC) is one of the revolutionary development in concrete construction in present time. **SCC** has proved **economical** specially in higher grades due to following factors

- Reduction in skilled labour;
- Speedy constructions;
- Smooth surface finishes;
- Longer life (durability);
- Freedom in design;
- Thinner sections;
- Better working environment.

SCC was first developed in **Japan** in 1988 and then in **Germany** and many other countries. Development of **super plasticizers** made it possible to develop SCC technology. **SCC**

technology is still based on short term investigations. After long term investigations **SCC** will become the most common construction material in near future. National standards are being developed in Japan, Germany, Europe and many other countries. These codes provide **specifications, compositions** and specific **applications** of **SCC**.

10.6.5 Terms In Reference To SCC

- **Additions** are finely divided inorganic materials used in **SCC** to improve or achieve certain specific properties. These **additions** are **nearly inert** (type I) or **pozzolanic/latent hydraulic** (type II).
- **Admixtures** are materials added in small quantities during mixing to modify the properties of **fresh** or **hardened** concrete.
- **Binders** are **cement** and **pozzolanic additions** in SCC.
- **Flowability** of SCC is the **ability to flow** through tight openings such as spaces between steel reinforcement **without segregation** or blockage.
- **Unconfined flowability (filling ability)** of SCC is characteristic to **flow** into and fill all the spaces within the **formwork** under its own weight.
- **Powder** (fines) material of particle sizes **smaller than 0.125 mm** (125 micron). This includes part of sand smaller than 125 microns.
- **Mortar** is the fraction of the concrete comprising of **paste plus** those aggregates having size smaller than 4.00 mm.
- **Paste** is the fraction of the concrete comprising of **powder plus water** and air.
- **Self compacting concrete** (SCC) is that which is able to flow under its own weight and completely fill the formwork, even in the presence of dense reinforcement, without the need of any external vibrations, whilst maintaining homogeneity.
- **Segregation resistance (stability)** is the ability of SCC to remain homogeneous in composition during transport, and placing.
- **Workability** is measure of the **ease** by which fresh concrete can be **placed and compacted**. It is a complex combination of aspects of **fluidity, cohesiveness**, transportability, compaction and stickiness.

10.6.6 Constituent Materials

The constituent materials, used for the production of **self compacting concrete** (SCC) shall comply with the standard specifications. The **material** shall be suitable for the **intended use** in SCC and **not contain any harmful ingredients** in such proportions that may be highly detrimental to the **quality** or the **durability** of the concrete or cause **corrosion** of reinforcement.

Cement for SCC

Cement conforming to standard specifications shall be considered **suitable for SCC**.

Aggregate for SCC

Aggregate conforming to standard specifications shall be considered **suitable** for SCC. Generally the maximum size shall be taken as **20 mm** unless and otherwise different sizes are specified for specific use. **Aggregate** size smaller than 125 micron (0.125 mm) will be considered as **part of powder** content. The **moisture** content in aggregate shall be closely monitored and must be taken into **account** for the desired **uniform quality**.

Mixing Water for SCC

Mixing water shall conform to the standard specifications and should not contain any harmful ingredients.

Admixture for SCC

Super plasticizers (conforming to standard specifications) will be **essentially** used for **workability** of SCC. Other admixtures such as **"Viscosity Modifying Agents (VMA)"** for stability, **"Air Entraining Agents (AEA)"** for freeze thaw resistance, **"Retarders"** for control of setting, etc. may be used as needed. It is essential to establish **suitability** of these admixtures and ensure **adequacy of supply** for the whole project.

Additions for SCC

Suitability (conformance to specifications) of **semi-inert** mineral filler (or pigments) and **pozzolanic** (or latent hydraulic) additions be established before use. Pozzolanic additions are **flyash, silica fume** and granulated blast **furnace slag**. Both **inert** and **pozzolanic** additions are commonly used in **SCC** to improve and maintain workability and regulate other properties. Pozzolanic additions improve long term **performance** in respect of **strength, impermeability** and **durability**. Crushed **lime stone, dolomite** or **granite** may be used to increase the powder content (fraction of size less than 0.125 mm). **Flyash** is a fine inorganic pozzolanic material (with reactive silica) to be added to **SCC** to improve its properties but may affect dimensional stability (shrinkage & creep). **Silica fume** gives very good improvement of the **rheological** as well as **mechanical** and chemical properties. It also improves the **durability** of SCC. Ground **blast furnace** slag is a fine **latent hydraulic binding** material, which also improves **rheological** properties of **SCC**.

Filler of SCC

Ground glass filler is usually obtained by finely **grinding** recycled glass with particle size less than 0.10 mm and **specific surface** area more than **2500 cm^2/gm**. Suitable **pigments** may be used in **SCC** after establishing its specifications.

Fibres for SCC

Fibres added to **SCC** will improve flexural and tensile strength of concrete. Commonly steel or polymer fibres are used. **Steel** fibres enhance mechanical properties such as flexural strength and toughness. **Polymer** fibres are used to **reduce segregation** and **plastic shrinkage** or increase **fire resistance**. Any fibres to be used in SCC shall **comply** with appropriate **standards** and shall be tested before use.

10.6.7 Requirements for SCC

Application Area

SCC can be used in **precast** or **site** constructions. SCC can be manufactured in a site batching plant or ready mix concrete plant. SCC can be **placed by pumping or pouring**. SCC **mix design** must consider the size of structure, formwork, reinforcement and cover, etc. These aspects will facilitate for decision on **workability requirement**. SCC has made it possible to construct structures not possible with conventional concrete.

Requirements

SCC mix can be designed to satisfy special requirements of **density, strength development, ultimate strength, durability** and **impermeability**. SCC may show **more plastic shrinkage** and creep due to **high powder** content. Knowledge available on this is limited and research is going on. Fibres may be used to limit negative characteristics. **Curing** should also be started at the earliest possible to avoid excessive shrinkage.

The workability requirement of **SCC** is very high and can be **characterized** by the following aspects:

- **Filling** ability;
- **Passing** ability, and
- **Segregation** resistance.

A concrete mix can only be classified as **self-compacting concrete** if all the above **three characteristics** are fulfilled.

Test Methods

There are many test methods but some combinations of tests are gradually getting accepted. No single method can characterize all relevant **workability aspects**. Each mix shall be tested by more than one test for each workability parameter for reliable results. Various alternative test methods for these workability parameters are listed in the Table 10.1. Performance of these tests will be briefly explained in the annexure. For initial mix design all the three tests are conducted to ensure all aspects of the workability of SCC.

Table 10.1 Alternative Tests for Workability and Acceptance Values for SCC

S. No.	Method	Property	Measurement Units	For 20 mm CA Range of Values for Acceptance Min. – Max.
1	Slump-Flow by Abram's cone	Filling Ability (Spread)	mm	650 – 800
2	T 50 cm Slump Flow (Time)	Filling Ability (Seconds)	Seconds	2 – 5
3	V-Funnel (Time)	Filling Ability	Seconds	6 – 12
4	V-Funnel Observation Increase at T_5 min	Segregation Resistance	Increase in Time – Seconds	0 – + 3
5	J-Ring	Passing Ability	Difference (mm)	0 – 10
6	Orimeter (Time)	Filling Ability	Seconds	0 – 10
7	L-Box	Passing Ability (h_1/h_2 Ratio)	Ratio	0.80 – 1.00
8	U-Box	Passing Ability (Difference $h_2 - h_1$)	Difference (mm)	0 – 30
9	Fill Box	Passing Ability	Percent (%)	90 – 100 %
10	GTM Screen Stability Test	Segregation Resistance	Percent (%)	0 – 15 %

The quality control of **SCC** at site can be monitored by usually conducting two test methods. Typical combinations of these tests are:

- Slump-flow and V-funnel test or
- Slump-flow and J-ring test.

Experienced technician can control quality of **SCC** even by conducting one set of tests and ensuring consistency in raw material quality.

The workability criteria must be satisfied just before placing **SCC**. For quality concrete tests may be carried out just before placement so as to ensure **workability retention** at the time of placing. Acceptance criteria specified in Table: 10.1 are valid for **20 mm** coarse aggregate and may be modified for **40 mm** or **10 mm** CA. The acceptance criteria is tentative and based on present level of knowledge and may be modified after long term experience in use of SCC. Special care must be taken to **avoid segregation**.

10.6.8 Mix Composition

General

The mix composition shall satisfy all the **performance criteria** both in fresh and hardened states. For the fresh concrete various requirements related to workability are already discussed in 10.6.7. In the hardened state the performance criteria are related to the **mechanical properties, durability** and **impermeability**. These criteria can be ensured by testing before use of SCC.

Initial Mix Compositions

The key components of SCC are proportioned by volume instead of mass. Tentative **initial** range of proportions are given. Further **modifications** may be introduced based on **fulfillment** of various **performance** criteria.

Initial mix compositions are considered on the basis of performance:

- **Admixture** as recommended type and proportion.
- **Water**/powder ratio by volume – **0.80 to 1.10**.
  (Trials 0.80, 0.90, 1.0 and 1.10 w/p).
- Total **powder** content – **160 to 240** litres (**400 to 600 kg**) per m^3.
- **Coarse** Aggregate content – **28 to 35** percent by volume of mix.
- **Water-content** as per codal provisions (water content shall not exceed **200 litres/m^3** concrete).
- **Sand** content equal to **balance of the volume** on the basis of other constituents.

Generally, **SCC** mix shall be designed to maintain its workability inspite of minor field variations in raw materials (such as moisture condition of aggregates). Viscosity modifying admixtures (**VMA**) shall be used to compensate fluctuations in sand grading and moisture conditions of aggregate.

Adjustments in Mix Compositions

Laboratory trials shall be made to verify the properties of **SCC** in fresh state and necessary cubes shall be cast for checking **mechanical properties** & durability. in hardened state. **Work-**

ability of SCC in **fresh state** shall be achieved with 4 – 5 water/powder ratios (**0.80**, 0.85, **0.90**, **1.0**, 1.05 and 1.10).

Even for each water/powder ratio, total powder content shall be varied (160, **180, 200**, 220 and 240 litres/m^3) to find optimum values.

Water cement ratio may also be varied on the basis of codal provisions keeping water content/m^3 less than **200 litres** (160, **180, 200** for trials).

Coarse aggregate content may be kept as 28, **30**, 33 and **35** % by volume of concrete. Balance volume shall be provided by **sand** content.

Percentage quantity and type of **super plasticizers** may be varied as per instructions of manufacturer (generally **2 to 5** %) by mass of cement). Viscosity Modifying Agent (VMA) may also be used to compensate certain variations in raw materials. Adjust the **dosage** of admixture (both super plasticizers and VMA) to modify **water content**, and the **water/powder** ratio.

10.6.9 Production and Placement

Production staff involved in **SCC** must be specially trained in quality control and various tests. Various raw materials must be properly stored so that the **variations** in moisture are **minimized**. All admixtures must also be stored properly with labels & other instructions for the users.

Mixing of raw materials of **SCC** is done in any conventional mixers with little more time than conventional concrete. A suitable **water-cement** ratio can be maintained with varying **water content** to make the necessary workability of **SCC** as desired.

Since **SCC** is more sensitive to variations in grading or moisture content in the aggregate, quality tests (**filling ability, passing ability** and segregation resistance) shall be regularly conducted to minimize variations.

SCC may be delivered and transported by appropriate method depending on the **size** of end structure to be casted, **distance** from manufacturing site to the construction site, etc. The **workability** shall be retained at the point of **placement**. **Placing conditions** (reinforcement, size, congestion, formwork, etc.) affect the **workability-retention** of SCC. Skip gates should close properly. **Full hydrostatic** pressure shall be considered for the design of forms in excess of 3 m depth. **Tendency of segregation** in SCC can be minimized by limiting free vertical fall below **5 m** and horizontal flow less than **10 m** from the point of discharge.

Cold joint formation with previously laid concrete may be avoided by laying SCC properly. SCC may be finished roughly at the time of laying and finally finished with steel trowel before the concrete surface stiffens.

SCC dries faster and therefore **initial curing** may be started at the earliest after final setting so as to **minimize** the **shrinkage** cracks for better **durability** and **impermeability**.

SCC shall be produced under strict control and shall follow the specific standards. SCC shall be accepted only if it complies with the agreed norms. In case of **non-compliance**, an appropriate action shall be taken as per the contract agreement. Acceptance test must be carried out by the competent, trained personnel.

Fresh **SCC** shall be tested for workability by measuring: Abram **slump flow** (filling ability), T_{50cm} **slump-flow** (filling ability), **V-funnel time** (filling ability), Increase in V-funnel at $T_{5minutes}$ (segregation resistance), **J-Ring test** (passing ability), L-box (passing ability ratio), U-box (passing ability difference), etc.

10.6.10 Workability Tests of SCC

Workability is measured by different tests of **filling ability, passing ability** and **segregation resistance**. **Filling ability** is generally measured by **slump flow, T$_{50cm}$ slump flow time, V-funnel time**, etc. **Passing ability** is measured by **J-Ring test, L-box** and **U-box**. **Segregation resistance** is measured by **V-funnel** at **T$_{5minutes}$**.

(a) Slump Flow Test

Slump flow test is conducted for **filling ability** by using Abram's truncated cone (top **100mm** and bottom **200 mm** diameters with **300 mm** height). About 6 litres concrete is required. The cone is placed with its centre on the centre of the base plate (Square or Circular). The base plate is marked with **200 mm, 500 mm** and **700 mm diameter circles**. The slump cone is filled with concrete upto the top and finished level. The cone is lifted vertically without any jerk or tilt. Stop watch is started when the cone is lifted. The concrete spreads. Time is noted when the concrete **crosses 500 mm diameter circle**. When the spread stops and reaches its final position, the spread is measured at **2 perpendicular diameters** and the average spread (mm)

$(d_0 = \dfrac{d_1 + d_2}{2})$ is registered as **slump-flow**. The spread of concrete may be observed for segregation tendency, if any. Average **time in seconds to reach 500mm** diameter circle is also registered as measure of Workability as **T$_{500mm}$** (seconds). **Greater** the time, **lesser** is the **workability**. Normal values of **slump-flow** should be **(650 ± 50) mm**. T$_{500}$ flow time may be accepted generally between **3 to 7** seconds and **2 to 5** seconds for housing projects.

$$\text{Slump Flow} = \frac{d_1 + d_2}{2} \text{ mm} \qquad\qquad T_{500mm} = t \text{ seconds}$$

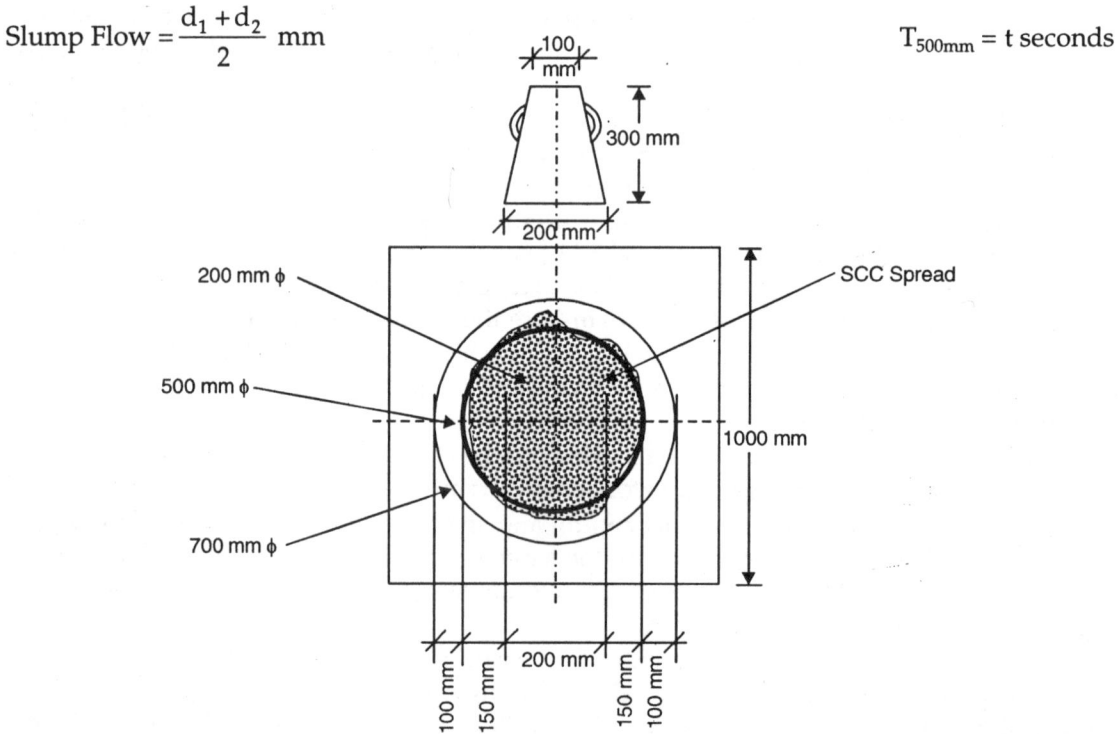

Fig. 10.5. Slump-Flow Cone

(b) J-Ring Test (Passing Ability)

J-ring test to determine **passing ability** can be conducted in **conjunction** with the **slump-flow** or **V-funnel test** for **flow-ability**. Concrete is allowed to pass through appropriate spaces (3 times maximum size of aggregate) between the reinforcement bars similar to actual situations. The **J-ring** comprises of **300 mm diameter** ring which can be placed **centrally** around **slump-flow** cone (bottom 200 mm diameter). The reinforcement bars of 100 mm heights are attached (or threaded) to the J-ring at a spacing appropriate to actual reinforcement situation.

Normal **slump-flow** is measured by spread of concrete after lifting slump-cone. In case of **J-ring test**, the 300 mm ring with 100 mm high reinforcement is placed **centrally around the slump cone**. The cone is lifted vertically and the concrete spreads with reinforcement bars resisting the **spread (flow)**. After the concrete has acquired **stable condition**, the height of concrete is measured at 3 to 4 places on **just inside** the ring (h_1) and on **just outside** the ring (h_2). The difference between h_1 and h_2 represents the **passing ability resistance**. **Lesser** the difference **greater** is the **passing ability** and vice versa.

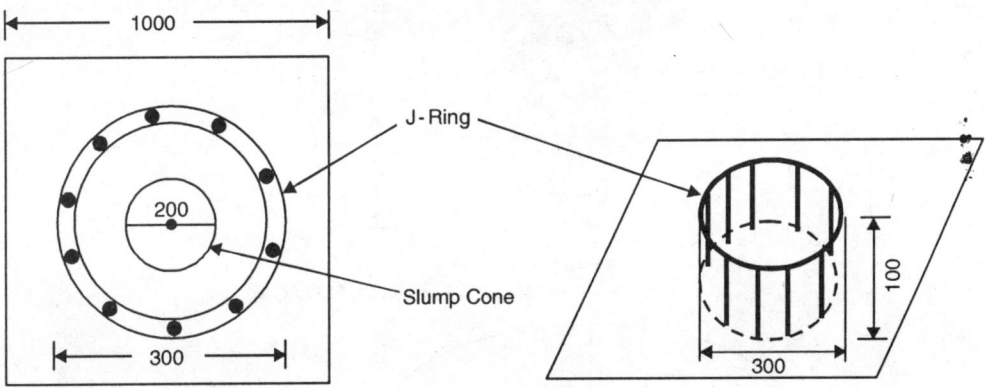

Fig. 10.6. J-Ring Apparatus (All Dimensions in mm)

(c) V-Funnel Test and V-Funnel at T5min Test

The test comprises of V-funnel of **490 mm × 75 mm** at top, **65 mm × 75 mm** at bottom and **425 mm height** with **150 mm** extension in height at the bottom (i.e. 65 × 75 × 150 mm tube). The bottom is fixed with a **trap door** which can be opened instantaneously with a start of stop-watch.

The trap door is closed and interior surfaces are lubricated and moistened. Prepare about **12 litres of SCC** in a bucket. Fill the V-funnel using **scoop** with no tamping or vibration. Finish top with trowel. **Open the trap door** instantly and **start the stop-watch** at the same time. Measure the **time in seconds (T_1)** when the V-funnel gets **fully emptied** by observing light through the top of funnel. Trap door must be opened within **10 seconds of its filling**. Whole test needs to be conducted **within 5 minutes** of adding water. Collect the V-funnel discharge in the same bucket.

Close the trap door again and refill.the V-funnel immediately after measuring the **flow time**. Keep the bucket underneath the V-funnel outlet. Finish the top surface level with the trowel without any tamping or vibration. Open the trap door instantly **after 5 minutes** after second filling and start the stop-watch simultaneously. The concrete **flows out** under gravity. Record the time of complete **flow-out** after refill. This time is recorded as the **flow time at** T_{5min}. This test indicates **ease of flow** of the concrete. A **flow time of 10 seconds** is considered appropriate for **SCC**. Inverted cone restricts flow and **prolonged flow times** may give rise to **susceptibility of the mix to blocking**. After **5 minutes of setting,** segregation of concrete will show a less continuous flow with an **increase in flow time** (T_2).

Check first **flow time** T_1 and second **flow time** T_2 after 5 minutes in seconds and compare. Observe the **tendency** of **segregation** if cement **paste** and **coarse aggregate** particles show separation, specially near the edges of spread concrete.

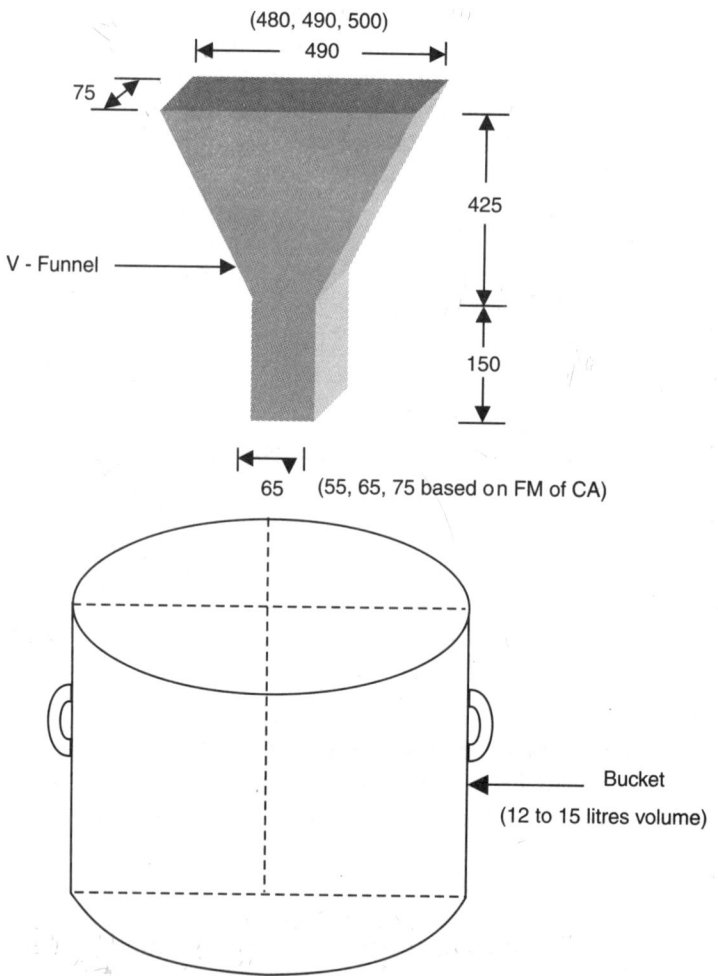

Fig. 10.7. V-Funnel (All Dimensions in mm)

There are many more tests being experimented in different countries but most commonly used tests are explained above. These tests are L-box test, U-box test, Fill box test, Orimeter test and GTM screen stability.

Workability Conformance

- Slump-Flow ———————— 650 to 800 mm
- T_{500mm} Slump-Flow (time) ———————— 2 to 5 seconds
- J-Ring (height difference) ———————— 0 to 10 mm
- V-Funnel (time of flow) ———————— 8 to 12 seconds
- V-Funnel (T_{5min}) ———————— + 3 seconds (Increase)
- Workability to be maintained for more than 1 hour
- Mechanical Strength ———————— Expected strength at 1d, 7d & 28d.
- Modulus of Elasticity ———————— As expected values
- Shrinkage ———————— As specified

Table 10.2 Test Results and Adjustments in SCC Mix

Test	Result	Observations/Cause(s)	Adjustment to Rectify
1. Slump Flow	650 mm	Viscosity High	• Increase water content • Increase paste volume • Increase super plasticizer
	750 mm	Viscosity Low	• Reduce water content • Reduce paste volume • Reduce super plasticizer
2. T_{500mm} Slump Flow	2 sec.	Viscosity Low	• Reduce super plasticizer • Increase viscosity modifying agent
	5 sec.	Viscosity High	• Increase water content • Increase paste volume • Increase super plasticizer
3. J-Ring (Difference $h_1 - h_2$)	10 mm	Viscosity High	• Increase water content • Increase paste volume • Increase super plasticizer
	2 mm	Viscosity Low & Segregation Tendency	• Reduce water content • Increase paste volume (cohesiveness) • Increase mortar volume (cohesiveness) • Use finer powder
4. V-Funnel (Time)	8 sec.	Viscosity Low & Segregation Tendency	• Reduce water content • Increase paste volume • Increase mortar volume • Use finer powder
	12 sec.	Viscosity High & Blockage Tendency	• Increase water content • Increase paste volume • Increase mortar volume
5. V-Funnel (at T_{5min})	0	Doubtful Result	NA
	+ 3 sec. (More)	Segregation observed, rapid loss of workability, blockage	• Use slow setting cements • Increase retarder • Replace part of cement by filler • Increase mortar volume • Increase paste volume

10.7 SUMMARY

Extreme temperatures and relative humidity affect the properties and quality of cement concrete construction and hence **hot weather, cold weather** and **underwater** concreting requires special precautionary measures. High temperatures accelerate hydration of cement and evaporation of surface water, thus resulting in shortage of water for adequate hydration affecting the quality of concrete. Though early strength is higher but ultimate strength is less in case of high temperatures. High temperatures cause greater shrinkage cracking and result in porous concrete. At **low temperatures** hydration of cement concrete slows down which increases the risk of **freezing and thawing.** Freezing and thawing of partially hardened concrete results in expansion of pore water spaces causing **disintegration** of weak concrete at low temperature.

During **extreme hot weather,** the concrete temperature can be controlled by adding **cold water** or **ice** for mixing, while workability is maintained by use of **retarding admixtures.** For better quality low heat cements should be used and concreting should be carried out in the **evening.**

In case of **extreme low temperatures** heated ingredients (except cement) should be used to maintain temperature of mixed concrete appropriate to hydration process. Rapid hardening cement and accelerating admixtures may be used to accelerate hardening process to eliminate freezing and thawing at an early age. Insulate and delay removal of formwork.

For **under water concreting** since no mechanical compaction is possible, cement concrete mix should be **more workable** which compacts on its own. It should be placed with **tremie** in its final position using about **10% extra cement to compensate** for the loss of cement during placement. Concrete should be **laid continuously** and the area to be concreted should be protected with **cofferdams.** High alumina cement or accelerating admixtures may be used for an early strength.

Polymer-impregnated special concrete can develop very high compressive strength ($140 \, N/mm^2$) and high elastic modulus with almost zero creep. Polymer impregnated concretes are suitable in highly corrosive environment viz distillation plants, marine structures and sewer lines. Synthetic or steel fibre reinforced concrete can also be used for structures requiring high tensile and flexural strengths such as pavements, blast resistant structures, marine structures, shell structures and refractory linings.

The **research** in concrete technology and development of **admixtures** and **pozzolanic** or inert fine materials (such as flyash, blast furnace slag, silica fumes, etc.) has resulted in **high performance concrete** (HPC). High workability concrete (slump-flow 650 – 750 mm) is being developed as **self compacting concrete** (SCC). With suitable mix design, **HPC** can achieve **strengths** as high as **200 N/mm².** HPC is manufactured by using high percentage of **super plasticizers, pozzolanic** or inert **additions** (flyash, silica fumes, furnace slag, etc.), **fibres** (steel or polymers) and ground glass **fillers** for specific properties. **HPC** leads to economy in construction and even **elimination of reinforcement** in some cases. SCC shall only be used after through **laboratory testing** for the desired properties.

PRACTICE QUESTIONS

10.1 Explain the effect of high temperature on quality of cement concrete in not more than 100 words.

10.2 Explain the effect of low temperature on the properties of cement concrete in not more than 100 words.

10.3 Describe special practices and precautions required to achieve good quality of cement concrete in extreme hot weather.

10.4 Calculate temperature of concrete mix of one bag batch. The concrete batch comprises of 110 kg sand, 450 kg of coarse aggregate, 20 kg mixing water at 20°C and 5 kg. ice. Sand contains 5% surface moisture and has temperature of 50°C. Cement temperature is 20°C.

Specific heats of cement, sand, and CA are the same at 0.22 k cal/kg °C, while that of water is 1.0 k cal/kg°C.

10.5 Describe precautions required to produce good quality concrete in extreme low temperatures in not more than 200 words.

10.6 Explain special features of cement concrete placed under water.

10.7 Describe special properties of polymer-impregnated concrete in not more than 100 words.

10.8 Describe special properties of fibre reinforced concrete (FRC) in not more than 100 words.

10.9 Explain the concept of **self compacting concrete**.

10.10 **List** the **materials** used in **SCC** giving very brief specifications of these materials.

10.11 **Define** the terms: **Additions, Admixtures, Flowability, Powder, Mortar, Paste** in relation to **SCC**.

10.12 **List alternative** methods of measurement of **flowability** & **passing ability**.

10.13 **Explain** briefly measurement of **workability** of SCC by any one method.

10.14 Explain J-ring test to measure passing ability.

10.15 Explain V-funnel filling ability test.

CHAPTER 11

Repair, Maintenance and Rehabilitation of Concrete Elements

LEARNING OBJECTIVES

After studying this chapter, the learner understands repair and maintenance of concrete elements and will be able to:

- Explain the need and importance of repair and maintenance of concrete elements;
- Explain symptoms and problems of concrete deterioration;
- Analyze the root causes of concrete deterioration;
- Describe preparatory steps required for concrete repair;
- Describe concrete repair materials;
- Describe main characteristics and specifications of concrete repair materials;
- Explain the concrete repair methods;
- Explain repair of concrete cracks and surface defects;
- Describe protective surface coating of concrete;
- Explain repair and rehabilitation of deteriorated concrete structures.

11.1 INTRODUCTION

Cement concrete elements are subjected to various forces and weathering actions and loose their serviceability with time. Generally deteriorations are rapid and failures take place due to lack of adoption of quality construction practices. Although concrete has proved to be durable material, it has not been completely durable due to adverse exposure or other physical conditions during or after the construction.

A very large number of modern structures are constructed using concrete in one or the other form. Scientific knowledge and skills are necessary for quality assurance and serviceability of concrete elements. **Serviceability** of concrete elements will require appropriate repair and maintenance to compensate or reduce the effects of chemical, physical or weathering forces. Depending on the basic quality of original construction and exposure conditions, the concrete deteriorates affecting its serviceability. To enhance durability and serviceability of concrete elements, **repair and maintenance** will be necessary at various stages. Rehabilitation of deteriorated concrete structure also becomes necessary.

Concrete has advantage of getting a set of properties by design during its construction by use of appropriate materials and workmanship so as to withstand service conditions. Certain deterioration and defects are inevitable even in concrete structures. The problem in concrete may range from **superficial to in-depth defects** including corrosion of steel reinforcement. Although there is no alternative to initial good quality concrete construction, suitable repair and maintenance practices can facilitate reduction in defects and retard deterioration to enhance serviceability and useful life. Repair and maintenance of concrete construction requires highly disciplined approach to select materials and techniques based on the nature and extent of defect and adverse ambient conditions. Inadequate and improper maintenance of structures may require **demolition** or **reconstruction of deteriorated concrete element** requiring high cost. **Repair and maintenance** of concrete structures is, therefore, as important as its original construction for maintaining original form, quality and serviceability.

Most of the concrete constructions in India are handled by skilled and semi skilled workers having inadequate or no knowledge of scientific principles underlying cement concrete. Inadequate supervision on smaller projects at various stages of concrete construction result in various types of defects causing rapid deterioration and reduction in quality of useful service life of such structures. Concrete deterioration may be due to unsound and porous concrete, or due to chemical, physical and weathering forces causing corrosion of structural steel. Steel in RCC elements may get corroded due to carbonation, presence of chlorides or other defects such as inadequate concrete cover to reinforcement.

11.2 DIAGNOSIS OF DEFECTS/PROBLEMS

11.2.1 Symptoms and Causes of Problem

Certain defects and decay in plain and reinforced concrete constructions are of common occurrence all over the world if proper construction and maintenance practices are not adopted. Deterioration in concrete usually results in some visible symptoms in the form of cracking or spalling with or without rust stains on the surface. Sometimes pitting corrosion may also occur with little or no visible effect on the concrete surface. The form of damage in concrete or steel reinforcement depends on many factors such as porosity of concrete, cover to steel, spacing of steel, size of bars, shape and size of the members and ambient conditions. Some defects and deteriorations may not always provide visible symptoms and such concealed problems need to be investigated thoroughly by observations of visible symptoms alongwith various types of tests. Important symptoms of defects and deterioration in concrete include:

- Cracking (surface and body of concrete);
- Spalling;
- Rust and dampness stains on the surface;
- Corrosion of reinforcement (exposed and hidden bars);
- Porous concrete near surface;
- Non conformity of surface, shape and size of member; and
- Surface defects.

Sometimes visible signs and symptoms represent deep-rooted serious concealed and hidden problems. Before undertaking any repair and maintenance of concrete, visible and concealed defects need to be thoroughly investigated and analyzed. It may be noted that some of the defects occur in almost any climatic condition, while others occur in certain specific adverse climate and ambient environment. The climatic and ambient conditions must also be assessed while investigating defects and problems. Logical and Scientific analysis of **visible symptoms** may lead to concealed defects, which can further be confirmed by **testing and detailed investigations.** The observation and analysis of ambient climatic and environmental conditions facilitate in correct **identification of forces causing deterioration** and decay in concrete elements.

Delamination in concrete tends to occur when cracks due to corrosion propagate in the plane of the reinforcement. Visible symptoms of cracking and spalling occurs after the corrosion has already taken place in wide range of the reinforcement. It may be noted that all cracks are not due to corrosion only but sometimes cracks do occur due to normal response of RCC to loading. It is, therefore, very important to investigate various visible signs and symptoms for **root causes.** In majority of cases the problem or defect occurs due to any one or combination of the following main causes:

- Physical forces and overstressing;
- Poor workmanship and inadequate concrete cover;
- Chemical attack;
- Ambient environment and biological factors; and
- Steel reinforcement corrosion;

Physical forces can attack concrete in many ways and cause defects in its serviceability. Water and wind borne sand causes erosion, moving machines and wheels cause abrasion, vibrations cause impact, overloading causes overstressing, fire cause high and differential temperatures, very low temperature cause freezing and thawing. Flowing water, beating by rain, beating by windborne sand particles, and flow of moisture through concrete body cause erosion. Moving machinery and wheels of carts cause severe wear and tear of floor surfaces by abrasion. Sudden impact and overloading may result in failure of concrete due to brittleness and inadequate tensile strength. Concrete may develop cracks and disintegrate due to fire, which results in large thermal expansions and contractions. Alternate freezing and thawing also disintegrates concrete due to **expansion of pore water due to freezing.** Thus, physical forces of nature are critical in causing concrete to deteriorate and fail.

Poor workmanship in original concrete construction results in poor quality of concrete, which facilitates and accelerates deterioration reducing serviceability of concrete structures. Poor workmanship may result from **poor mix design, incorrect placement** of reinforcement, **inadequate cover, inadequate vibration** and **compaction, poor curing** (failure to ensure adequate strength development) and use of inappropriate admixtures. Poor workmanship may result in **porous, weak** and **permeable** concrete prone to failure and defects. Poor workmanship results from **poor standards** and **inadequate supervision** coupled with lack of skill and knowledge. Poor workmanship may also lead to penetration of moisture due to surface defects and lack of surface drainage.

Chemical attack may be caused due to **external aggressive chemicals** in contact with concrete or due to those chemicals present or inducted internally during the original concrete constructions. Chemical attack, sometimes, result in crystal growth within concrete especially under high temperatures and high humidity. These crystal formation may cause disintegration of concrete due to expansion in volume. Certain ambient conditions and material composition of cement and aggregate may also cause **Alkali-Silica-Reaction** (ASR). The chemical attacks occur by carbonation and acid formation process due to presence of sulphates in ground water or soil. Direct alkali or acid attacks may also cause chemical attack.

Carbon dioxide and other gases alongwith moisture present in atmosphere may form acid, which react with cement hydrates. This reaction breaks chemical bonds in calcium silicates and cause disintegration of cement concrete. This damage depends on the density, porosity and composition of cement and aggregate in concrete. The sulphate ions in soil or ground water in contact with concrete elements react with calcium hydroxides to form calcium sulphate. Calcium sulphate so formed cause large increase in volumes. Sodium and potassium ions in alkali solutions attack the active silica, sometimes present in aggregate, resulting in increase of volume and consequent cracking. The rate of attack depends on the **presence of active silica** in aggregate, **porosity** of concrete, and **moisture movement** in concrete.

Chlorides in mixing water or admixtures of concrete promote corrosion of reinforcement. Chloride contents of concrete should not be in excess of permitted percentage at any stage. Chloride penetration can also occur from external sources in contact with concrete elements. Chloride attack is especially more dangerous in reinforced concrete as it **accelerates corrosion process in reinforcement.** Chloride ions in concrete disrupt the passivating film that form on the steel surface in highly alkaline conditions in uncarbonated concrete. The **ingress of chlorides** from the ambient environment can reach steel bars if concrete is permeable. The increase in concentration of chloride ions breaks the alkaline coating of reinforcement and promotes corrosion. Thus, repair of concrete has to ensure **control on chloride ingress** and flow upto reinforcement.

Corrosion occurs mainly due to **porosity** of concrete, **inadequate cover, carbonation** and presence of **chlorides.** Corrosion damage may be due to one or combination of these causes. Corrosion causes spalling or cracking of concrete due to increase in volume of corroded steel causing disintegration of concrete around its surface.

Carbonation of concrete occurs due to presence of carbondioxide and other gases in combination with moisture. Concrete in the cover first deteriorates. Carbonation in concrete makes the cover ineffective and promotes direct attack of atmospheric forces and chlorides on steel resulting in corrosion. Carbonation in concrete is also affected due to **porosity** of concrete. **Corrosion** of steel reinforcement also occurs if **chlorides** are present in concrete due to use of contaminated aggregate, sea water, chloride containing admixtures and ingress of chlorides from de-icing salts or from ambient environment. The degree of concentration of chloride ions in water directly affects the speed of corrosion in reinforcement, which has direct bearing on the life, and durability of concrete structure. Dense, impermeable and thick concrete cover enhances durability. The repair process must ensure **impermeable, dense** and **thick concrete cover** with use of appropriate repair materials and **impermeable protective coatings.**

Table 11.1 Symptoms and Possible Causes of Deterioration

	Symptoms	*Possible Causes of Deterioration*
a. During and Before Construction		
Cracking	Design Errors	- Incorrect joint spacing & locations - Restraints - Incorrect load calculations - Excessive slender designs
Cracking High permeability Increased Carbonation Surface Sanding Spalling due to Corrosion	Mix design	- Incorrect W/C ratio & Workability - Too high or too low cement content - Poor aggregate grading - Poor quality of cement and other ingredients
Pores, voids, Blemishes and cracks Low strength Surface Sanding Exposed Steel Premature Carbonation and Corrosion of Steel	PoorWorkmanship	- Incorrect W/C ratio and - Uncontrolled water addition, Bleeding - Poor compaction - Inadequate cover to steel - Poor shuttering - Inadequate mixing & - Finishing
b. Environment and Weather Effects		
Disintegration Cracking Sanding Wear and tear	Excessive MechanicalStresses	- Static or dynamic overloading e.g collision, explosion, abrasion, impact, etc.
Cracking Disintegration Corrosion in steel and breaking of concrete cover Surface Sanding	Thermal Stresses	- Temperature changes - Freezing/thawing cycles - Fire
Distintegration Surface Sanding Carbonation Corrosion	Chemical Attack	- Aggressive gases (CO_2, SO_2) - Corrosive Soils - Polluted groundwater - Acids and Salts
Cracking Flaking Expansion due to plant root growth Fungus appearance	Biological effects	- Plant growth - Algae - Micro-organisms

Carbonation in cover concrete can be tested with the help of **phenolphthalein test** while chlorides can be found by measurement of chloride ion concentration. Concrete samples are obtained from the cover at various depths and tested for carbonation and chloride ions to assess the durability. Chloride test can be performed to determine concentration of chlorides in concrete with the help of chloride ion sensitive electrode. Chloride content of 0.4 to 0.6% by weight of cement is considered quite critical for affecting corrosion of steel.

The continuous presence of moisture on the surface of concrete promotes **biological growth** such as moss, algae and small plants. The penetration and growth of roots of these plants also result in deterioration and cracking of concrete. Extreme conditions of temperatures, humidity, rains and alternate cycles of dry and wet conditions result in accelerated deterioration of concrete.

Table 11.1 provides a summary of possible causes and symptoms of deterioration in concrete. Figure 11.1 shows various types of deteriorations in concrete.

11.2.2 Method of Diagnosis

Reliable system of repair and maintenance requires a reliable system of observation and testing. A systematic programme of testing and observation facilitate correct assessment of defects, decay or deterioration in concrete elements. Generally the type of problems and causes can be related to the following one or combination of causes:

- Inadequate cover to steel;
- Porous or permeable concrete;
- High level of chlorides;
- Poor quality of concrete;
- Overloading, abrasion or erosion.

Reliable system of investigation follows a structured programme of preliminary inspections, detailed inspections and insitu and laboratory testing of affected concrete. Complete investigations should be carried out to identify the nature and extent of the problem before undertaking any repair and maintenance job. The extent of detailed investigations will depend on the nature and importance of structural elements and the problem. The diagnosis approach involves following stages:

- **Preliminary inspections** to assess the basic nature of problem and plan detailed inspections and collect samples for laboratory testing;
- **Detailed inspections** to measure and confirm the defects identified in preliminary inspections and collect samples for laboratory testing;
- Insitu and **laboratory tests** to confirm the defects and measure their quantities and extent.

Preliminary inspections include:

- Visual inspections for the whole structure using binoculars;
- Sounding hammer test;
- Recording of visual symptoms including photographs;
- Collection of spalled concrete samples for visual observations and confirmatory tests for carbonation and presence of chlorides;
- Recording depth of concrete cover using covermeter;
- Evaluation of safety hazards arising from the possible dislodging of damaged concrete.

During visual inspection location and areas under defect are measured and noted. Sounding hammer test is conducted to find spalling and cracking of concrete in the affected element due to corrosion of steel. Preliminary inspections are used to identify the defects as much as possible.

Specific detailed inspections are carried out for critical defects identified during preliminary inspection. The detailed inspections are carried out to determine:

- Whether the problem exists **locally** or **widely** in the structure?
- The range of **carbonation** depth in concrete cover;
- Percentage of **chlorides** present in concrete;
- Whether the chlorides present are due to **external source** or part of **original concrete**?
- General quality of concrete and extent of variation in quality at various locations;
- Whether the deterioration has affected **strength and load bearing capacity** of the structural element?
- Whether the symptoms of concrete degradation are caused by **chemical or frost attack**.

During detailed inspections, it is necessary to arrange **access to all affected elements** for close examination and investigation. Consistency in observations, possible causes of problem, measurements of defects and likely materials and methods to be used are assessed during detailed inspections. **Measurements**, preparation of **drawings** and **quantities of repairwork** are also assessed during detailed inspections. Details of existing reinforcement and deteriorated elements should also be shown by drawings and photographs for analyzing **load bearing capacity of distressed element** and repair approach. Samples are collected for laboratory tests or surface marked and prepared for following insitu tests:

- Depth of cover, using a covermeter;
- Permeability test by initial surface absorption equipment;
- Carbonation test for affected depth using chemical indicators;
- Strength, using **NDT** such as rebound hammer or ultrasonic techniques.

Samples are also collected from the affected portions for laboratory tests to determine and confirm regarding:

- Carbonation;
- Chloride content;
- Cement content;
- Density;
- Water absorption; and
- Concrete strength.

Insitu test is conducted to assess concrete cover to steel, using covermeter. Covermeter is calibrated and validated by exposing steel at one or two points by drilling and actually measuring the depth of concrete in cover. The results of covermeter are recorded to fall within 5 mm range.

Permeability results are found approximately by surface absorption test at site but reliable results are only obtained by laboratory tests.

The depth of **carbonation** can be obtained at site by using a chemical indicator (phenolphthalein) test. When freshly cut surface of concrete is sprinkled with phenolphthalein solution, strong alkaline **uncarbonated concrete will indicate pink colour** while **carbonated** concrete will indicate no colour. Fresh concrete surface is obtained by drilling or cutting concrete upto depths of concrete cover.

To avoid carbonation of concrete samples, tests must be conducted immediately. Samples are taken to a depth upto, which the phenolphthalein solution starts giving pink colour representing alkalinity of concrete in absence of carbonation. While collecting samples, contamination should be avoided with previous samples. Samples should be collected from different places in affected areas for getting the average result.

Insitu **strength data** can also be obtained by use of **concrete hammer** rebound test. Concrete hammer rebound test results are used to represent a **general pattern of strength** of concrete. More accurate strength results of concrete can be obtained by **core testing**.

Samples are obtained by drilling holes and collecting dust at various depths or cutting concrete core (lumps) from various points of affected area. Drilling holes of atleast 10 mm diameter should be used for full depth of concrete cover. Samples may be collected from about 10% of affected elements (or one **sample per 50 m²** of large affected area). Generally core samples of 50 mm diameter are collected for laboratory testing.

Problem of corrosion of reinforcement can be detected by electro-potential mapping using electro potential instruments. Electro chemical tests are also available for detecting corrosion of steel reinforcement.

Laboratory tests are conducted to determine total chloride contents by acid test and results are expressed as chloride ion content by weight of concrete. This chloride ion content is further expressed in terms of chloride ion by weight of cement from the known cement content of concrete. Chloride ion concentration covers all sources (aggregate, mixing water, admixtures, de-icing salts and ingress water salts). Chloride concentration is compared with permissible limits for concrete under the given ambient conditions in accordance with codal provisions.

Confirmatory laboratory tests can also be conducted on freshly cut concrete core samples for cabonation effect. Water absorption tests are conducted on fresh concrete core samples in laboratory by measuring increase in mass of sample after soaking in clean water for intervals of 10, 30 and 60 minutes, 6 and 24 hours. A graph of percent absorption can be prepared to study porosity and permeability of concrete. Concrete core samples of 75 mm diameter and 75 mm thickness are preferred for the absorption test. Sometimes petrographic examination of concrete may also be carried out to study cement gel formation and other micro level details.

Chloride tests can be carried by obtaining powder samples of concrete by drilling and mixing with special chloride extraction. Half-cell potential survey instrument can also be used to assess corrosion of steel and condition of concrete cover. Corrosion potentials are obtained by placing a suitable reference electrode against concrete surface ensuring good contact. An electric connection is made to reinforcement and reference electrode to give potential difference. Probability of corrosion is directly dependent on corrosion measured by Half Cell potential. A half-cell potential survey can be presented as a potential contour.

After correct investigation and diagnosis of the problem and identifying its basic causes, the defective concrete element is prepared for durable and effective repair.

11.3 CONCRETE REPAIR MATERIALS

The selection of repair materials is quite important for effective and durable repair and maintenance of concrete elements. The choice has to be made depending upon the **durability**, **strength** and **compatibility** with the original concrete, **extent of damage** and **speed** with which these can be applied within the available period for repair shut down and resources available for carrying out the repair. The repair materials can be (a) **Grout**, (b) **Mortar, Shotcrete** or **Concrete**, (c) **Patching materials** (d) **Bonding aids**, (e) **Anticorrosive** and **Protective coatings.**

These repair materials essentially comprise of basic binding ingredients such as **cement, polymer modified cement**, or **epoxy resins.** Selection of these materials depend on:

- type and extent of damage;
- type of structure;
- ambient environment;
- site conditions;
- serviceability conditions;
- economic considerations;
- time factor; and
- durability aspects.

Concrete repair materials must have the following basic characteristics:

- Adequate mechanical **strength** close to the substrate (compressive, tensile, shear and bond);
- **Matching modulus of elasticity (E)** and **coefficient of thermal expansion 'α'** with the original substrate material;
- A **low drying shrinkage;**
- **Chemical compatibility** with the original substrate and resistance to permeability of chloride, oxygen, and carbondioxide for durability; and
- Sufficient high rate of **gain of strength** after placement

The main thrust in repair and **rehabilitation** of deteriorated structure is to achieve the desired **integral action** of the **old and new materials** to bring back serviceability of the repaired structure. Further these repair materials must be **environment friendly** and capable of resisting damaging forces specially **ingress of moisture**, chlorides, oxygen and carbon dioxide. Thus **strength, adhesion, compatibility, durability** and **impermeability** are important considerations for selecting concrete repair materials.

Besides the above materials **protective coatings** are also used specially to provide barrier to ingress of chemicals, moisture or gases. These protective coatings can be applied on repaired surfaces as well as other old surfaces. These protective coatings shall be discussed in subsequent paragraphs.

Commonly used repair materials may be classified as under:

(i) **Patching Materials:** These materials are mainly used to reinstate damaged or worn out areas. The commonly used materials to patch up worn out areas are:

- Plain cement mortar
- Polymer/latex modified mortars
- Epoxy resin mortars
- Polyster resin mortars

(ii) **Bonding Aids:** These are used where a new layer of concrete is to be laid over the existing substrate. The function of this bonding aid is to develop adequate adhesive force between the old and new layers to enable them to act monolithically for transfer of load and deformations evenly. The commonly used bonding aids are:

- Cement slurry
- Polymer modified cement slurry
- **Polymers**/water borne **latexes** (polyesters, Acrylics, Polyurethanes, & Silicones)
- Epoxies

(iii) **Anti Corrosive Coatings:** These are used for protecting steel reinforcement and structural steel against corrosion. Variety of corrosion protection coatings are available but **zinc rich epoxy** coatings are most effective. Commonly used protective coatings against corrosion are:

- **Epoxies**-Passivating and nonpassivating,
- **Zinc rich epoxy coatings,**
- **Bitumens,**
- **Fusion bonded epoxies,** and
- **Interpenetrating polymer network** system (IPN) coatings.

(iv) **Special Concretes:** These are used for reinstatement of large damaged areas or for strengthening an existing structure or to provide some special property to the structure. Some of the special concretes used for repair jobs are:

- Cementitious concrete,
- Polymer concrete,
- Polymer modified concrete,
- Polymer impregnated concrete,
- Shrinkage compensated concrete,
- Heat resistant concrete,
- Under water concrete,
- Free flow micro concrete,
- Epoxy concrete,
- **Sulphur concrete,** and
- **Fibre reinforced concrete**

(v) **Grouts:** These are used to fill the voids and cracks in the concrete. The commonly used grouts are:

- Cement grout
- Cement sand grout
- Cement sand grout with admixtures
- Epoxy grout
- Polyurethanes
- **Polymer modified cement grout**

(vi) **Gunites and Shotcretes:** Gunite or shotcrete is mortar or concrete conveyed through a pressureline and applied pneumatically at a high velocity on to a damaged or worn out surface or interior locations. Several types of fibres and additives are added to the gunites or substrates to improve its properties and performance. These gunites and shotcretes use various types of admixtures for plasticity and adequate bonding characteristics.

(vii) **Overlays and Toppings:** These are applied over the concrete substrates either to improve its abrasion or impart properties of chemical resistance or enhance dustproofing and hygienic conditions. The materials normally used as overlays may be cementitious, epoxies and polyurethanes developing good adhesive properties by using appropriate admixtures.

(viii) **Protective Coatings:** Apart from use of various repair and maintenance materials for concrete, protective coatings have been tried in the recent past to save concrete from fast deterioration under adverse environmental and weather conditions. Protective coatings resist effect of deteriorating forces and **enhance concrete serviceability** and durability. The potential of these coatings to improve long-term performance of concrete has also been well recognized. This has created an opportunity to develop special coatings for concrete structures. Various types of coatings are now available for the protection of concrete structures. Some of commonly available protective coatings are:

- **Conventional Coatings** - such as Oleoresins, paints, alloys and phenolic modified alkyls.
- Bituminous coatings
- Vinyl coatings
- Chlorinated rubber coatings
- Epoxy coatings
- Coaltar epoxies
- Polyurethanes
- Inorganic zincrich coatings
- Silicone, and
- **Acrylics and methacrylic resins**

Most of these coatings have water resistant and alkali resistant properties, which are necessary for protecting the concrete from damage. Despite the improvements in concrete practices, many problems of **efflorescence, leaking, moisture vapour transmission, spalling, cracking** and **staining** in exterior faces of concrete occur. Application of protective coating is necessary to treat these defects and to improve durability and impermeability of concrete elements on long-term basis. Successful use of protective coatings has promoted development of many new coating materials during recent past.

Once damaged concrete has been reinstated, it is advisable to consider the long-term serviceability of the total structure including adjacent parts of the structure. These areas will remain vulnerable to possible future attack from the adverse environment. Further repairs could become necessary after a short time unless suitable preventive measures are taken. To avoid the need for further repair and slow down the process of deterioration, certain protective coatings are available in the market. Suitable protective coatings will arrest the corrosive process, facilitate moisture to escape from the concrete, and extend the service life of the structure.

Certain range of high performance joint sealants provide barrier to water ingress through joints and avoid the risk of moisture contact with steel reinforcement. These coatings are applied to concrete surfaces to prevent ingress of moisture and/or provide protection against aggressive environment. These coatings can be applied to both old and new surfaces. These protective coatings are thus useful for enhancing the effectiveness of maintenance of repaired portion. These protective coatings can further be used to slow down the process of deterioration. Different coatings can provide resistance to ingress of moisture, aggressive chemicals and gases and abrasion in addition to providing beautiful surface.

11.4 PREPARATION FOR REPAIR

For effective repair and maintenance preparation of affected surface is to be carried out systematically and with thorough knowledge of its implications. Reinforced concrete elements involve preparation of decayed concrete and corroded steel reinforcement.

Plain Concrete surface preparation includes:

- **Removal of damaged and loose concrete**;
- **Cleaning** of concrete surface **with water/air jet** or sand blasting;
- **Removal of all oil/grease** from the surface, if any;
- **Decayed surfaces to be cut** to a depth of 25 mm for shotcrete or concrete repair and to a depth of 15 mm for mortar repair;
- **Cleaning and sealing of cracks**, if any; and
- **Roughening of surface** to enhance bonding of repair material.

Corroded Steel surface preparation includes:

- **Cleaning of steel rust** and concrete by light hammering or sand blasting;
- **Washing of reinforcement** using Rustonel to remove rust completely;
- **Providing and welding additional steel** to compensate for reduction in steel area due to corrosion in original steel;
- Providing **welded wiremesh wherever** required specially in case of repair by guniting; and
- **Cleaning and sealing of cracks**, if any.

It is necessary to provide passive primer bond coat of cement slurry/mortar, latex mortar or epoxy resin to achieve adhesion of old and new concretes just prior to application of repair material. Bond coat is required both in plain and reinforced concrete. The purpose of primer bond coat is to achieve effective adhesion between old concrete and repair material. Where a hand applied resin or cementitious mortar repair system is used, the old concrete surface is to be coated with a primer bond coat and repair material is applied while this bond coat is still wet or tacky.

For resin system of repair either a recommended primer bond coat is applied or the resin mortar itself acts as a bond coat. But unless it is water tolerant, the bond coat should be applied on dry surface.

For cementitious systems, the primer bond coat may be either resin or polymer cement slurry. The material should be selected with the advice of technologist and supplier. If the polymer is to be incorporated in the repair mortar, it is usually incorporated in the bond coat. Generally, acrylic and copolymer emulsion and epoxy resins are used as bonding materials.

11.5 REPAIR METHODS AND SYSTEMS

The success of repair of concrete structure depends on proper preparation of damaged portion as described in previous section. The prepared surface of plain or reinforced concrete is repaired using appropriate materials and method depending on the nature and extent of deterioration. A primer-bonding coat is laid on the prepared surface just before starting repair.

Repair of concrete can be carried out by three methods:

   (i) Hand/Trowel Application
  (ii) Fluid Application
 (iii) Spray Application

   (i) **Hand/trowel application** of repair mortars is still the most commonly used approach and is quite suitable for small repairs. Many companies introduced prepackaged mortarts to simplify the process. Clean water is required to be added to this dry mortar at site just before application. Wet the substrate with clean water before applying mortar. Apply bonding aid using Nitobond AR or Nitobond HAR primer or any other bonding primer. Steps involved in this method are:

- Measure correct quantity of clean water;
- Mix chosen grade of ready-made dry Renderoc or other mortar according to instructions. Generally a full bag should be added to the measured quantity of water (part bags should not be used);
- Mix thoroughly using a forced action mixer;
- Apply mortar with trowel or by gloved hand on the prepared, wetted and primer coated area for repair; Remove and strike off surplus mortar with trowel;
- Finish with trowel, float or sponge; and
- Spray or paint immediately the repaired surface with Nitobond AR or other curing compound.

Curing compounds are useful for quick and effective strength gain in repaired portion. This method is simple and can be used for small repair jobs.

  (ii) **Fluid application** is better method in case of repairs involving large volumes of damaged portions, congested reinforcement, and difficult architectural/complicated features. The fluid treatment is based on self-compacting fluid micro concrete. The technique was originally developed for major repairs of bridges in UK but with the success of this technique, it became popular all over the world for large volume of concrete replacement in difficult situations.

For fluid micro-concrete application, follow the sequence as under:

- **Fabricate and erect loosely** grout tight form-work to contain the micro-concrete in repair zone after preliminary preparations;
- **Soak substrate** with clean water or **apply epoxy primer** coat prior to application of micro-concrete in accordance with the situation;
- **Erect grout-tight formwork in its final** position to support the fluid micro-concrete in its desired shape and size;
- **Measure clean water** of required quantity;
- **Add Renderoc** or other suitable micro-concrete material to water;
- **Mix fluid micro-concrete** for stated time in a forced action mixer just prior to application;

- Pour or pump the prepared **fluid micro-concrete** in the erected form work around the damaged area; and
- Leave the area for hardening after fully covering the repaired portion with suitable curing technique.

Fluid application techniques have following advantages over traditional hand/trowel application:

- Fluid application provides **self-compacting concrete** requiring no vibration;
- **Fully homogeneous** repaired surface even for large volume applications;
- **Full encapsulation** (coverage) of even the most congested reinforced concrete area;
- Thick sections and **complex architectural shapes** are achievable homogeneously;
- **Maximum density** is always achievable.

(iii) **Spray applications** of premixed mortars are superior to other techniques for repairing concrete elements due to enhanced bond.

Also for speed and enhanced physical characteristics, repair is carried out by spray application of pre-packaged mortar either dry or wet. In case of dry spray, dry mortar powder is driven by compressed air to a delivery nozzle where measured quantity of water is added immediately prior to application. In case of wet spray approach, premixed spray grade mortar is pumped to delivery nozzle where compressed air is introduced immediately prior to application. Special mixer and pump unit is used for spraying by wet spray techniques. Such spraying units are available with specialised agencies.

Spray applications of premixed mortars provide distinct advantages over other conventional approaches. These advantages are:

- **Higher speed** of application;
- **Enhanced bond** strength;
- **Better mortar compaction** and density;
- **Constructed both** horizontally and vertically; and
- **Reduced permeability** of sprayed mortar.

Depending on the extent and nature of damage, only one technique of cementitious long-term repair can be adopted. Hand/trowel application is adopted in case of small and isolated jobs while fluid or spray application techniques are adopted in case of important elements and involves large quantum of work. Repair and rehabilitation of concrete elements facilitate in strengthening and enhancing serviceability and durability of the deteriorated structures.

11.6 REPAIR OF CRACKS AND SURFACE DEFECTS

11.6.1 Introduction

For achieving quality of serviceability and enhancement of life of concrete elements, it is necessary to repair cracks and surface defects immediately. Cracks, if any, must be repaired before removing surface defects. The causes of crack should be determined first to identify the nature of crack viz. stable (dormant) or growing (active). Shrinkage cracks are stable while **cracks due to foundation settlement may be live and continue to grow.**

On removal of shuttering, the concrete surface sometimes have defects of honeycomb, voids, tie holes, air bubbles and surface cracks. All such defects need to be repaired before undertaking finishing. First stage of surface defect repair is on removal of shuttering and formwork. Repair of these defects increases the durability of the concrete element due to protection from environmental forces. Carryout repair of cracks in the concrete elements before repair or plastering of surface defects so as to stop passage of moisture using any of the methods described below.

11.6.2 Crack Repair Methods

Following methods are used to repair cracks in concrete members:

- **Epoxy Injection**;
- **Grooving and Sealing**;
- **Grouting**;
- **Stitching**;
- **Flexible Sealing**;
- **Adding Reinforcement in RCC**;
- **Polymer Impregnation**;
- **Dry Packing**;
- **Overlays and Surface Treatment** and
- **Autogenous Healing**.

a) Epoxy Injection

Epoxy injection method is an important modern method for crack repair. A successful epoxy injection requires evaluation, preparation and planning. Epoxy injection can be used to restore structural soundness of concrete elements. Cracks as narrow as 0.05 mm can be bonded by the injection of epoxy. The technique involves drilling of holes at close intervals along the crack for injecting epoxy under pressure. The cracks should be cleaned of pollutants by washing and drying before injecting epoxy for maximum effectiveness. The steps involved are:

- Preparation of crack;
- Drilling holes;
- Cleaning and drying of cracks;
- Sealing of crack surface;
- Fixing of injection ports in holes;
- Mixing of epoxy resins;
- Injection of mixed epoxy resins;
- Removal of ports and plugging the holes; and
- Removal of surface sealing and finishing the surface.

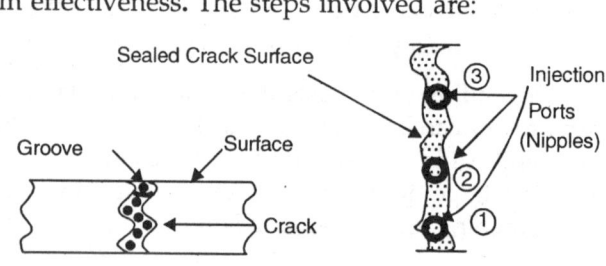

Epoxy injection is done at certain pressure using simple hand operated caulking guns or hydraulic pumps. The pressure of injection should be properly selected so that cracks are not damaged any further. Vertical cracks are injected from the **lowest port towards highest port in the last.** Remove the surface seal after completion of injection to allow the epoxy to cure. Remove nipples (ports) and plug the holes next day after completing injection of epoxy.

The simplest of the injection methods is the brushing method. The resin is brushed on the surface crack and the epoxy is absorbed in by capillary action. In case of pressureless injection, epoxy is poured into the nipples. For cracks of width between 0.2 to 1.0 mm, injection by low pressure may be adopted. Handguns can be used for low-pressure injection. High-pressure injection may be resorted for repairing **structural cracks** by using mechanical or **pneumatic pumps.**

Injection materials are mostly **synthetic resin** or **cement** based. The synthetic resins are usually two components comprising of epoxies and polyurethane. The cement-based materials are invariably modified with polymers to impart flowability, non-shrinking characteristics and better bonding. These injection materials are of variety of composition and characteristics. Polyurethane based two-component injection resin can be used for non-rigid and elastic sealing of cracks.

Polymer modified ready to use **cement grouts** at very **low W/C** ratio can be used for wet or dry cracks and joints above 2 mm width. These can be used for developing high compressive strengths (35 N/mm^2).

b) Grooving and Sealing

The cracks are grooved and sealed before injection. Grooving involves enlarging the crack along its exposed face for filling and sealing with suitable material. Grooving and sealing is not tried on active (movable) cracks. Clean the grooved surface with blow of air. Allow the crack surface to dry before placing the sealant. Sealing is done to avoid entry of moisture or water to protect the reinforcement. Sometimes epoxy compounds are used as sealants.

c) Grouting

Cracks in thick concrete walls and dams can be filled with cement grout. The crack surface must be cleaned by compressed air and water under pressure. Grout nipples are installed and the cracks are sealed between the nipples with cement paint or grout. Flush the crack to clean and test the seal and then pump the grout. The **grout** mix consists of **cement, admixture,** and **water** with or without sand. Plasticizers or water reducers are used as admixture to improve primary and secondary properties of the grout.

Narrow cracks may be filled with chemical grouts consisting of solutions of chemicals that combine to form gel. Cracks as narrow as **0.05 mm** have been filled with chemical grouts successfully.

d) Stitching

Continuity across major cracks in structural concrete elements can be restored by stitching across the cracks. Holes are drilled on both sides and along the crack at suitable interval. Reinforcement bars (known as dogs) are fixed in holes by means of non-shrink grout. These dogs comprise of variable size and placed along varying plane to distribute and transfer load on different planes. **Stitching** is a simple **cost-effective technique** for restoring **tensile and shear strengths** across the cracked section. The most common stitching method uses stitching dogs (U-shaped metal units) or **dowel** bars. The dowel bars or legs of stitching dogs are fixed on each side of the crack with the help of **epoxy mortar** or **cementitious grout.** The stitching dogs are surface mounted on both sides of the crack by fixing the doglegs in holes drilled on either side of the crack. Dogs are generally not capable of taking compressive loads across the crack. Stitching with dogs for repair of concrete or brick wall elements is quite effective for transferring tensile forces.

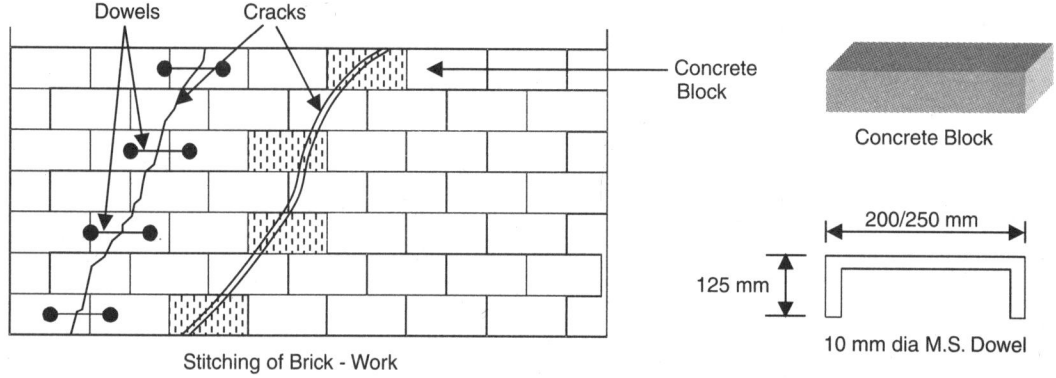

Fig. 11.1. Stitching of Crack with Dowels

e) Flexible Sealing

Active cracks can be cleaned by sand blast, air or water jet or both and filled with a suitable field moulded **flexible sealant.** The slot or groove for the sealant should be of suitable shape and width for expected movement of the crack. Where appearance is not important and active cracks are not subjected to traffic or mechanical abuse, these may be sealed with flexible surface seal using a bond breaker. The flexible joint sealant is trowelled at the top for bonding to the adjoining concrete but no bonding in the base of the joint or crack.

f) Adding Reinforcement

Cracked reinforced concrete can be successfully repaired by inserting additional reinforcing bars to supplement epoxy injection in the cracks. Holes are drilled across the crack plane at about 90 degrees. Reinforcing bars are inserted in the drilled holes after injecting epoxy in holes and the crack at low pressure. The epoxy should have a very low viscosity and modulus of elasticity for rebonding the crack.

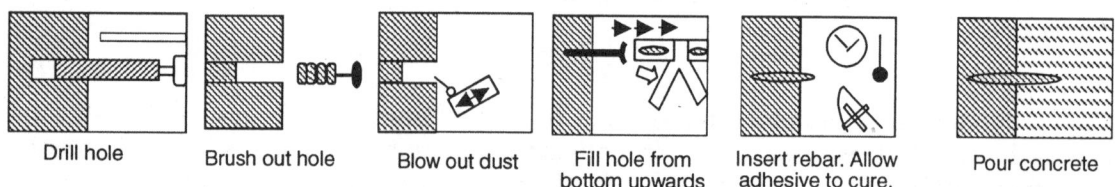

Fig. 11.2. Adding Reinforcement

g) Polymer impregnation

A monomer system consists of small organic molecules in fluid form and can be soaked by dry concrete to fill the crack in concrete element. These monomers vary in volatility, toxicity and flammability characteristics and are capable of forming solid plastic. The monomer systems used for impregnation contain a catalyst or initiator and the basic monomer (or combination of monomers). The monomers join together or polymerise on heating to make a tough-strong durable plastic that enhances the properties of concrete.

If cracked concrete surface is dried, **flooded with the monomer and polymerized** in place, the cracks will be filled and structurally repaired. The cracks must not contain any moisture, otherwise repair will be unsatisfactory. Polymer impregnation is difficult in existing structure and the repair may be unsuccessful.

h) Dry Packing

Dry pack is the hand placement of **low water content mortar** followed by tamping or ramming of mortar into place to produce tight contact between the surfaces of crack. There is little shrinkage and the mortar remains tight with good durability, water tightness and strength. Dry pack is used for repair of dormant cracks but is **not recommended** for filling or repairing of **active cracks.**

The cracks are widened at the surface with the help of power driven saw with a slot size of 25mm´25mm. Clean and dry the slot and then apply a bond coat of cement slurry or mortar paste. Place the dry pack mortar immediately and fill with force of tamping hammer. Surface may be finished with a rag or sponge float. Cure by applying curing compound or wet burlap.

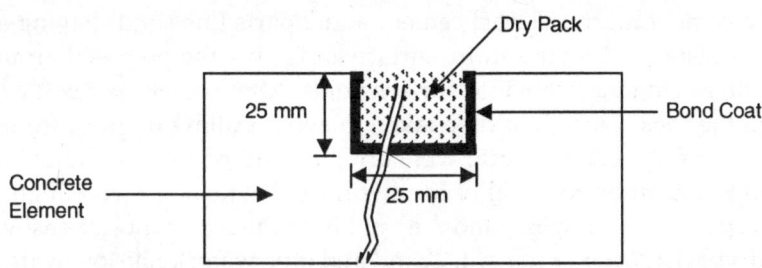

Fig. 11.3. Dry Packing

i) Overlays and Surface Treatment

Cracks subjected to variations in loading, temperature and moisture reoccur if bonded overlays are used to repair. **Moving cracks are repaired by unbonded overlays.**

Dormant cracks are repaired by applying **bonded overlays of latex** modified portland cement concrete or mortar after cleaning the crack surface. A bond coat of broomed latex mortar or an epoxy adhesive should be applied immediately before placing the overlay. Latex mortars solidify rapidly, continuous mixing and laying should be adopted.

j) Autogenous Healing

Dormant fine cracks close by continuous hardening by creating moist environment. This process of natural crack repair is called **autogenous** healing. This healing depends on the carbonation of calcium hydroxide in cement paste by carbondioxide present in the atmosphere in presence of moisture. **Calcium carbonate ($CaCO_3$) and calcium hydroxide crystals form and grow within the cracks.** The resulting chemical and mechanical bonding between the crystals and the surfaces of the paste and the aggregate helps in restoring some of the tensile strength of the concrete across the cracked section and the crack may get sealed. Healing does not occur if the crack is subjected to movement. Flow of water through crack may dissolve and wash away the lime deposit and may not allow healing of the crack. Saturation of the crack and adjoining concrete with water is essential for healing and development of strength.

11.6.3 Treatment of Surface Defects

Surface defects in concrete can occur even in new casting. These surface defects may sometimes lead to serious problem at a later stage. For getting optimum quality of serviceability, it is essential to repair the surface defects immediately within 24 hours on removing of shuttering and formwork. Surface defects can also occur at later stage due to occupational forces. Surface defects in new constructions are due to bugholes, form tieholes and honeycomb. Surface defects in old concrete can be due to spalling, cracking, uneven wear and surface voids. Surface repairs are thin and small patchwork. These thin and patch repair jobs may fail easily and requires a special attention. The mix water of repair concrete is absorbed by existing old concrete and also evaporates from the surface leaving little water for hydration to establish adequate bond strength. The repaired portion should be wet cured for atleast 2 weeks to prevent excessive shrinkage. Surface repair work sometimes require colour matching for better appearance.

Bugholes are air bubbles trapped at the surface of cast insitu form concrete during placement and compaction. These bug holes can be repaired by sack rubbing. Sack rubbing is carried out with cement mortar (1 part cement and 2 parts fine sand) having adequate water for thick paint consistency. Wet the entire surface and apply the prepared grout mortar with a rubber hand float, forcing the grout into surface voids. After sometime rub the excess grout off the surface with the sack. Care should be taken to avoid pulling of grout from inside the bug holes. After removing the excess grout, wet curing should be done for about 2 weeks.

Form tie holes are left on removal of formwork. These tie holes have high depth to width ratio (40 mm deep and 25 mm wide) and may not hold the repair mortar easily. Such tie holes are filled with drypack mortar (1 cement, 2 sand and little water). The low water content of dry pack results in minimal drying shrinkage, thereby improving durability. The inside surface of tieholes is first roughened and then saturated with water before filling drypack mortar for better bond. For stronger bond, epoxy-bonding agents can be added to dry pack mortar. Dry pack mortar is filled in layers of 10 mm and finally overfilling the hole slightly so as to level it after several beatings of flat float and mallet (hammer). Few light strokes of the mortar with a rag may help in blending repair mortar with the surrounding concrete surface.

Honeycomb and large voids use epoxy or cementitious grouts with bonding agents. First chip off surface voided concrete with light chipping hammers (< 5 kg mass). Apply sand blasting or water blasting to remove all loose and fractured concrete. Place repair mortar with trowel or a small pneumatic mortar gun. The repair mortar may consist of 1 part cement, 4 parts sand, adequate admixture and W/C ratio of about 0.35. For repair thickness more than 25 mm, apply mortar in **layers of 15-20 mm thickness** to avoid sagging and loss of bond. Each layer should be placed after 30 minutes after placing previous layer. Overfill the void slightly and finish by trimming and trowelling avoiding impairing of the bond. Removal of surface defects will not only provide better appearance but also enhance the service life of the concrete element.

11.6.4 Specific Cases of Crack Repair

Protection of concrete structures is the key to the **durability** and longetivity of structures. Decorative concrete protection uses **Emce Colour** – Flex Betonflair Elastic, Elastomeric **Crack**

Bridging System. Protection of Cooling Towers can be done by using **Ultra-violet Resistant, Crack Bridging, Carbonation Resistant, Breathable** Chloride Ion **Diffusion Resistant** Injection Systems.

Liquid Chemicals (e.g. Samafit VK) are used to repair DPC in case of rising dampness. Injection of these liquids depresses dampness by creating **chemical DPC**. These chemicals are available in one or two components. These chemicals preserve monuments and masonry from rising dampness.

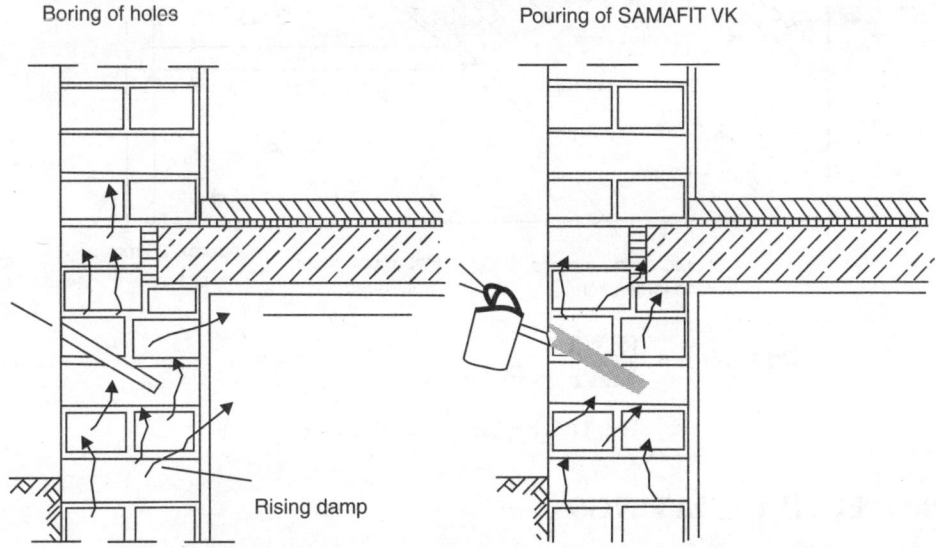

Fig. 11.4. Repair of Rising Dampness in DPC

Leakage due to cracks in water structures can be repaired by using **MC-Fix ST** Powder (Polymer based quick sealer). This stops leakage even under pressure. It allows surface to receive water proofing treatment by sealing water leakage.

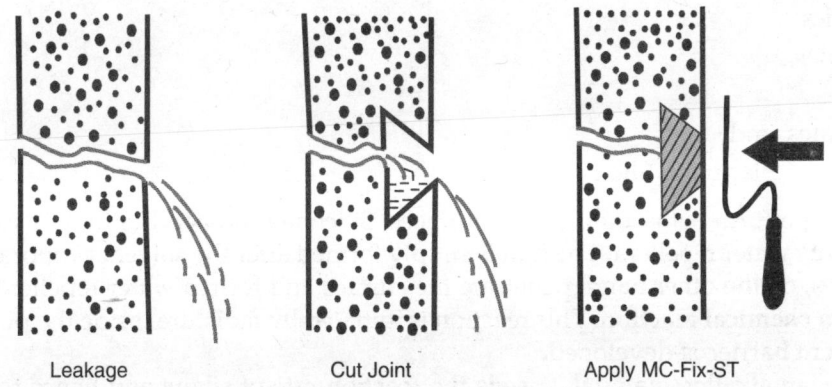

Fig. 11.5. Repair of Leakage through wall

Mc-Floor Patch Powder plus Nafafill liquid can be used for repair and protection. This comprises of two pack, polymer modified patching compound and mortar for floors. It develops rapid strengths and suited to cement substrate. It develops high compressive and bond strength.

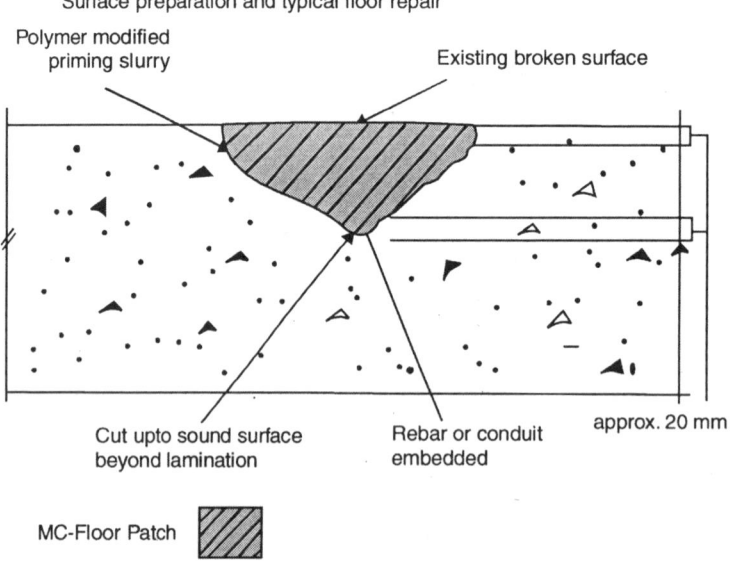

Fig. 11.6. Repair of Floor Patches

11.7 PROTECTIVE COATINGS

When water or moisture penetrates concrete, it can cause many problems such as efflorescence, **freeze-thaw** damage, carbonation, **corrosion of reinforcement**, mildew and loss of insulation value. A clear **water repellent coating** when applied with care helps to reduce moisture penetration in concrete to minimize the problems. Following are five main categories of water repellent coatings:

- Acrylics,
- Silicones,
- Silanes,
- Siloxanes, and
- Blends of Silane and Siloxanes.

Silicones perform better on concrete surfaces. Silicones and acrylics form a **surface film**, which is highly water repellent. The film is simply formed after the solvent evaporates. Silanes and Siloxanes, on the other hand, penetrate the surface and form a water-repellent barrier by undergoing a **chemical reaction.** This reaction is initiated by moisture inside the pores and the water repellent barrier is developed.

Concrete, an alkaline material, speeds the reaction rate of silane and hence reducing the loss of silane by evaporation. **Silanes are volatile** in nature. Avoid application of silanes on hot-windy days to minimize the loss of silanes by evaporation. Silanes generally have 20 to 40% solids content.

For effective protective coating on concrete surface, it is necessary to select and apply suitable type of coating based on the situation, weather conditions and the properties of the coating as determined from manufacturer's literature and trial testing. The **properties** to be considered are:

- Water repellency;
- Water vapour transmission;
- Surface Gloss;
- Weathering and ultravoilet stability;
- Resistance to efflorescence; and
- Water permeance.

The selected material for coating must ensure the material properties with reference to the concrete substrate. These protective coatings may not allow external weather forces (specially water) to penetrate and hence reduce the rate of deterioration. For optimum results the surface should be prepared, cleaned and dried properly for coating. Air temperature between 10 to 30°C is the best for applying protective coating.

Silicone materials are available as **water-soluble formulations** or mineral spirit soluble. Water-soluble silicones are preferred as these can be used both on dry or damp surfaces. Mineral solvent type silicones can be used in sub-zero temperatures. Water repellents fill the fine pores and seal the cracks less than 250 micron width only.

Protective coatings are required to be applied at certain interval (generally **3 to 5 years**) for maintaining water repellency of surface to enhance the life of the concrete element. Reapplication of these coatings is always done after preparing the surfaces properly.

Some of these protective coatings also serves the purpose of **decorative coatings.** These decorative coatings are: cement paints, oil free urethanes, Acrylics, Oil modified Urethanes, Catalysed epoxies, latex emulsions, Alkyd, two compound urethanes, epoxy esters, Polyesters, Vinyls, Chlorinated rubber and phenolics.

11.8 SUMMARY

Almost all modern constructions involve cement concrete construction. All constructions are subjected to environmental forces and loose their serviceability with time unless scientifically repaired and maintained. Due to lack of proper repair and maintenance of concrete structures, the service life reduces from **80-100 years to 30-40 years** resulting in great loss to the country and the owner. Repair and maintenance, therefore, plays most vital role in country's development.

Deterioration in concrete usually occurs in terms of **cracks, spalling, damp stains, corrosion** in reinforcement, **carbonation of concrete, delamination, porous surface** and other surface defects. Before undertaking any repair work, the root causes of deterioration must be assessed. Root causes of deterioration pertains to **physical overstressing, poor workmanship** (inadequate cover), **chemical attack, ambient environment,** and **corrosion of steel.**

The problem of concrete deterioration pertains to one or combination of inadequate cover to steel, porous or permeable concrete, high level of chlorides, poor workmanship, overloading, abrasion or erosion.

Reliable system of investigation requires preliminary **inspections**, detailed inspections, **laboratory testing** and insitu **NDT.** The extent of deterioration depth of carbonation and depth of cover, permeability of concrete, actual strength and chloride contents are determined during detailed investigation to confirm defects and causes identified during visual inspections. Depth of **carbonation can be determined by phenolphthalein test.** More accurate strength results can be obtained by core cutting. For approximate results NDT may be carried out.

Concrete repair materials are **grouts, mortars, concrete** or **shotcretes,** patching materials, **bonding aids, anticorrosive** and **protective coatings.** Repair materials essentially comprise of binding ingredients such as cement, polymer modified cement or expoxy resins. Repair materials must have adequate ultimate strength and rate of gain, matching modulus of elasticity and coefficient of expansion, low drying shrinkage, chemical compatibility, adequate **impermeability** for durability, and sufficient **bond with substrate** for monolithic action. Apart from basic repair materials, protective coatings are also used for enhancing the durability of repaired elements.

Commonly used concrete repair materials are: patching **mortars, bonding aids, anticorrosive coatings, special concretes, grouts, shotcretes,** overlays (toppings) and **protective coatings.** The selection of repair material depends on the extent of damage, situations and locations of placement, desired quality and availability of materials and skilled manpower.

Effective repair requires preparation of the concrete surface. Preparation includes removal of damaged concrete, cleaning the surface, cutting of decayed concrete to a depth of 25 mm for shotcrete or 15 mm for mortar repair, sealing of cracks, roughening of surface for adhesion, cleaning and removal of corrosion in RCC elements, and providing primer bond coat for adhesion.

Concrete repair methods include hand/**trowel application, fluid application** or **spray** application. Trowel application requires the surface to be prepared with primer coat and application of readymade mortar mixed at site with trowel and finished with float or sponge. Fluid application is better method in case of large areas or difficult architectural features. Repair fluid involved is micro-concrete which is poured or pumped after erection of fluid tight formwork and application of primer epoxy coat. The repaired element is appropriately cured. Fluid application provides homogeneous and self-compacting concrete. Spray of premixed mortars is quite effective due to superior bond, reduced permeability and better speed and compaction. Spray application can be either dry or wet with compressed air through delivery nozzle of a pump.

Repair of cracks form an essential part of any successful repair of concrete elements. Cracks must be investigated for its nature and the main causes before repairing them. Cracks must be prepared well by **cleaning, washing, drying** and **applying suitable bonding coat.** Variety of crack repair methods are: **epoxy injection, grooving and sealing, stitching, adding reinforcement, grouting, flexible sealing, dry packing, polymer impregnation, overlays** and **surface treatment,** and **autogenous healing.** Epoxy injection method is commonly used for structural repair of cracks. Epoxy injection is carried out by preparing the crack, drilling port holes, sealing the surface, inserting the ports, mixing and injecting epoxy, removing of surface seal and removing and plugging of ports. Variety of crack repair materials are: solvent free epoxy, solvent free **modified epox***y*, **epoxy injection resin, polyurethane, water compatible polyurethane, nonshrink grouts, polymer modified** ready to use **grouts** and **polymer modified cement grout.** Handguns, grease guns, foot pumps, and automatic mixing machines can be used for repairing cracks.

Stitching of cracks can be done by anchoring **stitching dogs** and **dowel bars** across the crack with suitable grout or epoxy mortars. In certain cases gel forming chemical grouting is also adapted to repair crack.

Surface defects such as **bugholes**, form **tie holes** and honeycomb are immediately treated within 24 hours of removing forms. This is necessary for protection and appearance of concrete elements. **Sack rubbing** with cement mortar paste is a simple technique for treating surface defects.

Use of **protective coatings** protect concrete elements from the environmental forces and enhance their service life by **slowing down deteriorating forces.** These protective coatings form water repellent film or barrier to ingress of moisture. The surface of concrete is prepared by cleaning, washing, drying and applying primer coat to receive suitable protective coat. Various protective coatings are: cementitious coats, silicone treatments, linseed oil, bituminous, surface hardeners, plastics and epoxy resins. Some of these protective coatings also serve as decorative coatings. Most of the **coatings** need replacement after **3 to 5 years** for their effectiveness.

PRACTICE QUESTIONS

11.1 Describe the **importance** of repair and maintenance of concrete structures in not more than 200 words.

11.2 Explain briefly various **symptoms and causes** of deterioration of concrete elements.

11.3 Explain briefly in about 100 words each: **carbonation** of concrete and **corrosion** of reinforcement.

11.4 Explain the method of preliminary inspections for diagnosing defects in concrete.

11.5 Explain main **causes of cracks** in concrete.

11.6 List methods of **repair of cracks** in concrete.

11.7 Explain **effect of permeability** of concrete on its serviceability in not more than 100 words.

11.8 **List factors** on which selection of concrete repair material depends.

11.9 **List** concrete repair materials.

11.10 Write short notes on (not more than 100 words each): **Anticorrosive coatings, Special concretes, Grouts,** and **Protecting coatings.**

11.11 Describe steps in **preparation of concrete** for repair.

11.12 List **repair methods** for concrete.

11.13 Explain the following in not more than 100 words: **Epoxy injection in cracks, Stitching, Polymer inpregnation,** and **Dry packing.**

11.14 Explain two techniques of treating **surface defects.**

11.15 Explain importance and characteristics of **protective coatings** for concrete.

11.16 Explain repair of dampness at DPC.

11.17 Explain repair of leakage through walls.

11.18 Explain repair of concrete floors.

CHAPTER 12

Testing of Materials for Quality Control

LEARNING OBJECTIVES

After studying this chapter, the learner understands testing procedure to check quality of various materials for ensuring **quality of cement concrete construction** and will be able to:

- Explain **importance of tests** in quality management of concrete construction;
- Describe **procedure of various tests** for cement;
- Interpret results of cement tests for quality assurance of concrete;
- Describe **procedure of various tests** for different aggregates;
- **Interpret results** of aggregate tests for quality assurance of concrete;
- Describe **procedure of testing water** quality for concrete;
- Describe **procedure of testing fresh** (plastic) **concrete**;
- **Interpret results** of fresh concrete tests for quality;
- Describe **procedure of testing hardened concrete** for desired quality;
- Explain **procedure of non destructive tests** (NDT) for the desired quality of concrete;
- Explain **field tests** and **adjustments** in proportions for achieving desired quality.

12.1 TESTS ON CEMENT

Experiment No. 1 : DETERMINATION OF WATER FOR CEMENT PASTE OF NORMAL CONSISTENCY (Refer IS: 4031-1967)

Objective

To determine the percentage of water for normal consistency for a given sample of cement.

Principle

Cement paste of normal consistency is defined as percentage of water by mass of cement to produce a consistency, which permits a plunger of **10mm diameter** to penetrate upto a depth of **5 to 7 mm** above the bottom of the Vicat Mould. It is necessary to determine the quantity of water to be mixed to prepare a cement paste of standard consistency for performing the tests for setting times, compressive strength and soundness. The quantity of water to be added in each of the above mentioned experiments bear a definite relation with the percentage of water for normal consistency.

Equipment

Vicat's apparatus with the plunger of **10 mm** diameter and **50 mm** length, weighing **300 gm** and Vicat's mould with non porous plate.

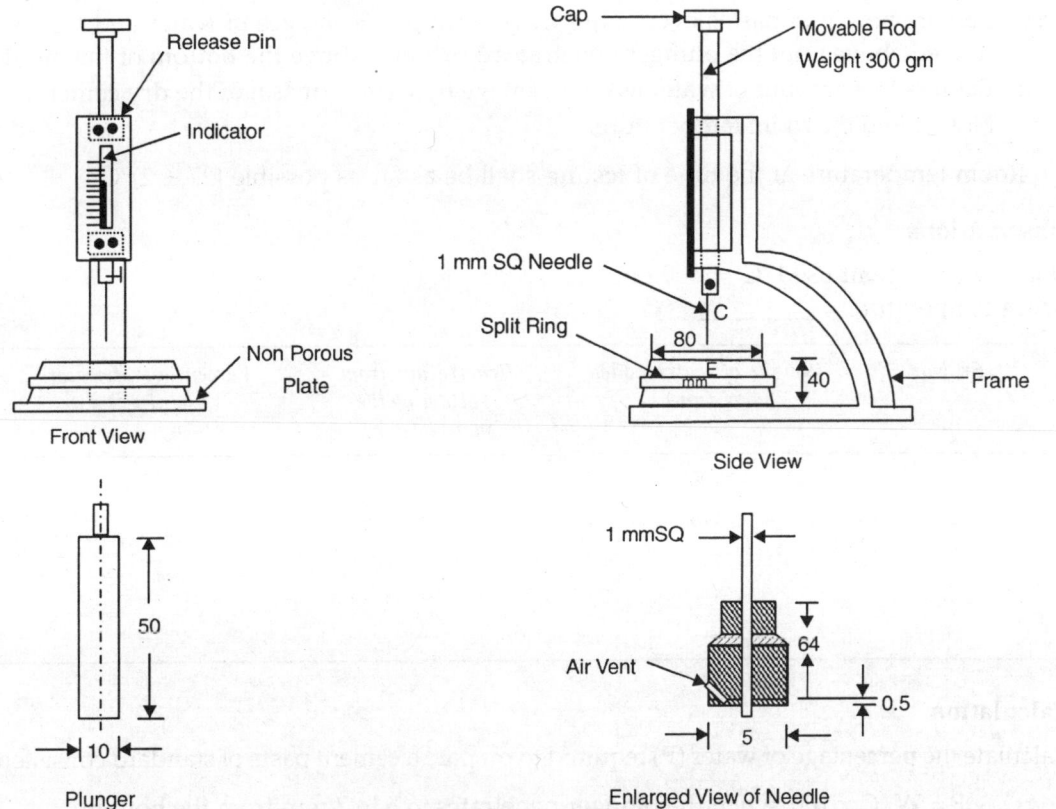

Fig. 12.1. Vicat's Mould (Dimensions in mm)

- Balance with masses of 1 kg, graduated measuring cylinder.
- Trowel, enamelled trays, standard spatula.

Thermometer, stopwatch.

Procedure

i. Weigh about 400 gm of cement, accurately and place it in the enamelled tray.

ii. To start with add about 25% of clean water and mix it by means of a spatula. Care should be taken that the time of gauging is **not less than 3 minutes** and **not more than 5 minutes**. The gauging time shall be counted from the time of adding water to the dry cement until commencing to fill the mould.

iii. Place the mould together with the non-porous glass plate under the plunger of Vicat's apparatus. Adjust the indicator to show zero reading when the plunger just touches the bottom surface of the test mould. Fix the plunger assembly near its top position.

iv. Fill the Vicat mould with the prepared paste, the mould resting on non-porous glass plate.

v. Make the surface of the cement paste level with the top of the mould with a trowel weighing 210 gm. The mould should be slightly shaken or tapped to expel air. Place the filled mould under the plunger with the surface of the paste just in contact with the plunger bottom.

vi. Release the plunger quickly, allowing it to sink into the paste.

vii. Prepare new trial pastes every time with varying percentages of water and test as described above until the plunger penetrates **5 to 7 mm above the bottom** of the mould.

viii. Express this amount of water as a percentage by weight (mass) of the dry cement.

ix. Note down the room temperature.

Room temperature at the time of testing shall be as far as possible $(27 \pm 2)°C$.

Observations

Quantity of cement used 'C' = 400 gm
Room temperature = _____ °C

Sr. No.	Weight of water added W (gm)	Penetration from bottom of the mould (mm)	Percentage of water (P)
1.	100	15	25 %
2.	115	05	28.75 %
3.	—	—	—
4.	—	—	—
5.	—	—	—

Calculation

Calculate the percentage of water (P) required to prepare a cement paste of standard consistency.

$P = W/C \times 100$, when the plunger penetrates to 5 to 7 mm from the bottom.

Conclusions:

_____

_____

Precautions

  i. Clean appliances should be used for gauging.
 ii. The temperature of cement and water and that of the test room at the time of test should be as far as possible from 25 to 29 [(27 ± 2)°C].
iii. In filling the mould, the operator's hands and the blade of the gauging trowel alone are used.

Experiment No. 2 : DETERMINATION OF INITIAL AND FINAL SETTING TIME OF CEMENT (Refer IS:4031-1967)

Objective

To determine the initial and final setting times of cement for a given sample of cement.

Principle

When water is mixed with cement to form a paste, hydration reaction starts. Out of the three active compounds viz. C_3A, C_3S, C_2S, react quickly with water to produce a Jelly like compound which starts soldifying. This action of changing from a fluid state to a solid state is called setting (**loss of plasticity**) and should not be confused with 'Hardening'. This point of starting the process of solidifying is termed as **initial setting** and is arbitarily judged by needle (1 mm^2 cross-section) when it penetrates in test block upto **5 to 7 mm** from the bottom. Final setting is assumed to have occured when the annular ring fails to make the impression on the surface of the test block.

Equipment

  - Vicat's apparatus with mould and non-porous plate.
  - Needle 1 mm^2 and needle with annular attachment of 5 mm diameter.
  - Balance (with weight box) of capacity 1 kg.
  - Graduated measuring cylinder 100 ml.
  - Trowel of about 210 gm weight, enamelled Trays, standard spatula.
  - Stop watch, thermometer (0 to 100°C). (Refer figure 12.1)

Procedure

  i. Weigh about 400 gms of neat cement.
 ii. Prepare a neat cement paste by adding **0.85 times** the percentage of water required for standard consistency.
iii. Start the stop watch at the instant when water is added to the cement.
 iv. Fill the Vicat's mould (E) with the cement paste prepared with the mould resting on the non-porous plate. Gauging time should **not be less than 3 minutes** and should **not exceed 5 minutes**.
  v. Fill the mould completely and smoothen off the surface of the paste, making it level with the top of the mould.

vi. Place the test block confined in the mould and resting on the non-porous plate, under the rod, bearing 1 mm² needle (C).

vii. Lower the needle gently till it comes in contact with the surface of the test block and quickly release, allowing it to penetrate into the test block and note penetration, after every two minutes.

viii. Repeat this procedure until the **needle** fails to pierce the block for about 5 to 7 mm, measured from the **bottom of the mould**. Stop the stopwatch and note the time, which is the initial setting time.

Determination of final setting time

i. Replace the needle (C) of the Vicat apparatus by the **needle with an annular attachment**.

ii. Release the needle with annular ring attachment as described earlier till the **needle point makes an impression** thereon the surface, while the **attachment fails to do so**.

iii. The time that elapses from the moment water is added to the cement till the needle only makes an impression is considered as **final setting** time for the cement under test.

Observations

(i) Quantity of Cement (C) = _____ (gm)

(ii) Water for standard consistency percent (P) = _____

(iii) Water to be added $0.85 \ P \times C = $ _____

(iv) Temperature at the time of testing

Cement _____ °C

Water _____ °C

Room _____ °C

Initial Setting Time

Sr. No.	Readings of penetration from bottom (mm)	Time at which water first added (T₁)		Time at which penetration readings are taken (T₂)		Initial Setting Time	
		Hr.	Min.	Hr.	Min.	Hr.	Min.
1.	—	—	—	—	—	—	—
2.	—	—	—	—	—	—	—
3.	—	—	—	—	—	—	—
4.	—	—	—	—	—	—	—
5.	—	—	—	—	—	—	—
6.	5 to 7	—	—	—	—	—	—

(v) Time T_2' needle fails to penetrate 5 to 7 mm from the bottom of the mould = _____.

(vi) Initial setting time $(T_2' - T_1) = $ _____ minutes.

(vii) Time when the **needle makes an impression** but the **attachment fails to make one** $(T_3') = $ _____ .

(viii) Final setting time $\left(T_3' - T_1\right)$ = _____ minutes.

Final Setting Time

Sr. No.	Record of Impression	Time of recording impression (T_3)		Final Setting Time	
		Hr.	Min.	Hr.	Min.
1.	—	—	—	—	—
2.	—	—	—	—	—
3.	—	—	—	—	—
4.	—	—	—	—	—
5.	—	—	—	—	—
6.	—	—	—	—	—

Conclusions

_____
_____

Precautions

   i. Needle must be cleaned with spirit each time before use.
   ii. Shift the position of the mould after recording the penetration reading so that the penetration may not be at the same place.
   iii. Check up the stopwatch for accuracy.
   iv. Clean appliances shall be used for gauging.
   v. Test Mould should be kept in 90% relative humidity and at $(27 \pm 2)°C$ and away from direct draught of air.

Experiment No. 3 : TO VERIFY THE SOUNDNESS OF CEMENT (Refer IS: 4031 - 1967)

Objective

To test the soundness of the given sample of cement.

Principle

The presence of free lime (CaO) and magnesia (MgO) in cement tends to increase the volume of the hardened concrete thus causing disintegration. During manufacture of cement, free lime is produced which **reacts with water to increase its volume** considerably. Magnesia also has the same effect but its rate of reaction is slow. Cement having such a property of expansion during hardening is classified as an unsound cement. Unsoundness is measured with the help of the **Le-Chatlier** mould.

Equipment

   - Le-Chatelier's apparatus, two glass plates.
   - Trowel, enamelled tray.

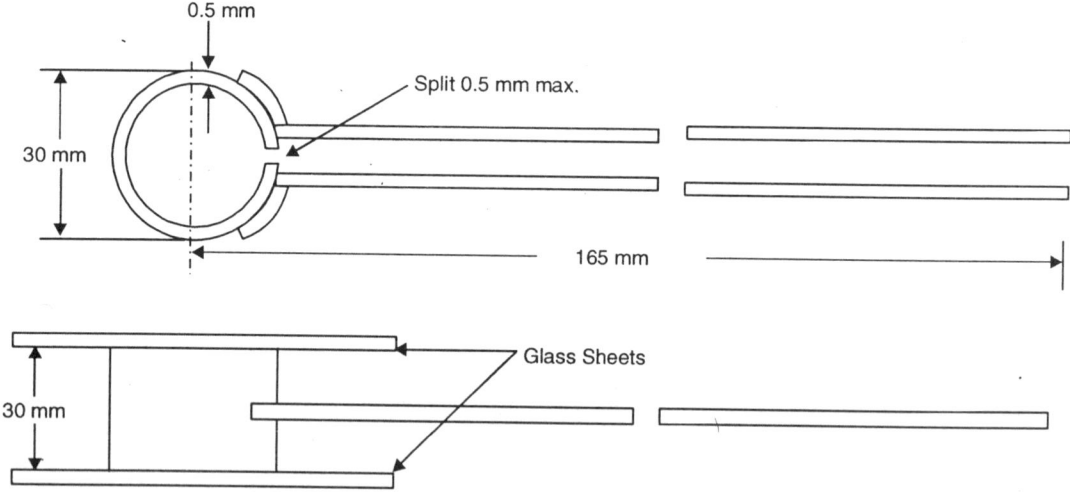

Fig. 12.2. Lechatelier Mould

- Measuring cylinder, blow lamp, a stove or water bath with electric heating arrangement.
- Thermometer, stop watch, balance and weight box.

Procedure

i. Place the mould on a glass plate and fill it with cement paste prepared by adding water equal to 0.78 P of cement mass. Care should be exercised to keep the edges of the mould gently together, while this operation is being performed. 'P' refers to percentage of water for normal consistency of the cement.
ii. Cover the mould with another piece of glass plate.
iii. Place a small weight on this, covering glass plate and immediately submerge the whole assembly in water at a temperature of 29°C to 32°C and keep it there for 24 hours.
iv. Measure the distance between separating indicator points to an accuracy of 'a' mm.
v. Submerge the mould again in water at a temperature prescribed above.
vi. Bring the water to boiling in 25 to 30 minutes, with the mould kept submerged and keep it boiling for **three hours**.
vii. Remove the mould from the water, allow it to cool and measure the distance between the indicator points.
viii. The difference between these two measurements, represents the expansion of the cement.

Observations

Room temperature (t°C) = _____.
Weight of cement sample (C) gm = _____.
Water required for standard consistency (P) = _____.
Water added to the sample (0.78 P × C) = _____.
Distance between the pointer ends before heating (D_1) = _____.
Time at which the sample is put in water = _____.
Time when water is brought to boiling point = _____.

Time of heating (to be kept 3 hours) = _____.
Distance between the pointer ends after heating (D_2) = _____.
Difference $(D_2 - D_1)$ = Expansion.

Conclusions

_____

_____

Precautions

i. All the measurements of quantity of cement and water should be done accurately by mass.
ii. The edges of the split mould should be kept together gently while filling the mould with the paste. Even a fine thread, can be wound very gently so as to avoid the splitting up of the brass mould due to filling of the paste.
iii. Water should be brought to boiling point gradually in the specified time.
iv. Le-Chatelier apparatus should be handled carefully by pressing glass plates without disturbing the pointer arms.

Experiment No. 4 : FIELD TEST TO VERIFY QUALITY OF CEMENT

Objective

To verify in the field if the cement is adulterated, unsound or has its setting and hardening action improved.

Equipment

Stone, enamelled trays, test tube, glass tumbler and measuring steel rule.

Procedure

Setting and Hardening Action

i. Prepare three small pats, each 75 mm × 75 mm × 25 mm in size from the given sample mixed with 28 % water by weight.
ii. Prepare similar number of pats with good quality standard cement.
iii. Cover the pats with moist cloth for 24 hours.
iv. Make thumb nail impression or scratch. Good quality cement will resist this impression.
v. If the given cement sample does not resist this impression then continue curing it up to total 48 hours, after which try to break it with pressure of thumb. Bad quality cement will easily break under the pressure.
vi. If 48 hours-test show improvement in hardening but does not attain hardness comparable with genuine cement, further trial should be made after 72 hours of curing. If the only defect in the cement under test is its **slow-setting quality**, it will become as strong as the genuine cement in this third test.

Detection of Adulteration

i. Place a small sample of doubtful cement on a steel hot plate and heat it thoroughly for 20 minutes on a stove. **Adulterated cement** will **change its colour on heating.** In genuine variety there will be no change in colour.

ii. To detect adulteration with coal ash take a small quantity of doubtful cement in test-tube or a glass tumbler and add water till the glass container is half full.

iii. Shake the container thoroughly and allow it to settle for a few minutes.

iv. Cement particles will settle down and ash particles will either be found floating on the surface or held in suspension, because of their lightness.

Ascertaining soundness of Cement

i. Make a pat of cement 75 mm in diameter and 15 mm thick and cure it with moist cloth for 24 hrs and then boil in water for a period of 6 hrs.

ii. Observe the surface of the pat. If the cement is sound, the surface will not develop a pattern of cracks as shown in the figure (8.11). In sound cement, cracks are thin and uniformly distributed all over the surface.

Conclusions

Precautions

i. In a test for soundness of cement, the cracking of unsound cement should not be confused with shrinkage cracks.

ii. Shrinkage cracks develop during boiling where the test pats might have been exposed to heat or drying winds.

iii. Shrinkage cracks are well-defined cracks running from edge to edge as shown in figure 8.11 and these do not indicate anything wrong with the sample.

Experiment No. 5 : DETERMINATION OF FINENESS OF CEMENT BY (i) SIEVING (ii) AIR-PERMEABILITY (Ref. IS: 4031 pt 2:1999 & IS: 5516-1996)

Objective

To determine the fineness of a given sample of cement.

Principle

Strength development of cement concrete is the result of the reaction of water with cement particles. The reaction always starts with the cement available at the surface of the particles in contact with water. Thus, larger the surface area available for reaction, greater is the rate of hydration and development of strength. Rapid development of strength requires greater degree of fineness. However, **too much fineness** is also considered to **accelerate deterioration** of cement during storage when exposed to air. Fine cements are likely to cause **more shrinkage**, but **less prone to bleeding** in concrete during placing. Greater fineness also requires greater quantity of gypsum for appropriate retardation and setting action.

Fineness can be measured by (i) **Sieving** through **90 micron** sieve or (ii) **Air permeability** apparatus (**Lea and Nurse** or **Blaine**) to assess surface area.

Procedure

I Sieving

Greater the fineness of cement, greater will be the **percentage (by mass) passing** through 90 micron sieve or lesser will be the percentage **(by mass) retained on the 90 micron sieve.**

Steps

    (i) Weigh accurately 100 gm of cement and place it on a standard **90 micron IS sieve**.

    (ii) Break down any air set lumps in the sample with fingers but do not rub on the sieve.

    (iii) Continuously sieve the sample by holding the sieve in both hands and giving a **gentle wrist motion** or alternately mechanical sieve shaker may be used for this purpose of sieving.

    (iv) Weigh the **retained cement** on 90 micron sieve after **continuous sieving** for 15 minutes.

    (v) Repeat the procedure for at least two more such samples.

    (vi) Determine the average percent retained.

Precautions

    (i) Check weighing machine for **accuracy** before using.

    (ii) The sieve should be **cleaned gently** with the help of a 25 mm or 40 mm bristle brush with 250 mm handle.

    (iii) Sieving must be carried out **continuously**.

    (iv) **Remove the cement** sample from the **bottom surface** of the sieve gently after every sieving.

Observations

S. No.	Sample No.	Mass of Sample (gm)	Mass of Cement Retained w (gm)	Percent Retained (w/100)	Remarks
1	—	100	—	—	
2	—	100	—	—	
3	—	100	—	—	
4	—	100	—	—	
5	—	100	—	—	

Average Percent Retained

Conclusions

Percent retained represents coarseness and **lesser percent retained** represents **finer material** i.e. greater the percent retained, lesser is the fineness of cement.

II Air Permeability Test

The air permeability test consists of a means to draw a stream of dry air at a **constant** or **variable velocity** through a bed of powdered cement contained in a **permeability cell**. There are two types of air permeability apparatus.

    (a) Lea and Nurse Type   (Fig.: 12.3 (a))

    (b) Blaine's Type   (Fig.: 12.3 (b))

In an air permeability apparatus, the **fineness** of cement particles is expressed in terms of **specific surface area** in cm^2/gm (also as m^2/kg). Finer the particles, greater is the specific surface area per gm. Greater surface area in case of finer particles offer more **resistance to air flow** through the bed of cement in the cell. The reaction of cement and water (hydration) will be **rapid** in case of **greater specific surface** due to better contact with water and also the **reaction starts first from the surface** of the particles.

Let us study both these methods of air permeability and respective apparatus to determine specific surface area based on **airflow** through the bed of cement.

(a) Lea and Nurse

·Lea and Nurse air permeability apparatus is shown in Fig. 12.3 (a). This apparatus is used to measure the specific surface area of cement in terms of the measured head (h_1), in the manometer and the head (h_2) in the flow meter. The **principle** is based on the relation between the **flow of air** through the **cement bed** and the **surface area** of the **particles comprising the cement bed.** From this, the surface area per unit mass of the material can be related to the permeability of a bed of a given **porosity** (e). The cement bed in the permeability cell is **1 cm high and 2.5 cm in diameter.** Knowing the density of cement the mass required to make a cement bed of **porosity of 0.475** can be calculated. This quantity of cement is placed in the permeability cell in a standard manner. Pass on air slowly through the prepared cement bed at a **constant velocity.** Adjust the rate of air flow until the **flow meter** shows a **difference in level of 30 – 50 cm.** Read the **difference in level (h_1) of the manometer** and the **difference in level (h_2) of the flow meter.** Repeat these observations to ensure that **steady conditions** have been obtained as shown by a **constant value of (h_1/h_2).** Specific surface S_w is calculated from following formula:

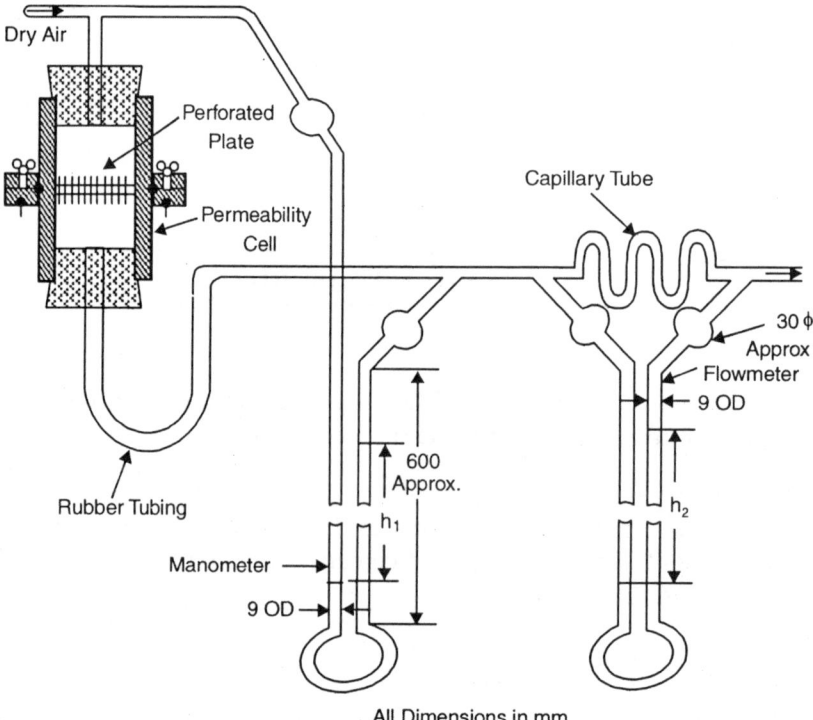

All Dimensions in mm

Fig. 12.3 (a). Permeability Apparatus with Manometer and Flowmeter

$$S_w = K\sqrt{(h_1/h_2)}, \text{ and } K = \frac{14}{d(1-e)}\sqrt{\frac{e^3 A}{CL}}$$

Where,

e	=	porosity, i.e. 0.475,
A	=	Area of the cement bed in the cell,
L	=	Length (cm) of the cement bed
d	=	density of cement,
C	=	flow meter constant,
h_1	=	difference of manometer levels when steady, and
h_2	=	difference of flow meter levels when steady.

(b) Blaines' Apparatus (Refer: IS 5516:1996)

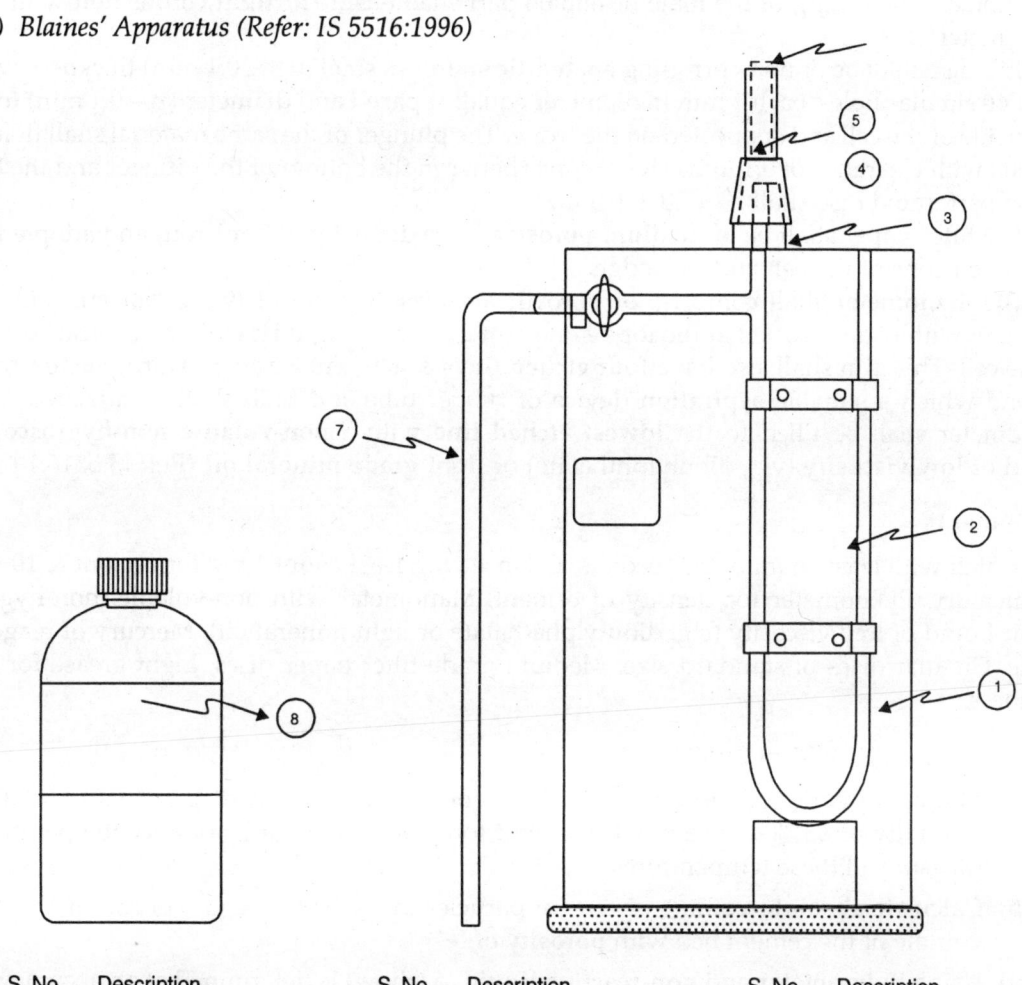

S. No.	Description	S. No.	Description	S. No.	Description
1.	Wooden Stand	4.	Permeability Cell	7.	Tube
2.	Manometer	5.	Perforated Disc	8.	Dibutylphthalate
3.	Rubber Cork	6.	Plunger		

Fig. 12.3 (b). Air Permeability Apparatus (Blaine Type)

Description

The working principle of Blaines' apparatus is based on the **variable flow** of air. It shall essentially consist of a means for **drawing a definite quantity of air** under a **falling pressure head** through a prepared bed of powder of a **definite porosity**. The **number and size** of the pores in a prepared bed of cement of **definite porosity** is a **function of the size of the particles** and determine the **rate of air flow through the bed**.

In Blaines' apparatus the **permeability** cell consists of a rigid right cylinder 12.7 ± 0.1 mm inside **diameter 'A'** of non-corroding **austenitic stainless steel**. The bottom of the cell shall form an **airtight connection** with the top of the **manometer**. The internal walls of the cell shall be smooth and truly vertical with a integral ledge to support the **perforated disc**. The top of the cell shall be fitted with a **protruding collar** to facilitate the **removal of the cell** from the manometer. The flaring of the male or female part shall ensure **airtight connection** with the manometer.

The disc shall be of non-corroding **austenitic stainless steel** (0.9 ± 0.1 mm) thickness with **30 to 40** circular holes, each 1 mm in diameter equally spaced and diameter (A – 0.1 mm) to fit the inside of the cell and supported on the ledge. The plunger of the same material shall fit into the cell with clearance of 0.1 mm. The distance between the bottom of the plunger and the top of the perforated disc shall be 15.0 ± 1 mm.

The filter paper shall be of **medium porosity** (pore diameter of 7 micron) and adopted to the dimensions of the cell and the ledge.

The **manometer** shall comprise of **borosilicate glass U-tube of (9.0 ± 0.4) mm** outside diameter with conical socket at the top of one arm to form **airtight fit** with the conical surface of the cell. This arm shall also have **four etched lines** & a T-joint leading to airtight **stop cock** beyond which a suitable **aspiration** device of rubber tube and bulb shall be attached. The manometer shall be filled to the **lowest etched line** with a **non-volatile non-hygroscopic** liquid of **low viscosity** (viz. dibutylphthalate) or **light grade mineral oil** (Ref: IS 5516:1996).

Equipments

Stopwatch with least count of 0.2 seconds, Balance with least count 1 mg for cement & 10 mg for mercury, Pyknometer for density of cement, Manometer with non-volatile, non-hygroscopic liquid of low viscosity (e.g. dibutylphathalate or light mineral oil), Mercury of reagent grade, Circular discs of standard size, Medium grade filter paper discs, Light grease for air tightness.

Procedure (Ref: IS 4031 pt 2:1999)

(i) Maintain the test laboratory at **27 ± 2 °C (or alternatively at 20 ± 2 °C)** and relative humidity of **65 %**. All test materials and tools need to be at laboratory temperature. Measure all these temperatures.

(ii) Calculate **absolute volume** of cement particles in the bed as **V (1 – e) cm^3**, if V is the volume of the cement bed with **porosity of 'e'**.

(iii) Using **Pyknometer** and **non-reactive liquid**, ρ **(rhow)** is determined at an accuracy of **0.01 gm/cm^3**. Repeat the determination of ρ to check accuracy and consistency of measurement. The volume V of cement bed can be determined and checked by using mercury and its data from the Table 12.1.

$$V = \left(\frac{m_2 - m_3}{D} \right) cm^3 ,$$

where

    D = density of mercury at the test temperature taken from the **Table 12.1**

    m_2 = **mass of mercury** to fill the cell **without cement bed**

    m_3 = **mass of mercury** to fill the cell **with cement bed**

Repeat the measurement of **V** until the values do not differ more than **0.005 cm³**. Record the mean value of **V**.

(iv) Weigh cement mass m_1 to prepare **cement bed** of **porosity 'e'** in the cell calculate mass **m_1 = ρ (1 – e) V gm**

Where

    ρ = density of cement particles (gm/cm³)

    e = porosity of cement bed, and

    V = volume of the cement bed (cm³).

For **e = 0.50, m_1 = 0.50 ρV gm**

Knowing values of ρ, e, and V, sample mass m_1 can be determined and weighed.

Prepare cement of specified volume based on **specified** porosity 'e' of cement mass.

(v) Place the **perforated disc** on the **ledge at the bottom of the cell** and place on it a **new filter paper disc** fully covering the perforated disc and keeping the **filter paper flat**.

(vi) Place the weighed mass m_1 of cement in the cell above the prepared cell base of perforated disc and filter paper disc taking care to avoid loss of cement from measured mass m_1. Tap the cell to level the cement bed.

(vii) Place a second **new filter paper disc** on the levelled cement bed. Insert the plunger to make contact with the filter paper disc. Press the plunger gently but firmly until the lower face of the cap is in contact with the cell. Slowly withdraw the plunger above 5 mm, rotate it through 90° and gently but firmly **press the bed** once again until the plunger **cap is in contact** with the cell. The **bed** is now compacted to the volume **V** with the porosity of 'e' and ready for air permeability test. Slowly withdraw the plunger.

(viii) **Insert** the conical surface of the **cell** into the **socket** at the top of the **manometer** using light grease if necessary to ensure an airtight joint. Take care not to disturb the cement bed.

(ix) **Close the top** of the cylinder with a suitable **plug. Open the stopcock** and **raise the level** of the manometer liquid with **gentle aspiration**. The manometer liquid should be brought to the **highest etched line** and close the stopcock so that the manometer **liquid remains constant**. If the liquid level falls, check all the **connections for air tightness**. Repeat the **leakage test** until the sealing produces a constant liquid level. Open the stopcock and adjust the liquid level to highest etched line with gentle aspiration. Close the stopcock.

(x) **Remove the plug** from the top of the cylinder. The manometer **liquid starts flowing**. **Start the stopwatch** as the liquid reaches the **second etched line. Stop** the watch the moment liquid reaches the **third etched line**. Record the time 't', to the nearest 0.20 (seconds) and the temperature to the nearest 1°C.

(xi) Repeat the procedure on the **same cement bed** and record the additional values of time and temperature.

(xii) **Prepare the fresh bed** of the same cement for the second sample, repeating the total procedure. Carry out the permeability test twice on the second bed, **recording** the values of time 't' and temperature as before.

(xiii) Calibrate the apparatus to calculate the **apparatus constant 'K'** by use of **standard material** of **known fineness**. Constant **'K'** for the apparatus is calculated as **average of at least three values**.

Prepare a cement bed of reference/standard sample as usual and conduct the permeability test and record the time 't_0' and temperature. Repeat the whole procedure on two further samples of cement of known specific surface area. Calculate mean time t_0 and mean temperature.

Calculate $K = \dfrac{S_0 \rho_0 (1-e) \sqrt{0.1\eta_0}}{\sqrt{e^3} \cdot \sqrt{t_0}}$

where

K = apparatus constant
S_0 = specific surface area of the reference (standard) sample
ρ_0 = density of the reference cement (gm/cm^3)
t_0 = mean of the three measurement times (seconds), and
η_0 = air viscosity at the mean of the three temperatures (PaS).

For specified porosity e = 0.500, $K = \dfrac{1.414 S_0 \rho_0 (1-e) \sqrt{0.1\eta_0}}{\sqrt{t_0}}$

Take mean of 3 values of K as the apparatus constant.

(xiv) Specific surface **'S'** of any cement sample can be calculate as:

$$S = \frac{K\sqrt{e^3}}{\rho(1-e)} \frac{\sqrt{t}}{\sqrt{0.1\eta}},$$

where

K = apparatus constant
e = porosity of the cement bed
t = measured time in seconds
ρ = density of cement (gm/cm^3), and
η = viscosity of air at the test temperature taken from the table 12.1 in Pascal Seconds

Precautions

(i) All joints and plugs shall be airtight and if required use light grease to ensure air tightness.

(ii) Manometer liquid shall be in level with the etched lines as accurately as possible.

(iii) Time of flow of air through the cement bed must be measured accurately to the nearest of 0.2 Seconds.

(iv) Cement sample shall be measured accurately taking care to avoid any loss while filling the cell.

Table 12.1 Density of Mercury D, Viscosity of Air (η) and $\sqrt{0.1\eta}$ as Function of Temperature

S. No.	Temperature (°C)	Mass Density of Mercury D (gm/cm³)	Viscosity of Air η Pascal-Second (PaS)	$\sqrt{0.1\eta}$
1	16	13.56	0.00001788	0.001337
2	18	13.55	0.00001798	0.001341
3	20	13.55	0.00001808	0.001345
4	22	13.54	0.00001818	0.001348
5	24	13.54	0.00001828	0.001352
6	26	13.53	0.00001837	0.001355
7	28	13.53	0.00001847	0.001359
8	30	13.52	0.00001857	0.001863
9	32	13.52	0.00001867	0.001366
10	34	13.51	0.00001876	0.001370

Note: Intermediate values by linear interpolation.

The equation for various standard temperatures can be simplified as under:

Porosity e = 0.500 and temperatures

(a) 27 ± 2 °C, $\boxed{S = \dfrac{521.08K\sqrt{t}}{\rho}\left(cm^2/g\right)}$, where t = time in seconds, ρ = Density gm/cm³

(b) 20 ± 2 °C, $\boxed{S = \dfrac{524.2K\sqrt{t}}{\rho}\left(cm^2/g\right)}$

with usual notations.

Observations

Porosity of cement bed e = 0.500

	I	II	III	Average
(a) Temperature of the room:	_	_	_	_
(b) Temperature of the sample:	_	_	_	_

(c) Apparatus Constant $K = \dfrac{S_0\rho_0\,(1-e)\sqrt{0.1\eta_0}}{\sqrt{e^3}\cdot\sqrt{t_0}}$ = _____

(d) Density of cement bed 'ρ_0' average = _____ (gm/cm³)

(e) Viscosity of air at the test temperature 'η' average = _____ (Pascal-Seconds)

(f) Time of flow of air from
     one etched line to the other (t):   _   _   _   _ (seconds)

Specific Surface at 0.500 porosity and at 27 ± 2°C test temperature (Refer: IS 4031 pt 2:1999)

$\boxed{S = \dfrac{521.08K\sqrt{t}}{\rho}\left(cm^2/g\right)}$

Specific Surface at 0.500 porosity and at 20 ± 2°C test temperature

$$S = \frac{524.2K\sqrt{t}}{\rho}\left(cm^2/g\right)$$

Conclusions

The cement sample tested **satisfies/does not satisfy** the requirements of fineness, as the observed specific surface area is **more/less** than that specified by the Indian Standard.

Experiment No. 6 : TO DETERMINE THE COMPRESSIVE STRENGTH OF CEMENT (Refer IS: 4031 - 1967)

Objective

To determine the compressive strength for the given sample of cement.

Principle

The compressive strength test is the final check on the quality of cement. The compressive strength of cement is measured by determining the compressive strength of **cement sand mortar cubes of 1:3 proportions**. The compressive strength test enables us to distinguish between rapid hardening and low heat portland cements.

Equipment

- Cube vibration machine, cube moulds 70.7 mm size,
- Trowel, enamelled tray, non-porous steel plates,
- Measuring cylinder 1000 ml, balance, and thermometer.

Procedure

i. Oil the interior faces of the mould.
ii. The materials for each cube shall be measured and mixed separately,
   Cement : 200 gm, standard sand = 600 gm (200 gm of each size).
   Water : (p/4+3.0) percent of combined **mass of cement and sand**, where p is percent of water for normal consistency of cement.
iii. Place a mixture of cement and standard sand in proportion of 1:3 by mass in an enamelled tray.
iv. Mix it dry with a trowel for one minute and then with water until the mix is of uniform colour.
v. Gauging time should not be less than 3 minutes and should not exceed 5 minutes. If it exceeds 5 min. the mixture is rejected and operation is repeated with new sample.
vi. Place the assembled mould on the vibrating machine and firmly hold it in position by means of suitable clamps.
vii. Securely attach a hopper of suitable size and shape at the top of the mould to facilitate filling and this hopper shall not be removed until completion of the vibration period.
viii. Immediately after mixing the mortar as specified above, place the entire quantity of the mortar in the hopper of the cube mould and **compact the same by vibration for a period of about 2 minutes** at a speed of 12000 ± 400 vibration/min.
ix. Keep the test cubes in the mould at a temperature of 27 °C ± 2 °C in an atmosphere of at least 90% relative humidity for 24 hours after completion of vibration.

x. At the end of this period remove the test cubes from the mould and immediately submerge in clean fresh water at 27 °C ± 2 °C and keep there until taken out just prior to testing. The water in which cubes are submerged should be renewed after every 7 days. The cubes should not be allowed to dry up before testing.

xi. Test three cubes for compressive strength at the periods mentioned below:
   a. Ordinary portland cement – 3, 7, and 28 days (optional)
   b. Rapid hardening cement - 1,3 and 7 days (optional)
   c. Low heat portland cement - 3,7 and 28 days.

xii. Place the test cube **centrally** on the platform of the compression-testing machine without any packing between the cube and the steel plattens of the testing machine.

xiii. Apply the load steadily and uniformly starting from zero at a rate of 35 N/mm^2 per minute till the cube fails.

xiv. Record the load at failure (in Newtons) and calculate the compressive strength as unit stress (N/mm^2)

Observations

Temperature:
   i. Room _____ °C
   ii. Water _____ °C

S. No.	Specimen Identification No.	Date & Time of Casting	Date & Time of Testing	Age (days)	L	B	H	Crushing Load at Failure P (N)	Strength P/LB (N/mm^2)	Remarks
1.										
2.										
3.										
4.										
5.										
6.										

Average Compressive Strength (**N/mm^2**) at _____ days is = _____.

Conclusions

_____

_____

Precautions

   i. All appliances should be clean.
   ii. The mixture which takes more than 5 minutes of mixing should be rejected.
   iii. In assembling the mould, cover the joints between the halves of the mould with a thin film of petroleum jelly in order to ensure that no water escapes during vibration.
   iv. Apply the load on the specimen gradually and centrally on the specimen.
   v. The specimen should be tested just after it is removed from the curing tank and should not be allowed to dry or contain extra surface water.

12.2 TESTS ON AGGREGATES

Experiment No. 1 : PARTICLE SIZE DISTRIBUTION AND FINENESS MODULUS (Refer IS: 2386) (Part 1) - 1963

Objective

To determine the **particle size distribution** and **fineness modulus** of coarse and fine aggregates.

Equipment

- Balance or scale shall be such that it is readable and accurate to 0.1% of the weight of the test sample.
- Sieves of sizes (80mm, 40mm, 20mm, 10mm, 4.75mm, 2.36mm, 1.18mm, 600micron, 300micron, 150micron) conforming to **IS 460-1962** shall be used.

Procedure

i. The sample shall be brought to an air-dry condition before weighing and sieving. This may be achieved either by drying at room temperature or by heating at a temperature of 100 °C to 110 °C.

ii. The air-dry sample shall be weighed and sieved successfully on the appropriate sieves starting with the largest size sieve.

iii. Each sieve shall be shaken separately over a clean tray for a period of not less than 2 minutes. The shaking is done with a varied motion backwards and forwards, left and right, circular clock-wise and anticlockwise, and with frequent jarring so that the material is kept moving over the sieve surface in frequently changing directions. If sieving is carried out with a mechanical sieve shaker **not less than 10 minutes sieving** shall be required for each test.

iv. Find the **weight** of aggregates **retained** on each sieve taken in order.

Observations

(A) Coarse aggregate Mass of each sample

$$I = \underline{\hspace{2cm}} (gm)$$
$$II = \underline{\hspace{2cm}} (gm)$$
$$III = \underline{\hspace{2cm}} (gm)$$
$$IV = \underline{\hspace{2cm}} (gm)$$
$$Total = \underline{\hspace{2cm}} (gm)$$

IS Sieve Size	Mass Retained (gm)				Percent Mass Retained Total	Percent Passing Total	Cumulative Percent Passing	Remarks
	I	II	III	Total				
80 mm								
63 mm								
40 mm								
20 mm								
16 mm								
10 mm								
4.75 mm								
Pan								

Sum of Cumulative Percent Passing (F) = _____ .

Fineness Modulus $= \dfrac{F}{100} =$ _____ .

(B) Fine Aggregate
Mass of each sample
I = _____ (gm)
II = _____ (gm)
III = _____ (gm)
Total = _____ (gm)

IS Sieve Size	Mass Retained (gm)				Percent Mass Retained Total	Percent Passing Total	Cumulative Percent Passing	Remarks
	I	II	III	Total				
10 mm								
4.75 mm								
2.36 mm								
1.18 mm								
600 micron								
300 micron								
150 micron								
75 micron								
Pan								

Sum of Cumulative Percent Passing (F) = _____ .

Fineness Modulus $= \dfrac{F}{100} =$ _____ .

Conclusions

_____

_____

Precautions

i. Each sieve shall be shaken for a period of atleast 2 minutes if hand sieving is used.

ii. Sieving should be in a circular clock-wise and anticlock-wise directions, forward and backward and left and right.

iii. If sieving is done in a shaker, atleast **10 minutes sieving** per test must be carried out.

iv. The sample to be sieved should be in **air-dry** condition.

Experiment No. 2 : SPECIFIC GRAVITY AND WATER ABSORPTION OF FINE AGGREGATE (refer IS: 2386 - 1963-Part II)

Objective

To determine **specific gravity** and water absorption of a fine aggregate sample.

Equipment

- A pycnometer or jar of 1 litre capacity fitted with a glass disc.
- A balance having a capacity of 2 kg and sensitive to 0.1 gm.
- A filter pump with the necessary pipe connection and gauge for removing the entrapped air.
- A water storage jar of about 20 litre capacity for maintaining the water at room temperature.
- A sample splitter and a thermometer. (Refer figure 12.4)

Procedure

i. The sand sample shall be immersed in clean water for 24 hours and dried to saturated surface dry condition. A representative sample of 1.2 to 1.5 kg mass (12 to 15 Newtons) of the above saturated surface-dry material is obtained by a sample splitter or other satisfactory means. The sample is divided into approximately equal parts from which two samples are weighed having identical masses (weights) from 0.6 to 0.75 kg (6 to 7.5 N). This mass (weight) of saturated surface-dry materials is recorded as '**B**'.

ii. One sample is placed in an oven and dried to a constant mass and the mass of the sample is recorded as **A**.

iii. The pycnometer jar is filled about three-quarters full of water having known temperature. The saturated surface dry sand sample '**B**' is added in the pycnometer. Entrapped air is removed either by a vacuum applied to the top of the pycnometer (jar) or by rolling the pycnometer (jar) or otherwise agitating the sand. The pycnometer (jar) is then filled with water and jar is covered with a glass disc by sliding the disc across the top of the jar. The pycnometer (jar) is shaken vigorously to remove all remaining entrapped air, after which the disc on the jar is removed and the pycnometer (jar) is carefully **refilled with water**. Replace the disc on the jar making sure that no air voids remain, and **wipe the outside surface**. The pycnometer filled with water is weighed and let this mass (weight) be recorded as '**W_1**'.

iv. Remove the sand sample and clean the pycnometer. Refill the pycnometer with water upto top and cover the jar with disc after ensuring complete **removal of air voids**. Now weigh the mass (weight) of the filled pycnometer and record the mass (weight) as '**W_2**'.

Observations and Calculations

A $\quad$ = Weight of Oven dry sample = (N)
B $\quad$ = Weight of Saturated Surface dry sample (N)
W_2 = Weight of Pycnometer filled with water (N)
W_1 = Weight of Pycnometer with water and sand (N)

$$\text{Specific Gravity on dry basis} = \frac{A}{W_2 + B - W_1}$$

$$\text{Specific Gravity on Saturated dry basis} = \frac{B}{W_2 + B - W_1}$$

$$\text{Water absorption percent by mass} = \frac{B - A}{A} \times 100$$

Conclusions

_____

_____

Precautions

  i. Remove the entrapped air completely while covering the jar with disc or filling the pycnometer.
 ii. Clean the Pycnometer from outside before weighing.
iii. All the weighings should be done accurately.

Fig. 12.4. Pycnometer For Determining Specific Gravity Of Aggregate Smaller Than 10 mm

Experiment No. 3 : SPECIFIC GRAVITY AND WATER ABSORPTION OF COARSE AGGREGATE (Refer IS:2386 - 1963 Part III)

Objective

To determine the **specific gravity** and **water absorption** of a coarse aggregate sample.

Equipment

- A balance of capacity not less than 3 kg, readable and accurate to 0.5 gm, and of such a type as to permit the weighing of the vessel containing the aggregte and water.
- A well ventilated oven, thermostatically controlled to maintain a temperature of 100 °C to 110 °C.
- A wide-mouthed glass vessel such as a jar of about 1.5 litres capacity, with flat ground lip and a plane ground disc of plate glass to cover it, giving a virtually watertight fit.
- Two dry soft absorbent cloths, each not less than 750 mm × 450 mm.
- A shallow tray of area not less than 32500 mm².
- An airtight container large enough to take the sample.

Procedure

  i. The sample shall be screened on a 10 mm IS Sieve, thoroughly washed to remove the particles of dust, and immersed in distilled water in the glass vessel. It shall remain immersed at a temperature of 22 °C to 32 °C for 24 ½ hours. Soon after immersion and again at the end of the soaking period, air entrapped in or bubbles on the surface of the aggregate shall be removed by gentle agitation. This may be achieved by rapid clock-wise and anti-clockwise rotation of the vessel between the operator's hands. Put the sample of soaked aggregate in a clean vessel.

ii. The vessel shall be overfilled by adding distilled water and covering with the plate ground-glass disc over the mouth so as to ensure that no air is trapped in the vessel. The vessel with sample and water shall be wiped on the outside and weighed (mass A).

iii. The vessel shall be emptied, cleaned and the aggregate allowed to drain. Refill the empty vessel with distilled water; Slide the glass disc cover in position as before. The vessel shall be wiped on the outside and weighed (mass B).

iv. The difference in the temperature of water in the vessel during the first and second weighings shall not exceed 2 °C.

v. The aggregate shall be placed on a dry cloth and gently surface dried with the cloth, transferring it to a second dry cloth when the first will remove no further moisture. It shall then be spread out not more than one stone deep on the second cloth and left exposed to the atmosphere away from direct sunlight or any other source of heat for not less than 10 minutes or until it appears to be completely surface dry. The aggregate shall be turned over atleast once during this period and a gentle current of unheated air may be used during the first ten minutes to accelerate the drying of difficult aggregates. The aggregate sample shall then be weighed in saturated surface dry condition (mass C).

vi. The aggregate sample shall then be placed in the oven in the shallow tray, at a tempera-ture of 100 °C to 110 °C for 24 ½ hours. It shall then be cooled in airtight container and weighed (mass D).

Observations

Weight of Saturated Surface dry aggregate = C
Weight of Jar filled with water and aggregate = A
Weight of Jar filled with water = B
Weight of Oven dried sample of aggregate = D

$$\text{Bulk Specific Gravity} = \frac{C}{C-(A-B)}, \text{ where } (A-B) = \text{wt. of aggregate in water}$$

$$\text{and } C - (A - B) = \text{wt. of water replaced by aggregate}$$

$$\text{Apparent Specific Gravity} = \frac{D}{D-(A-B)}$$

$$\text{Water Absorption (Percentage of Dry Weight)} = \frac{C-D}{D} \times 100$$

Conclusions

_____

_____

Precautions

i. The saturated surface dry sample is obtained when all the visible films of water are removed from the sample.
ii. All the weighings must be done carefully and accurately.
iii. Wipe and dry outside of the vessel every time before weighing.
iv. Ensure removal of entrapped air from the sample before weighing.

Experiment No. 4 : DETERMINATION OF BULK DENSITY AND VOIDS OF AGGREGATES (Refer IS:2386 - 1963 - Part III)

Objective

To determine the bulk density and percentage voids in aggregate.

Equipment

- A balance sensitive to 0.5% of the weight of sample to be taken.
- Cylindrical containers (having capacities of 3, 15 and 30 litres).
- Tamping rod 16mm in dia and 600 mm long rounded at one end.
- Glass plate for use in calibration of container.
- Trough, steel rule and measuring cylinder 250 ml.

Procedure

i. Condition of specimen: The test shall be carried out on **dry material** for determining the voids, but for bulk density the material with moisture may be used.
ii. Rodded or compacted mass: The measure shall be filled about 1/3rd full with thoroughly mixed aggregate and tamped with 25 strokes of the rounded end of the tamping rod. A further similar quantity of aggregate shall be added and a further tamping of 25 strokes given. The measure shall finally be filled to over-flowing, tamped 25 times and the surplus aggregate struck off, using the tamping rod as a straight edge. The net mass (weight) of the aggregate in the measure shall be determined and the bulk mass density (weight density) calculated in Kg/litre (Newton/litre).
iii. Loose Unit mass (weight): The measure shall be filled to over flowing by means of a shovel, the aggregate being discharged from a height **not exceeding 5 cm** above the top of the measure. Care shall be taken to prevent, as far as possible, segregation of the particle size of which the sample is composed. The surface of the aggregate shall then be leveled with a straight edge. The net mass (weight) of the aggregate in the measure shall then be determined and the lose bulk density calculated in kg/litre (Newton per litre).

Precautions

i. All the weighings should be done very accurately.
ii. While tamping the sample in bulk density test, there should be uniform rodding of the sample.
iii. Risk of segregation of particles should be avoided while performing loose bulk density test.
iv. Take container of appropriate volume (3, 15, 30 litres) depending on the particle size.

Observations

Condition of aggregate: Air dry/surface dry/moist _____

   i. Capacity of measure V = _____ 1
  ii. Weight of measure W_1 = _____ kg
 iii. Weight of measure + compacted aggregate W_2 = _____ kg
  iv. Weight of measure + loose aggregate W_3 = _____ kg
   v. Rodded bulk density $(W_2 - W_1)/V$ = a. _____ kg/1
$\qquad\qquad\qquad\qquad\qquad$ b. _____ kg/1
$\qquad\qquad\qquad\qquad\qquad$ c. _____ kg/1

  vi. Average bulk density (Rodded) $\left(\dfrac{a+b+c}{3}\right)$ kg/1

 vii. Loose bulk density $(W_3 - W_1)/V$ = a'. _____ kg/1
$\qquad\qquad\qquad\qquad\qquad$ b'. _____ kg/1
$\qquad\qquad\qquad\qquad\qquad$ c'. _____ kg/1

viii. Average bulk density (loose) = $\left(\dfrac{a'+b'+c'}{3}\right)$ kg/1

  ix. Specific Gravity (G_s) = _____
   x. Voids, $(G_s - \gamma)/(G_s) \times 100\%$ _____
  xi. Voids by measurement = _____

Conclusions

_____

_____

Experiment No. 5 : BULKING OF FINE AGGREGATES

Objective

To determine the bulking of fine aggregates in the filed.

Equipment

Graduated glass cylindrical measure and tamping rod.

Procedure

   i. Take the true representative sample of sand from the available lot at site.
  ii. Fill up the graduated jar with the sand upto certain height without compacting it exter-nally.
 iii. Level the sand surface by gentle motion and note down this height (h_1).
  iv. Now pour water in the graduated jar containing sand till the sample is completely submerged.
   v. Cover the jar with disc and give some motion, so that all the entnrapped air is released. The tamping rod should be moved throughout the sample in the jar so as to ensure removal of entrapped air completely.

vi. Now allow the contents to settle. After the sand has settled, note down the new height of the sample (h_2).

vii. Bulking factor can then be determined as
B.F = ($h_1 - h_2/h_2$)

Precautions

i. Sand should be filled lossely in the jar without carrying out any compaction.

ii. After adding water, the jar is thoroughly shaken to ensure that no entrapped air is left behind.

iii. Sand should be allowed to settle after shaking before noting down the final height (h_2).

iv. Take representative sample from the stack of sand.

Observations

a. Height of loose sand (h_1) = _____mm

b. Height of saturated sand (h_2) = _____mm

c. Percentage Bulking ($h_1 - h_2/h_2$) = _____%

d. Average Value = _____%

Conclusions

_____
_____

Experiment No. 6 : SILT CONTENT OF FINE AGGREGATE

Objective

To determine the silt content of a fine aggregate sample.

Equipment

- Measuring cylinder of 200 ml capacity
- salt

Procedure

i. Fill a measuring cylinder with sand upto 100 ml mark and add water upto 150 ml, to perform this test. For better result, dissolve a little salt in water (1 tea spoonful in 250 ml, water).

ii. Shake the sample vigorously for one minute and the last few shakings being in a side wise direction to level off the sand.

iii. Allow the cylinder to stand for three hours during which time any silt present will settle in a layer on the top of the sand and its thickness can be read in the graduated cylinder itself. There should not be more silt than about **6 to 10 %** of the amount of the sand.

Precautions

i. Shaking of the sample be upside down and sidewise so that water reaches every location of the sample.

ii. Allow the cylinder to stand away from shocks and vibrations.

iii. Read the total height of sand-silt layer after settlement (h).

iv. Read thickness of silt layers accurately after settlement (t).

v. Calculate silt percent $\left(\dfrac{t}{h} \times 100\right)$.

Conclusions

_____

_____

12.3 TESTS FOR WATER

Experiment No. 1 : SUITABILITY OF MIXING WATER (Refer IS:3025 - 1964)

Objective

To assess suitability of water on Concrete Mix.

Equipment

- Measuring cylinders, cube moulds, trowel, tamping rod, vibrating table.
- Testing machine.

Procedure

i. Analyze water for its chemical compositions as per IS:3025-1964 and compare these with permissible limits given in IS:456-2000

ii. In case water done not satisfy the permissible limits or it is not possible to obtain chemical analysis data readily, the suitability of water is tested by making concrete cubes. Concrete of desired grade is prepared with available water with designed proportions and cubes casted for testing.

iii. Prepare the same concrete mix with distilled water and cast the cubes in exactly the similar manner as with water of unknown nature.

iv. Cure both the samples in respective water under exactly the same conditions.

v. Test both the cube specimens after 7 and 28 days in exactly the same manner. Compare the test results of strength and conditions of the surface of cubes in respect of mix prepared with available water and distilled water.

vi. Available water shall be considered suitable if the strengths are not much less and no deterioration observed in case of concrete prepared with available water of unknown quality.

Precautions

i. For assessing the suitability of water for mixing and curing the concrete cube specimens should be prepared and tested under similar set of conditions with the available water of unknown quality to be assessed and distilled water.

ii. Mix proportions should be kept exactly the same in both the cases.

iii. All appliances should be clean.

iv. Cement, sand, and coarse aggregate should be the same as likely to be used at site.

Observations

Water	Specimen	Mix Proportions					Area	Crushing Load	Compressive Strength	Conditions
Type	Nos.	W	C	FA	CA	W/C	mm^2	N	N/mm^2	of Cubes
	1									
A	2									
Unknown	3									
Water	4									
	5									
	6									
							Average			
	1									
B	2									
Pure-distilled	3									
Water	4									
	5									
	6									
							Average			

Conclusions

Ratio of average compressive strength of concrete prepared with unknown water and distilled water = _____

12.4 TESTS FOR FRESH CONCRETE

Experiment No. 1 : CONSISTENCY OF FRESHLY MIXED CONCRETE BY SLUMP TEST (Ref. IS:1199-1959)

Objective

To determine the consistency of fresh concrete of given proportions by slump test where the nominal maximum size of the aggregate does not exceed 38 mm.

Equipment

- Mould in the form of a frustum of a cone with bottom diameter = 200 mm, Top diameter = 100mm, Height = 300 mm
- Tamping rod, 16mm diameter and 0.6 m length rounded at one end.
- Trough, trowel, G.I. plain sheets.
- Steel scale, stopwatch, etc. (Refer figure 12.5).

Procedure

   i. The internal surface of the mould shall be thoroughly cleaned and freed from moisture and any set concrete before commencing the test.

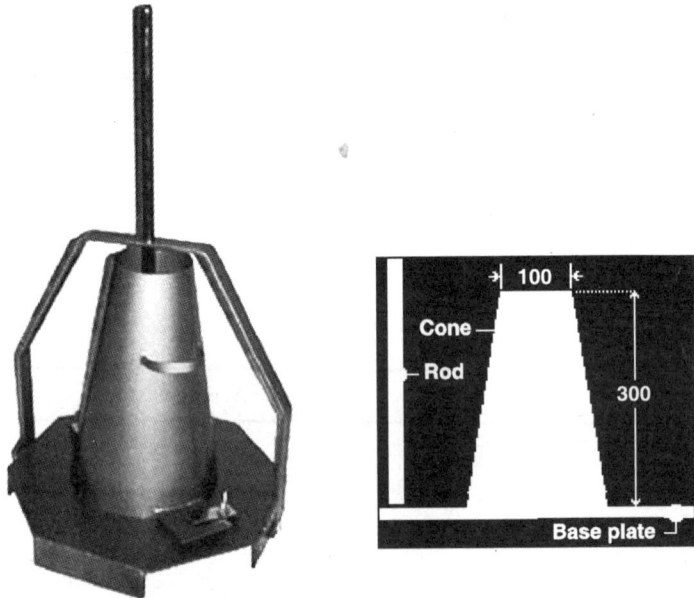

Fig. 12.5 (a). Slump Test Apparatus

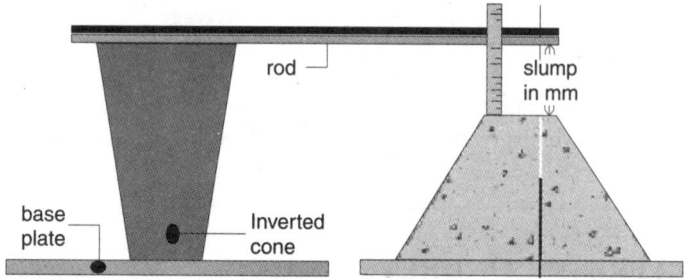

Fig. 12.5 (b).

   ii. The mould shall be placed on a smooth horizontal, rigid and non-absorbent surface, such as a carefully levelled metal plate.

  iii. Prepare the concrete mix as per design of proportions taking batch of about 15-20 kg total. Mix them thoroughly first, dry and then with water.

  iv. The mould shall be filled in **four layers,** each approximately one quarter of the height of the mould. Each layer shall be **tamped with 25 strokes** of the rounded end of the tamping rod. The strokes shall be distributed in a uniform manner over the cross-section of the mould and for the second and subsequent layers shall penetrate into the underlying layers.

   v. The concrete shall be struck off level with a trowel or the tamping rod, after the last layer has been rodded. Any mortar which may have leaked out between the mould and the base plates shall be cleaned away.

vi. The mould shall be immediately removed from the concrete by raising it slowly and carefully in a vertical direction. This allows the concrete to subside.

vii. The slump shall be immediately determined as the difference between the height of the mould and that of the highest surface of the subsided concrete specimen. The above test shall be carried out at a place free from vibration or shock and within a period of five minutes after mixing of water.

Precautions

i. Apply the strokes with the tamping rod uniformly through the full depth of the concrete of previous layer.

ii. Remove the mould very slowly by lifting it vertically upwards so that concrete within mould does not get disturbed.

iii. The base plate used in the experiment should be clean and smooth so that there is complete contact between base plate and the bottom of the mould.

iv. Test should be performed away from the ground vibrations produced due to machinery or some other sources.

v. Test should be completed in a minimum time say 2 to 5 minutes after mixing water.

Observations

S. No.	Proportions					Slump (mm)
	W/C	W	C	FA	CA	
1						
2						
3						
4						
5						

Conclusions

_____

_____

_____

Experiment No. 2 : WORKABILITY OF FRESHLY MIXED CONCRETE BY COMPACTING FACTOR TEST (Ref. IS: 1199-1959)

Objective

To determine the workability of freshly mixed concrete by the compacting factor test.

Equipment

- Compacting factor Apparatus (Refer Fig 12.6)
- Two trowels, Hand scoop, tamping rod.
- Platform weighing machine.

Procedure

i.   Prepare the concrete mix as per designed proportions and mix it thoroughly.

ii. Place the sample of concrete mix gently in the upper hopper with a hand scoop. Fill the concrete in level with the brim of the hopper.

Fig. 12.6. Compacting Factor Apparatus

iii. Open the trap door so that the concrete falls into the lower hopper.

iv. If the concrete sticks to the sides of the hopper, push it gently with the help of rod from top.

v.   Open the trap door of the lower hopper and allow concrete to fall into the cylinder.

vi. With the trowel in each hand and move them simultaneously from each side across the top of the cylinder and at the same time keeping them pressed on the top edge of the cylinder, remove the excess concrete remaining above the level of the top of the cylinder.

vii. Determine the weight of the cylinder to the nearest 10 gm.

viii. Refill the cylinder from the same samples of concrete in 4 layers of approximately 75.0 mm depth. Layers being heavily rammed or preferably vibrated so as to obtain full compaction.

ix. Finish the top of fully compacted concrete in level with the top of the cylinder. Clean the outside of the cylinder and weigh it again.

x.   Calculate the compacting factor as the ratio of the masses of partially compacted concrete with standard effort to the masses of fully compacted concrete in layers.

Precautions

i.   Oil the inner surface of the hoppers and cylinder before starting the experiment.

ii. Perform the test on a level ground.

iii. Fill the top hopper gently and to the same extent each time.
iv. The time between the end of mixing and release of concrete from top hopper must be constant say 2 to 5 minutes.
v. The hoppers and the cylinder must be washed clean and wiped off before use.

Observations

Mass (weight) of empty cylinder (W_1) = —————————Kg (N)

S. No.	Proportions					Wt. of Cylinder with Partially Compacted Concrete $W2$ Kg (N)	Wt. of Cylinder with Fully Compacted Concrete $W3$ Kg (N)	Wt. of Partial Compacted Concrete $(W2 - W1)$ Kg (N)	Wt. of Fully Compacted Concrete $(W3 - W1)$ Kg (N)	Compacting Factor CF = $=\dfrac{W_2 - W_1}{W_3 - W_1}$
	W/C	W	C	FA	CA					
1	0.50	—	—	—	—	—	—	—	—	—
2	0.55	—	—	—	—	—	—	—	—	—
3	0.60	—	—	—	—	—	—	—	—	—
4	0.65	—	—	—	—	—	—	—	—	—
5	0.70	—	—	—	—	—	—	—	—	—

Conclusions

———————————————————

———————————————————

Experiment No. 3 : WORKABILITY OF FRESHLY MIXED CONCRETE BY VEE-BEE CONSISTOMETER METHOD (Refer IS: 1199-1959)

Objective

To determine the workability of freshly mixed concrete by Vee-Bee Consistometer method.

Equipment

i. The Vee-Bee consistometer comprising (Refer Fig.: 12.7) of
  - a vibrating table resting upon elastic supports, a cylindrical metal pot,
  - a sheet metal frustum of cone, open at both ends and
  - a standard tamping rod (600 mm long and 16 mm diameter with one end rounded).
ii. Stopwatch, weighing machine, scale and trowel.

Procedure

i. Concrete mix shall be prepared with the given design mix Proportions.
ii. First a slump test shall be performed in the cylindrical pot by filling the conical mould placed in the cylindrical pot in the same manner as for slump test.
iii. The glass disc (C) attached to the swivel arm shall be moved and adjusted just touching the top of the slump cone concrete in the pot. Swing back the glass disc to clear the cone. The cone shall then be lifted up and the slump noted on the graduated rod by

lowering the glass disc to touch the top of the concrete cone. The electrical vibrator shall be switched on and stop watch simultaneously started.

iv. The vibration shall then be continued to level off the concrete surface and the time taken for this (conversion of conical to cylindrical mould) shall be noted with the stop watch in seconds.

Fig. 12.7. Vee-Bee Apparatus

Precautions

i. The remoulding is assumed to be complete when the glass plate rider is completely covered with concrete and all cavities in the surface of the concrete have disappeared. This is judged visually and may introduce error in establishing the end point.

ii. The slump test to be performed before giving vibrations to the table, should be performed as per standard practice.

iii. The test should be performed away from vibrating machinery.

iv. The electric switch and stopwatch should be started at the same instant.

Observations

The input of energy required for compaction is a measure of workability of the mix and this is expressed in Vee-Bee seconds i.e. the time required for the concrete remoulding from conical specimen to cylindrical specimen.

S. No.	Proportions of Mix					Slump (mm)	Vee-Bee Time (Seconds)
	W/C	W	C	FA	CA		
1.	—	—	—	—	—	—	—
2.	—	—	—	—	—	—	—
3.	—	—	—	—	—	—	—

Conclusion

_____
_____

12.5 TESTS FOR HARDENED CONCRETE

Experiment No. 1 : COMPRESSIVE STRENGTH OF CONCRETE (Ref. IS: 516 - 1959)

Objective
To determine the cube strength of concrete of given design mix proportions.

Equipment
- Cube moulds 150 mm and 100 mm (machined to a tolerance of ± 0.025 mm and angles 90 ± 0.5 on internal faces) in size.
- Weighing machine, mixer, tamping rod and compression testing machine.

Procedure
i. Weigh ingredients as per design mix proportions including water.
ii. Mix them thoroughly in the mechanical mixer until uniform colour concrete is obtained. The materials should be sufficient for casting of 6 cubes of 150 mm size or 6 cubes of 100 mm size. The concrete can also be mixed by hand in manner so as to avoid loss of water. In mixing by hand the cement and fine aggregate shall be first mixed dry to uniform colour and then the coarse aggregate is added and mixed until uniformity throughout the batch is achieved. Next water shall be added to dry materials and mixed until uniform concrete is obtained.
iii. Pour the concrete so prepared in the moulds, which have been oiled with medium viscosity oil. Fill concrete in cube moulds in **three layers**, each of approximately **50 mm** and ramming each layer 25 times with standard tamping rod, or by suitable vibrators.
iv. Finish off surplus concrete from the top of moulds with trowel.
v. Cover the moulds with wet mats or gunny bags and mark them after about 3 to 4 hours.
vi. Specimens are removed from the moulds after 24 hours and cube specimens are submerged in clean water at 27 ± 2°C for curing.

Precautions
i. All the materials should be weighed to an accuracy of 1 to 1000.
ii. The mould and the base plate must be oiled lightly before use to prevent the concrete from sticking to the moulds.
iii. During compaction the blows should be evenly distributed over the surface of each layer.
iv. Excess vibration should be avoided when compaction is done by vibration to avoid segregation and loss of water from the spaces between moulds and their base plates.
v. The cubes should not be allowed to dry and they must be tested after taking out from water.

  vi. At least three specimens should be tested for each test period and mean crushing strength should be taken as crushing strength of concrete for the given age. While calculating the average strength, if any individual variation from the average is more than 15%, that result should be rejected and the test may be repeated or more than three cubes tested for each age so that at least three cubes are within 15% of average.

  vii. Cubes should be placed in the testing machine **centrally** on plattens and load increased gradually at the rate of **14 N/mm² per minute** till crushing occurs. Testing is carried out according to IS:516-1959. Smooth parallel faces must be kept towards top and bottom between the plattens keeping the identification marked face vertical in front (Fig. 12.8).

Observations

S. No.	Identification Mark	Date of Casting	Date of Testing	Actual Age of Specimen	Dimensions of Specimen (mm)			X-Section Area	Max. Load	Comp. Strength (N/mm²)	Remarks
					L	B	H				
1											
2											
3											
4											
5											
6											

Conclusions

    _____

    _____

Experiment No. 2 : FLEXURAL STRENGTH OF CONCRETE (Ref. IS:516- 1959)

Objective

To determine the flexural strength (modulus of rupture) of concrete of given proportions.

Equipment

- Steel rectangular prismoidal (beam) moulds of the following sizes:
  - a. 100 mm × 100 mm × 500 mm, when maximum size of aggregate is less than 20 mm.
  - b. 150 mm × 150 mm × 700 mm, when maximum size of aggregate is upto 30 mm.
- Standard tamping rod, mixing tools and appliances, needle or table vibrator.
- The testing machine capable of applying the load at suitable rate (4 KN/minute for 150 mm specimens and 1.8 KN/minute for 100 mm specimens).
- Beam bed of testing machine with two similar steel roller supports (38 mm in diameter), mounted at 600 mm or 400 mm C/C distance. (Refer figure 12.9).

Procedure

  i. Prepare the concrete mix with designed proportions and record these proportions.

ii. Cast three beams of size 100 × 100 × 500 mm when the maximum size of aggregate is upto 20mm. Compact these by standard method tamping and filling in two layers, each layer being rammed more than 100 times by standard tamping rod or vibrated to ensure full compaction. Ramming rod is of 25 mm square type, 400 mm long and has 2 Kg mass (20 N Weight). Finish the surface of the concrete.

iii. After casting, the moulds shall be covered with wet cloths or gunny bags. Specimens shall be removed by opening the moulds after 24 hours and wet cured for 6 or 27 days. The dimensions of each specimen shall be noted before testing. Specimen shall be tested immediately on removal from water.

iv. The bearing surfaces of the supporting and loading rollers shall be wiped clean and any loose sand or other material removed from the surfaces of the specimens where they are to make contact with the rollers. The specimen shall then be placed in the machine in such a manner that the load shall be applied to the upper most surface as cast in the

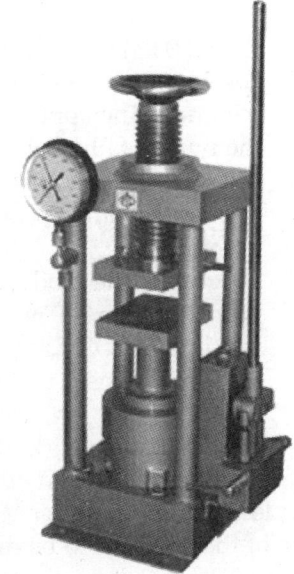

Fig. 12.8. Compressive Strength Testing Machine

mould, along middle third points spaced 200mm or 133 mm apart. The axis of the specimen shall be carefully aligned with the axis of the loading device. No packing shall be used between the bearing surfaces of the specimen and the rollers. The load shall be applied without shock and increasing continuously at a rate such that the extreme fibre stress increases at approximately 0.7 N/mm² (7kgf/cm²) minute that is at a

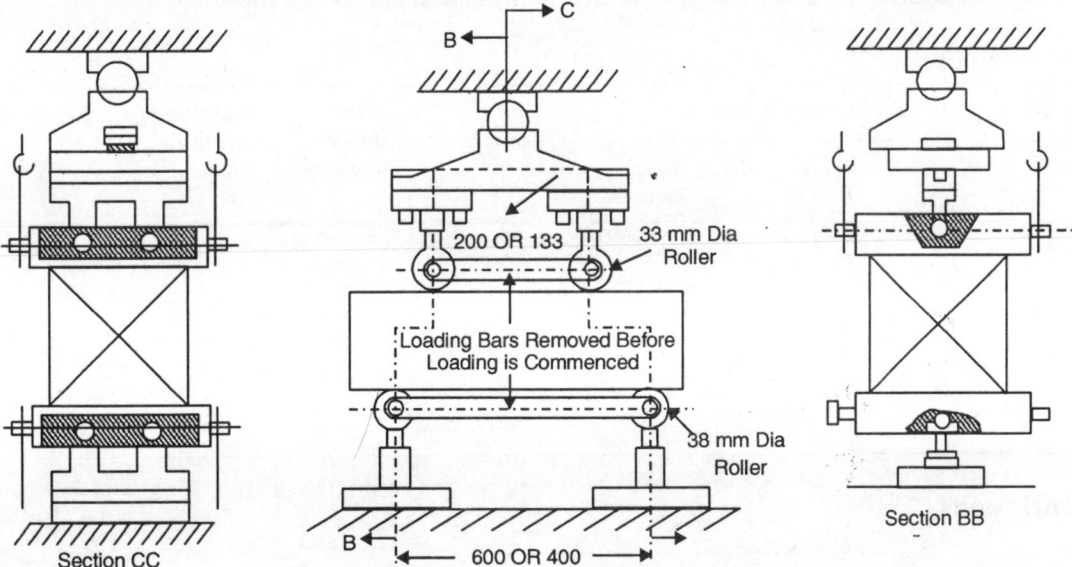

Fig. 12.9. Adaptor for Testing Beams in Flexure

rate of loading of 4KN (400kg) per minute for the 150 mm specimen and at a rate of 1.8 KN (180 kg) per minute for the 100 mm specimens. The load shall be increased until the specimen fails and the maximum load applied to the specimen during the test shall be recorded. The appearance of the fractured faces of concrete and any unique features in the type of failure shall be recorded.

v. Measure the distance between the line of fracture and the nearer support 'a'. Note if 'a' is greater than 200 mm for 150 mm specimen or greater than 133 mm for 100 mm specimens.

vi. Determine the modulus of rupture(flexural strength) by formula $f_b = pxl/bd^2$, when 'a' is greater than 200 mm for 150 mm specimens (or 'a' is greater than 133 mm for 100 mm specimens). OR $f_b^3 pxa/bd^2$ N/mm^2. When 'a' is less than 200 mm but more than 150 mm specimens (or 'a' is lesser than 133 mm for 100 mm specimens).

Where:

p is maximum load in N (kgf).

b,d,l are width, depth, and span of the beam respectively (all in mm).

If 'a' is less than 170 mm for 150 mm specimen or less than 110 mm for 100 mm specimens, the results of the tests are to be discarded.

Precautions

i. The loading should be applied centrally and without subjecting the specimen to any torsional stresses or restraints.

ii. The axis of the specimen shall be carefully aligned with the axis of the loading device.

iii. No packing shall be used between the bearing surfaces of the specimen and the rollers.

iv. The load shall be applied without shock and increasing continuously at the specified rate.

v. The load shall be correctly measured at the time of development of cracks.

vi. The location of cracks shall be correctly measured from the supports.

Observations

S. No.	Proportions of Mix					Size of Specimen	Max. Load N (Kgf)	Modulus of Rupture (N/mm²)	Remarks
	W	C	FA	CA	W/C Ratio				
1.									
2.									
3.									
4.									
5.									
6.									

Conclusions

Experiment No. 3: INFORMATION ON ACCELERATED CURING TEST FOR COMPRESSIVE STRENGTH OF CONCRETE (Reference IS: 10262-1982)

Concrete is usually placed in a structure in stages or lifts, one on top of another. Thus by the time the results of the 28 day test, or even the 7 day test results are available a considerable amount of concrete has been laid and completed. It is then rather too late for remedial measures if the concrete is too weak or if it is too strong. It is clear that it would be a tremendous advantage to be able to **predict the 28 day strength** within a few hours of casting. The strength of concrete at 24 hours is an unreliable guide in this respect, not only because even small variation in temperature during the first few hours after casting have a considerable effect on the early strength. It is, therefore, necessary for the concrete to have achieved a great proportion of its potential strength before 28 days, testing and a successful **test based on accelerated curing** has been recently developed.

In this test, standard concrete cubes are made but the moulds are immediately covered by top plates, sealed with grease on the metal surfaces of contact in order to prevent drying. **Within 30 minutes of adding the mixing water**, the cubes in their covered moulds, are placed in an airtight oven which is then switched on. The oven temperature should reach **95° ± 5°C in about an hour** and the cubes are kept at this temperature for a further **period of 5 hours**, making a **total of 6 hours in the oven**. At the end of this period the cubes are removed from the oven, stripped, allowed to cool, and tested in compression in the standard manner, the time allowed for these operations being **30 minutes**.

Thus the strength of the concrete is determined within **7 hours of casting**, and this accelerated **strength shows good correlation with the 7 and 28 day strengths of normally** cured concrete (**See figure 8.1 & 8.2**). The reliability of the results so obtained is high but, because of the variation in the rate of gain of strength of different cements, it is not possible to **use the 7-hour** strengths to **predict the 28-day values**. If the 7-hour period is inconvenient, because, for instance, it may fall outside the normal working hours, alternative schemes can be used; the same **accelerated strength is achieved if the cubes are kept in a closed tank over water at** 60°F for 18 to 24 hours and are then heated **for about 4 hours instead of 7 hours**.

The curves of figure 12.5 can be used as a general guide, but with unknown aggregates under unfamiliar condition it is preferable to obtain empirical curves relating the accelerated strength to the strength of normally cured cubes. In this manner a rapid and reliable means of strength control is achieved.

12.6 INFORMATION ON NON-DESTRUCTIVE METHODS OF TESTING

12.6.1 Schmidt Rebound Hammer Test

Principle

The test is based on the principle that the rebound of an elastic mass depends on the hardness of the surface against which the mass impinges.

Description

This is a useful and practical instrument for site work, but the results must be interpreted with caution taking into account its limitations.

The instrument measures the surface hardness of the concrete but experience based on tests has shown that the rebound number can be related empirically to the compressive strength, provided the following points are taken into consideration:

i. The result is affected by the position in which the concrete is struck whether it be over a stone, or the mortar paste, or near an air bubble inside the concrete.

ii. Floated or troweled surfaces give higher results than moulded surfaces.

iii. If a repeat test is made on the same spot or near to it a lower reading is obtained.

iv. Low readings are obtained within about 50 mm of the edge of the concrete.

v. If the object tested is small, it may be jerked by the blow and a low result will be obtained. All concrete elements tested should have a minimum mass or be suitably held. The solidity of the backing to the object tested also influences the result even if the object is held.

vi. The surface hardness of the concrete is the property measured and the hardness varies with the depth from the surface of the concrete. An accurate indication of the concrete compressive strength can be obtained only by calibration and other observations.

vii. The accuracy is greater if a separate calibration is carried out for each particular type of concrete to be tested.

The test determines in reality the **hardness** of the concrete **surface** and, although there is **no unique relation** between hardness and strength of concrete, empirical relationships can be determined for similar concretes cured in such a manner that both the surfaces tested by the concrete hammer and the central regions have the same strength.

Changes affecting only the surface of the concrete, such as the degree of saturation at the surface, would be misleading as far as the properties of the concrete within the structure are concerned (figure 12.10). The type of aggregate used affects the rebound number so that the relation between the rebound number and the strength should be determined experimentally for every type of concrete used on a site.

This test is mainly used for comparative study only. However, the hardness of concrete depends on the elastic properties of the aggregate used, and may also be affected by large differences in mix proportions. The test is useful as a **measure of uniformity** of concrete and is of great value in checking the quality of the existing material throughout a structure.

Procedure

In the rebound hammer test (See figure 12.10) a spring loaded mass has a fixed amount of energy imparted to it by extending a spring to a fixed position say (h_1). This is achieved by pressing the plunger against the surface of the concrete under test. Upon release the mass rebounds from the plunger, still in contact with the concrete surface and the distance traveled by the mass is h_2. The ratio of h_2 to h_1 expressed as a percentage gives us the rebound number. The distance traveled by the mass is measured by a rider moving along a graduated scale. The rebound number is arbitrary measure since it depends on the energy stored in the given spring and on the size of the mass.

Limitations

i. The hammer test has to be used only against smooth surface, preferably a prepared one.

ii. Troweled surfaces have to be rubbed with a carborundum stone to make them smooth if they are to be subjected to a hammer test.

iii. Small objects, if subjected to this test should have solid backing which otherwise due to jerking effect give erroneous readings.

iv. Test on concrete cannot be repeated on the same spot or near to it as it would give a lower reading.

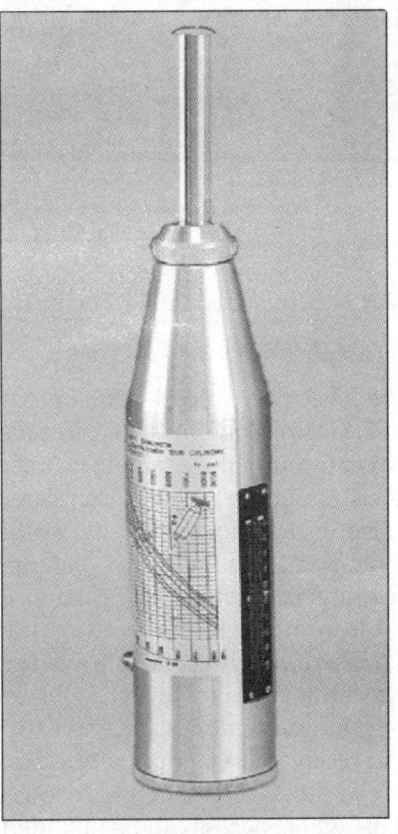

Fig. 12.10. Rebound Hammer

1. Plunger 2. Concrete 3. Tubular Housing 4. Rider 5. Scale 6. Mass 7. Release Button 8. Spring 9. Catch

Observations

S. No.	Location	Rebound Number Expressed as a %age $(h_2/h_1) \times 100$	Compressive Strength from Curves	Remarks
1.				
2.				
3.				
4.				
5.				

Note: Portable rebound hammer is easily available in India.

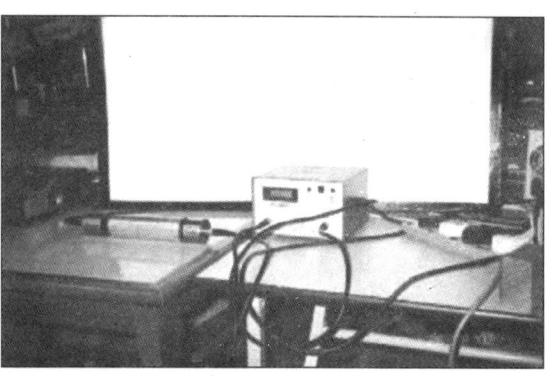 

Fig. 12.11. Layout of the Ultrasonic Pulse Apparatus

12.6.2 Ultrasonic Pulse Technique

Principle

The basic principle of ultrasonic testing of concrete is that concrete is an elastic material and will transmit longitudinal, compression, and shear waves. The velocity with which these waves travel through the concrete is determined by its properties, which control the elastic modulus. These properties are in turn related to the strength of the concrete. The apparatus generates a pulse in the concrete by the application of a mechanical impulse. It collects the impulse at some point at a measured distance from the point of generation. It contains a timing mechanism, which accurately measures the time taken for the pulse to pass from the loading edge of the transmitter to the receiver (transducers). Refer figure 12.11.

There are three basic ways in which the pulse can be transmitted and recorded. There is direct transmission, which is the most satisfactory. In this case the time measured is that for the longitudinal compression wave to pass between the transmitter and the receiver. The transmitter and receiver are on **opposite sides** of the structural unit concerned. The next method, which is not so satisfactory as the previous one, is **semi direct**, and is used for such units as thick floor slabs where access can be obtained to the top surface of the unit and to the side, but not to the underside. The third method, which is the least satisfactory, is **indirect**, where transmission and receiving have to be carried out from **one side only** as in ground floor slabs and basement walls after back filling has been carried out.

Description

Ultrasonic pulse velocity testing is an extremely useful method of testing concrete in a structure. Tests were conducted to study the effect of **density, age, water/cement ratio, aggregate/cement ratio**, type of **aggregate, moisture** content, type of cement and **presence of reinforcement**, and the results obtained are reproduced. The effect of density on the longitudinal wave velocity is illustrated from which it will be seen that there is a linear relation in which the velocity increases by 1.2% for every 1% increase in density. The strength of concrete depends in some way on its density and also on its age. It would be expected therefore, that definite relationship would exist between the **longitudinal wave velocity** and **the strength** of concrete. These curves were obtained by testing the longitudinal wave velocity in test cubes and immediately after wards crushing the cubes in a cube-crushing machine. It will be seen that there is initially a rapid increase of the wave velocity at early ages and then a more gradual increase at the later ages in exactly the same manner as the strength of concrete increases with age.

It will be seen that curve X is similar in shape to the curve for an identical mix. Curve 'Y' represents the results for M15 mix with high Alumina cement and water/cement ratio of 0.52 to 0.76 and the same aggregate. All the points irrespective of the kind of cement or water/ cement ratio fall closely on the curve. It would appear from this that the relationship between the **longitudinal wave velocity** and the **compressive strength** is **independent of the age**, type of **cement or water/cement ratio** and that the **water/cement ratio affects the longitudinal wave velocity only through the change in the compressive strength**, the flexural strength and the density.

Although the type of aggregate is normally assumed to have no appreciable effect on the strength of concrete it is found to produce an effect on the longitudinal wave velocity. The relationship between the longitudinal wave velocity and the modulus of rupture is however affected little by the type of aggregate. It would be expected that the longitudinal wave velocity would be greater for the leaner mix (or with higher proportion of aggregate in the mix). This is found to be so and is illustrated, where an existing structure has to be investigated, the calibration of wave velocity in terms of strength can be carried out on cores cut from the structure. A classification of concrete according to the longitudinal wave velocity is given as follows:

	Longitudinal Wave Velocity	Condition of Concrete
ft/sec	*(m/sec)*	
More than 15000	More than (4500)	Excellent
12000 to 15000	(3500 to 4500)	Good
10000 to 12000	(3000 to 3500)	Fair or questionable
7000 to 10000	(2000 to 3000)	Poor
Less than 7000	Less than (2000)	Very Poor

The modulus of elasticity can be computed from the formulae:

$E = V^2 \rho$ for laboratory beams

$E = V^2 \rho (1 - \mu)$ for flat slabs

$E = V^2 \rho (1 - \mu)(1 - 2\mu)/(1 - \mu)$ for mass concrete

Where E = modulus of elasticity

V = Velocity of wave transmission

ρ = Density of concrete

μ = Poisson's ratio (assumed as 1/6)

In addition to the control of the quality of concrete the ultrasonic pulse measurements can be used to detect the development of cracks in structures such as dams, and to check deterioration due to frost or chemical action. The ultrasonic pulse apparatus can also be used to determine the thickness of a concrete pavement or runway, provided the under side of the slab is fairly even.

Procedure

The apparatus for high precision measurement of the velocity of ultrasonic pulse in concrete is shown in figure 12.11. The wave velocity is not determined direct but is calculated from the time taken by a pulse to travel a measured distance. This ultrasonic pulse is produced by applying a rapid change of potential from a transmitter driver to a piezo electric crystal trans-ducer emitting vibrations at its fundamental frequency. Barium titan ate transducers have been found to be most suitable. The transducer is in contact with the concrete, so that the vibrations travel through it and are picked up by another transducer in contact with the oppo-site face of the specimen under test. The transducers generate an electrical signal, which marks at fixed intervals. Thus from the measurement of the displacement of the pulse signal relative to the position when the transducers are in direct contact with one another, the time taken by the pulse to travel through the concrete can be measured with an accuracy of ± 0.1 microsec-onds.

Limitations

i. The increase in the length of the path to the traveled decreases the sharpness of the onset waveform, so that there is no gain in accuracy.
ii. This technique cannot be employed for the determination of strength of concretes made of different materials in unknown proportions.

Observations

Sr. No.	Part of the Structure	Distance of Path (m)	Time of Travel (mic.s)	Velocity of Ultrasonic Wave (m/s)	Quality of Concrete	Range of Strength of the Concrete (N/mm²)
1.	Face 1					
	Face 2					
2.	Face 1					
	Face 2					

12.6.3 Core Testing

Core test of the concrete in a structure is generally the last resort, and adopted when cube results are significantly below the specification and the concrete is seriously suspected of being below the required strength. It is expensive and time consuming. The taking of cores in a liquid retaining structure should not be lightly undertaken, as it means boring holes 100 mm or 150 mm diameter through the structure and it is not easy to ensure a watertight joint afterwards.

Usually, a core is cut by means of a rotary cutting tool with diamond bits. In this manner a cylindrical specimen is obtained, sometimes containing embedded fragments of reinforcement, and usually with end surfaces far from the plane of reinforcement. The core should be soaked in water, capped, and tested in compression in moist condition.

Height/diameter ratio greatly influences the recorded strength of the core and usually this ratio should be nearly equal to two as height/diameter ratio lower than one yields unreliable results. Cores are cut to determine the strength of concrete and can also be used to detect segregation or honey combing or to check the bond at construction joints. In some cases, beam specimens can be sawn from road or airfield slabs, using a diamond or carborundum saw. Such specimens are tested in flexure. If the siliceous aggregate is used, sawn specimens give appreciably lower strengths than comparable moulded beams. Cutting of beams is not much used.

12.7 SUMMARY

To ensure expected properties and characteristics, cement concrete is required to be tested both in its fresh and hardened states. Properties and characteristics of ingredients are also tested to ensure the expected properties of cement concrete. Testing of various ingredients and concrete forms an essential part of quality management of cement concrete construction.

Correct test procedures, accurate measurements and proper interpretation play a critical role in achieving and managing specified and desired quality of cement concrete construction. Cement is tested to satisfy its standard specifications in respect of its fineness, setting and hardening, soundness and strength characteristics. Water is also tested to satisfy its suitability in respect of chemical reaction and hardening process without ill effects. Both fine and coarse aggregates are tested to determine suitability and basic design parameters (**size, shape, grading** or fineness modulus, **specific gravity, bulk density**, chemical and silt composition, and its moisture state).

According to the desired and expected values, workability of fresh concrete shall be measured by appropriate method (slump, compacting factor, Vee Bee degrees, and % flow.

Observance of certain precautions are essential for achieving correct and accurate test results. Determination of correct characteristics of concrete ingredients is necessary for the design of cement concrete mix for the quality management of construction.

PRACTICE QUESTIONS

12.1   Describe importance of **laboratory tests** of concrete ingredients

12.2   Describe importance of **field tests** of ingredients and fresh concrete for quality management of cement concrete construction.

12.3   Describe steps in testing of **compressive strength of cement**.

12.4   Describe procedure and precautions in conducting test for **soundness of cement**.

12.5   Describe the procedure of determining aggregate grading and fineness modulus.

12.6   Describe **field method** of assessing **suitability** of available **water** for cement concrete.

12.7   Explain the **influence of bulking** of sand on the quality of concrete.

12.8   Explain the importance of **assessing moisture condition** of aggregate in the field for quality construction.

12.9   Describe procedure of measuring **compacting factor** of fresh concrete.

12.10 State precautions necessary in the conduct of concrete **slump test** in the field.

12.11 Explain the procedure of conducting **concrete hammer test** to assess the strength of hardened concrete.

12.12 Explain the relationship of **ultrasound wave velocity and strength** of hardened concrete.

12.13 Explain the **need for accelerated tests** on concrete for quality construction.

12.14 List **precautions in non-destructive tests**.

12.15 List simple **field-tests for managing quality** of cement concrete construction.

12.16 Explain adjustment of 20% **bulking of sand** in case of volume proportion of 1:1.8:3.5. Assume data as required.

Glossary of Related Terms

Absolute volume	The volume of the solid particles of a loose granular material **excluding voids**.
Accelerator	A material that increases the rate of a chemical reaction
Acrylic	Polymer and copolymers of the esters of acrylic and methacrylic acids. One of group formed by polymerizing the esters or amides of acrylic acid.
Adhesives	The group of materials used to join or bond similar or dissimilar materials; for example, in concrete work, the epoxy resins.
Admixture	A material other than coarse or fine aggregate, cement or water, added in small quantities to the concrete during mixing to produce some desired modification in one or more of its properties.
Air entraining agent	A material, which introduces **tiny air** (micro size) bubbles into a concrete mixture.
Air-entrained Concrete	Concrete in which **minute bubbles of air** are deliberately introduced to improve resistance to freezing and thawing.
Bleeding	The migration of mixing water to the surface of freshly poured concrete (oozing of water to surface of concrete).
Calcine	To become quick powedery lime by the action of heat.
Cast stone	Simulated stone made from concrete cast in unit with artificially stone coloured surface.
Catalyst	A substance whose presence increase the rate of a chemical reaction (In some cases the catalyst is consumed and regenerated, in other cases the catalyst seems not to enter into the reaction, but functions by virtue of some other characteristic).
Cellular concrete	**Lightweight** concrete made by **introducing large numbers of air cells** into the mix.
Cement lime mortar	Mortar made with a proportion of slaked lime added to the cement also known as (gauge mortar).
Clinker	A stage in the manufacture of cement in which the **ingredients are fused** into small pieces by heat.
Colloidal Concrete (Prepacked)	Concrete made by injecting a cement grout under pressure into the void of a dry compacted mass of graded aggregate using micro level aggregate generated by chemical reaction.

Compressive mortar	Capacity of a material to withstand compressive load. The maximum compressive load (in newtons) that a masonry unit (in square mm) can carry. The resulting strength is expressed in (Newton) per square mm.
Concrete bond	The adhesion of two concrete surfaces together.
Co-polymerization	Polymerization of two or more dissimilar monomers.
Decorative Concrete	Concrete, which is given a special surface finish for architectural effect.
Dolomite	A limestone rich in magnesium carbonate.
E.J. Sealant	A compressive material used to exclude water and soiled foreign materials from joints.
Epoxy Concrete	A surface coating made with an epoxy resin and hardener, which stiffens at room temperature and is suitable for repair jobs.
Epoxy resins	A class of organic chemical bonding system used in the preparation of special coatings or adhesive for concrete or as binder in epoxy resin mortar and concrete.
Expansive Concrete	Concrete in which drying shrinkage is fully compensated.
Fibre-reinforced Concrete	Concrete in which fibres of glass, steel or other materials are introduced to improve certain properties.
Fineness modulus	It is a measure of relative fineness or **coarseness** of aggregate.
Flexural strength	Tensile strength in bending (modulus of rupture).
Float	A flat-faced wood (or metal) hand tool for spreading or smoothening concrete or Mortar.
Flux	A substance used to promote fusion at low heat.
Foamed or aerated concrete	Lightweight concrete in which the low density is obtained by the chemical reaction of an admixture with the cement resulting in the formation of a **cellular structure with bubbles of gas**.
Granolithic Concrete	Concrete made with specially selected hard aggregates and used for the wearing course of floors and pavements.
Granolithic finish	A surface layer of granolithic concrete, which may be laid on a base of either fresh, green, or hardened concrete.
Grout	A fluid concrete made with **small sized aggregate**, such as **sand** and **pea gravel**. Grouts used to fill wall cavities or cores of hollow masonry units. It must be fluid enough to fill all voids and completely encase the reinforcement but without segregation.
Gunited Concrete (Shotcrete)	Concrete made by spraying under pressure a mixture of cement, micro aggregates and water on to a surface.
Heavy-Weight Concrete	Concrete made with specially selected heavy aggregates to give a density exceeding 3000 kg/m^3.
HPC	High Performance Concrete.
Hydration	The **chemical reaction between cement and water**, which results in the hardening of concrete.
Hydraulic cement	Cement, which will set under water.
Impact load	An imposed load whose effect is increased due to its sudden application.
Initial set	The beginning of stiffening of the cement paste, which results in the concrete losing its plasticity.
Insitu	Literally means 'inplace', referring to material or components that are cast or assembled in their permanent positions in a building or structure as distinct from being cast or assembled before installation.

In-situ Concrete	Concrete which, whilst in its plastic state, is deposited in the location where it is required to form a part of the structure.
Latex	Organic polymer particles dispersed in water.
Lean mix	A concrete mix with a low proportion of cement and high proportion of aggregates.
Load bearing concrete masonry	Concrete block masonry made for load bearing applications.
Modular size	A size, which allows a material to a given module of standard measurement.
Moisture content	The amount of moisture in any material expressed as a percentage of its oven dry weight.
Monomer	An organic liquid, of relatively low molecular weight, that creates a solid polymer by reacting with itself or other compounds of low molecular weight.
Mortar	A plastic mixture of cementitious materials, fine aggregate, and water. Generally made up of the portland cement, lime, sand, and water.
No-fines concrete	Concrete, which contains **little or no fine** aggregates.
Partially reinforced masonry	Masonry that is reinforced only where design analysis indicates likely development of tensile stress. Partially reinforced walls have no minimum steel area requirements, where as fully reinforced masonry must have a minimum area and maximum spacing of steel limitations.
Permeability	A measure of the rate of the **passage of water or moisture through concrete** or other materials.
Pile	A support driven into or cast insitu in the ground for bearing or transferring load to substrate of foundation.
Plain Concrete	A hardened mass of cement and aggregates without reinforcement.
Plasticizer	(1) A substance added to **polymer or copolymer to reduce its minimum film** forming temperature and/or its glass transition temperature. (2) A substance added to **concrete mix to improve its workability** in plastic stage.
Plum Concrete	Plain concrete in which stones exceeding 15 cm in size are embedded.
Polymer	Polymers are long molecules of simple units called monomers. Monomers are generally organic compounds. Conversion of monomers into polymers is called polymerization, which is effected either by heat radiation (Gamma or ultraviolet) catalysts.
Polymer Concrete	**Concrete, which is impregnated** with a polymer compound to improve certain properties.
Polymerization	The reaction in which two or more molecules of the same substance (monomer) combine to form a compound containing the same element, but of high molecular weight.
Pozzolana cement	Cement made from volcanic rock containing proportion of **reactive silica**.
Pre-cast Concrete	Concrete, which is cast in a place away from its final position and then placed in position.
Pre-stressed Concrete	Concrete in which compressive stresses are deliberately induced internally, usually by means of tensioned steel, prior to loading of the structure to **compensate tensile stresses** likely to develop on actual loading.
Prestressed light-Weight Aggregate Concrete	Structurally lightweight aggregate concrete which is pre-stressed and density is not exceeding 1800 Kg/m^3 and capable of developing compressive strength of 30 N/mm^2 to 50 N/mm^2.

Pumped Concrete	Concrete, which is transported from the mixer or delivery point of ready mixed concrete to the placing position, being pumped through a pipeline.
Ready-mixed Concrete	Concrete made at a plant away from the construction site and conveyed in special vehicles.
Refractory Concrete	Concrete made with high-alumina or calcium aluminate cements and heat-resistant aggregate to **withstand high temperatures**.
Reinforced Concrete	Concrete in which rods, bars of fabric usually of steel are embedded in such a manner that the two materials act together under load.
Reinforced masonry	Masonry that is strengthened by the addition of reinforcing steel. Rebar grouted into the cores of masonry units or into the cavity between masonry to increase the wall's ability to resist flexural tensile stresses.
Rodding	Consolidation of grout in a cavity or core with the help of rod.
Rolled Concrete	Concrete having a high aggregate cement ratio, which is compacted by a road-roller and used as a base for road construction.
SCC	Self Compacting Concrete.
Screed	(1) A layer of mortar on a hard surface (also as verb, to screed). (2) A strip, usually of wood or metal, used as guide for striking off or finishing a surface. (3) A strip moved over a guide to strike off or finish a surface.
Shock Concrete	Concrete compacted by dropping the fresh concrete in a mould through a predetermined height several times.
Slurry	A very sloppy mixture of water and mortar or cement ingredients.
Spun	Concrete compacted by centrifugal action.
Terrazzo Concrete	Concrete made with marble aggregates and used as a surface finish of floors and walls.
Tremie Concrete	Concrete placed under water through a vertical steel pipe.
Vacuum Concrete	Concrete in which excess water added to make the mix workable is extracted from the plastic concrete by a vacuum process before the cement has set.
Vibrated Concrete	Concrete compacted by a vibratory effect, introduced either internally or externally in plastic (fresh) concrete.
Water gain	Absorption of water by hardened concrete.
Water retentivity	Ability of a mortar to retain the mixing water for hydration purpose.

References

1. **Kulkarni, P.D;** Ghosh, R.K; And Phull, Y.R.'*Text Book of Concrete Technology*' Oxford & IBH Publishing Co. – 1983,New Delhi.

2. Neville, A.M. '*Properties of Concrete*'Sir Issac Pitman & Sons Ltd. London.

3. Troxell, G.E. And Davis H.E. '*Composition and Properties of Concrete*' Mc Graw Hill Book Company Ltd. New York.

4. Orchand, D.F. '*Concrete Technology*' Vol. I and II Applied Science Publishers Ltd. London.

5. Portland Cement Association '*Principles of Quality Control*' John Wiley & Sons, Inc London.

6. Murdock, L.J. '*Concrete Materials and Practices*' Edwardd Arnold London.

7. Kulkarni, P.D; Gahlot, P.S; And Subramanian, R.S. '*Handbook of Concrete Technology*' T.T.T.I. Sector – 26, Chandigarh – 1985

8. Kulkarni, P.D; And Mittal, L.N. '*Laboratory Manual for Concrete Technology*' T.T.T.I. Sector – 26, Chandigarh – 1985

9. Kulkarni, P.D; Mittal, L.N; Gahlot, P.S.; Choudhary, M.R.; Jain, B.L.; And Inderchand '*Concrete Technology Ki Prayog Pustika*' – Hindi T.T.T.I. Sector – 26 Chandigarh – 1985

10. HMSO '*Design of Normal Concrete Mixes*' – 1975.

11. Bureau of Reclamation, U.S.A. '*Concrete Manual*' – 1981.

12. SP 23-1982 '*Hand Book on Concrete Mixes*'

13. IS: 456-2000 '*Code of Practice for Plain and Reinforced Concrete*'

14. IS: 10262-1982 '*Recommended Guide Lines for Concrete Mix Design*'

15. IRC: 44-1976 '*Tentative Guidelines for Cement Concrete Mix Design*'

16. IS: 269-1976 '*Specifications for Ordinary and Low Heat Portland Cement*'

17. IS: 383-1970 '*Specifications for Coarse and Fine Aggregates from Natural Sources for Concrete*'

18. IS: 2386 (Part I – VIII) '*Method of Test for Aggregates for Concrete*'

19. IS: 553-1969 '*Specifications for Constant Flow Type Air-Permeability Apparatus*' (Lea and Nurse Type)

20. IS: 650-1966 '*Specifications for Standard Sand*'

21. IS: 1199-1959 '*Method of Sampling and Analysis of Concrete*'

22. IS: 516-1959 '*Method of testing concrete*'.

23. IS: 7861 (Part I) – 1975 '*Code of Practice for Extreme Weather Concreting*' (Hot Weather)

24. IS: 7861 (Part II) – 1981 'Code of Practice for Extreme Weather Concreting' (Cold Weather)
25. IS: 8041-1978 'Specification for Rapid Hardening Cement'
26. IS: 8112-1976 'Specification for High Strength Ordinary Portland Cement'
27. IS: 1489-1967 'Specification for Portland-Pozzolana Cement'
28. IS: 445-1976 'Specifications for Portland Slag Cement'
29. IS: 6452-1972 'Specifications for High Alumina Cement for Structural Use'
30. IS: 6461 (Part VII) – 1973 'Glossary of Terms Selecting to Cement Concrete'
31. Unwalla, B.T. "Concrete Technology" an overview Vol. 61, Nov. 1980, Journal Institution of Engineers (India)
32. EFNARC, Association House, 99 West Street, Farnham, Surrey GU97EN, UK

Index